BIODIVERSITY
Monitoring, Management and Utilization

The Editor

Dr. Surendra Singh is working as Professor in the Department of Biological Science, Rani Durgawati University, Jabalpur (M.P.), India. He has done M.Sc. and Ph.D. in Botany from Centre of Advanced Study in Botany, Banaras Hindu University, Varanasi, Uttar Pradesh. Dr. Singh has more than 26 years of PG teaching and research experience in Environmental Microbiology and Biotechnology. He was awarded with Common Wealth Post Doctoral Scholarship by Govt. of UK and Visiting Scientist Fellowship by Govt. of Israel and PR China. Dr. Singh has supervised major research projects sponsored by Council of Science and Technology, Bhopal, Council of Scientific and Industrial Research, New Delhi, University Grants Commission, New Delhi and Department of Science and Technology, New Delhi. Dr. Singh has rare distinction to work in a Biotech Industry for about 4 years as Chief Technology Officer followed by Director Research and Development. Twelve students have been awarded Ph.D. degree and many more are in pipeline under his supervision apart from M.Phil. and M.Sc. dissertations. He is Fellow of Indian Botanical Society and Life member of several Scientific Societies and Association. Dr. Singh has published more than 45 research papers in National and International Journals of repute and contributed several Chapters in various books. He has participated and delivered lectures in several National and International Conference/ Seminar/ Symposium in India and Abroad. Presently Dr. Singh is Director, Distance Education; Director, UGC Community College; Director, University Institute of Vocational Studies & Skill Development; Chairman Board of Studies in Botany; Coordinator Career Guidance, Counselling, Training & Placement Cell; Nodal Officer, Skill Development Center in Rani Durgawati University, Jabalpur.

BIODIVERSITY
Monitoring, Management and Utilization

– Editor –
Dr. Surendra Singh
Department of Biological Science,
Rani Durgavati University,
Jabalpur-482001 (M.P.), India

2018

Daya Publishing House®
A Division of
Astral International Pvt. Ltd.
New Delhi – 110 002

ISBN: **9789387057555 (International Edition)**

Publisher's Note:

Every possible effort has been made to ensure that the information contained in this book is accurate at the time of going to press, and the publisher and author cannot accept responsibility for any errors or omissions, however caused. No responsibility for loss or damage occasioned to any person acting, or refraining from action, as a result of the material in this publication can be accepted by the editor, the publisher or the author. The Publisher is not associated with any product or vendor mentioned in the book. The contents of this work are intended to further general scientific research, understanding and discussion only. Readers should consult with a specialist where appropriate.

Every effort has been made to trace the owners of copyright material used in this book, if any. The author and the publisher will be grateful for any omission brought to their notice for acknowledgement in the future editions of the book.

Published by   :   **Daya Publishing House®**
A Division of
Astral International Pvt. Ltd.
– ISO 9001:2015 Certified Company –
4736/23, Ansari Road, Darya Ganj
New Delhi-110 002
Ph. 011-43549197, 23278134
E-mail: info@astralint.com
Website: www.astralint.com

Prof. S .M. Singh (Retd.)
M.Sc., Ph.D., FPSI, FGIIAP

Principal Investigator
Major Research Project
UGC, New Delhi
(2009-2012)

Department of Biological Science
Rani Durgavati University
Jabalpur - 482001 (M.P.) India
Mobile : 09425325756
Phone : 0761-2678821
E-mail : smsingh_rdvv@rediffmail.com

Foreword

Biodiversity constitutes the most important working component of a natural ecosystem. It helps maintain ecological processes, creates soils, recycles nutrients, has a moderating effect on the climate, degrades waste, controls diseases and above all, provides an index of health of an ecosystem. Providing food, medicines and a wide range of useful products, it is the natural wealth that exists on land, in freshwater and in the marine environment. Plant diversity alone offers more than just food security and healthcare for the one-quarter of humanity who live their lives at or near subsistence levels; it provides them with a roof over their heads and fuel to cook, and, on average, meets 90% of their material needs. The diversity of life on earth is so great that if we use it sustainably we can go on developing new products from biodiversity for many generations. This can only happen if we plan and manage biodiversity as a precious resource and prevent the extinction of species. The book before you offers an overall view of this biological diversity and carries the urgent warning that we are rapidly altering and destroying the environments that have fostered the diversity of life forms. It helps the reader quantify and interpret patterns of ecological diversity, highlights advances in measurement, application of measures of diversity to conservation and environmental management and addressing sampling issues. This book is entirely unique in its treatment of the topic. This book expertly combines biodiversity monitoring, management and utilization. Although there have been many volumes on biodiversity, this book is a unique contribution because it is able to unite these vast and complex arenas of scientific literature and then eloquently represents key findings in a way that is accessible to a broad readership. The editor has well compiled the practicalities of conservation management, economic, social and political issues and has examined conflicts between sustainable development and human dependence on biodiversity in agriculture, environmental management and biotechnology, and encourages contributions from developing countries to promote broad global perspectives on matters of biodiversity and conservation. Biodiversity knows no political boundaries and its conservation is therefore, a collective responsibility of all nations. Everybody should come together and take appropriate measures for conservation of biodiversity and sustainable utilisation of its benefits. I congratulate Dr. Singh for his successful effort in presenting relevant and clear information on Biodiversity and wish him the best in his future endeavour.

Prof. S.M. Singh (Retd.)
Department of Biological Science,
Rani Durgawati University, Jabalpur (M.P.)

Preface

This book "Biodiversity: Monitoring, Management & Utilization" explores biodiversity as measurement, management and utilization of resources in satisfying human needs in multi-sectional areas including agriculture, forestry, fisheries, wildlife and other exhaustible and inexhaustible resources. The term biodiversity defines not only the variety of life on Earth but also their complex interactions. The subject of biodiversity is huge in scope and significance. Under the current scenario of biodiversity loss, and in order to preserve it, it is essential to achieve a deep understanding on all the aspects related to the biological interactions, including their functioning and significance. This book brings together an international group of experts on the subject, each with a distinctive focus, giving an overview of its many different aspects and combining academic rigour with a concern to make the topic intelligible to the non-specialist reader. This book concise introductory text provides a complete overview of biodiversity to what it is, how it arose, its distribution, why it is important, human impact upon it and what should be done to maintain it. The contents of the book provide access to the latest scientific knowledge on biodiversity in an engaging and clear format. It describes the ancient origin and unique features giving an overview on diversity of animals, plants and microorganisms. It outlines tools for measuring, monitoring, planning and managing biodiversity. It highlights indigenous perspectives on biodiversity, its aesthetic values and explains the relation of biodiversity with respect to agriculture. It also explores the policies and legal status of our biodiversity. This book is an accessible guide to inform readers about biodiversity science and underpin decisions for its conservation and sustainable management. It provides a broad view of the many threats to global biodiversity and of the policy responses required to combat them. It explores business opportunities related to biodiversity and how it can be utilized for human welfare. The integration of biodiversity into national decision-making processes will enable parties to appropriately assess the consequences of biodiversity loss, possible trade-offs and increase coordination among government ministries and different level of government. Improving the international governance of biodiversity is very necessary and challenging, this book provide a rich and diverse source of informed perspectives on how to reconcile global economic development with the preservation of our biological and cultural diversity. This book is a great resource for higher secondary, undergraduate, postgraduate students, researchers and academicians worldwide. Also it will be of value to conservation biologists lead by the NGOs, wildlife managers, ecologists, environmental scientists, policy makers, literary figures, entrepreneurs, wildlife activists, social activists, certain organizations like PETA etc. The book is a must read for anyone seeking to understand the far reaching policy, legal, economic and social implications of current discussions over such issues.

Dr. Surendra Singh

Acknowledgement

I would like to express my eloquent gratitude to all the authors, academicians, researchers and reviewers, who provided their detail research and views for "Biodiversity: Monitoring, Management & Utilization". I would like to thank my family for their constant support and encouragement. This book could see the light of day due to generous support from the Astral International Pvt. Ltd. This volume is wholly a collective venture. I hope it will serve to increase understanding, awareness and appreciation of the diverse values of biodiversity, necessary to create the willingness to undertake the behavioural changes required to conserve and sustainably use biodiversity. This cause would not have been possible without the great efforts paid by all the authors and I am sure their valuable contributions increased the significance of the book. The book should leave its leverage on higher secondary, undergraduate, postgraduate students, researchers, conservation biologists, ecologists, wildlife managers, policy makers, entrepreneurs, wildlife activists etc. worldwide.

Dr. Surendra Singh

Contents

List of Contributors

Abhay Singh Yadav[1]* and Ankita Yadav[2]

[1]*Department of Zoology, Kurukshetra University, Kurukshetra-136119, Haryana, India*

[2]*University Law College, University of Rajasthan, Jaipur-302004, Rajasthan, India*

Abhay Singh Yadav* and Shikha Jaggi

Department of Zoology, Kurukshetra University, Kurukshetra-136119, Haryana, India

Anand Kar* and Lata Sunhre

School of Life Science, Devi Ahilya Vishwavidyalaya, Indore-452001, Madhya Pradesh, India

Anjana Pant

World wildlife fund, SATS Integrated Conservation Solutions Pvt. Ltd., 716 Imperial, Supertech Estate, Vaishali 9, Ghaziabad-201010, Uttar Pradesh, India

Ashok K. Jain

Department of Botany, Institute of Ethnobiology, Jiwaji University, Gwalior-473995, Madhya Pradesh, India

Bhumesh Kumar* and Disha Jaggi

Directorate of Weed Science Research (DWSR), Jabalpur-482004, Madhya Pradesh, India

Deepak Rahi

Department of Microbiology, Panjab University, Chandigarh-160014, India

Dolly Wattal Dhar*, N. K. Singh, D. B. Patel and B. G. Morad

Indian Agricultural Research Institute, New Delhi-110012, India

Gopal Shankar Singh*, Rinku, A. Sharma and S. K. Patel

Environment & Sustainable Development, Banaras Hindu University, Varanasi-221005, Uttar Pradesh, India

Harbansh Kaur Kehri* and Ovaid Akhtar

Department of Botany, University of Allahabad, Allahabad-211002, Uttar Pradesh, India

Hemant Rathore*, Dev Narayan Gautam, S. Marmat and Taj N. Qureshi
School of Studies in Zoology & Biotechnology, Vikram University,
Ujjain-456010, Madhya Pradesh, India

Himangini Singh
University Institute of Management and Commerce,
Rani Durgavati University, Jabalpur-482001, Madhya Pradesh, India

Jagdish Khobragade, P.P. Singh and Vikash Agrawal*
School of Law, Dr. Harisingh Gour Central University,
Sagar-470003, Madhya Pradesh, India

Manali Datta and S. L. Kothari*
Amity Institute of Biotechnology, Amity University,
Rajasthan, Jaipur-303007, Rajasthan, India

Meenakshi Banerjee
Biosciences Rice University, Houston, Texas-77005, United States

Nair Bindu Vijay
Gujarat National Law University, Koba,
Gandhinagar-382007, Gujarat, India

P. Girish Kumar* and P. M. Sureshan
Western Ghat Regional Centre, Zoological Survey of India,
Kozhikode-673004, Kerala, India

Pawan Kumar Bharti
Antarctica Laboratory, R&D Division,
Shriram Institute for Industrial Research, Delhi-110007, India

R. Ramanibai* K. Vikram and M. Gomathi Jeyam
Department of Zoology, University of Madras,
Guindy Campus, Chennai-600025, Tamiol Nadu, India

Sadhna Chaurasiya* and Ravindra Singh
Department of Energy and Environment,
Mahatma Gandhi Chitrakoot Gramodaya Vishwavidhyalaya,
Chitrakoot, Satna-485334, Madhya Pradesh, India

Shivesh Sharma*, Kanchan Vishwakarma, Neha Upadhyay, Nitin Kumar, Ashish Tiwari and Vivek Kumar
Department of Biotechnology, IPR Cell,
Motilal Nehru National Institute of Technology,
Allahabad-211004, Uttar Pradesh, India

Surendra Singh*, Ankita Kachhwaha and Sameer Choudhary
Algal Biotechnology Laboratory,
Department of Post Graduate Studies and Research in Biological Science,
Rani Durgavati University, Jabalpur-482001, Madhya Pradesh, India

1

Biodiversity: Leisure, Cultural and Aesthetic Values

Anand Kar* and Lata Sunhre

School of Life Sciences, Devi Ahilya University,
Takhshila Campus, Indore-452001, Madhya Pradesh, India
E-mail: karlife@rediffmail.com

Abstract

All living forms in the earth can be considered together as biodiversity. Importance of biodiversity is a globally debated issue. However, the number of biodiversity, particularly of wild types is gradually declining and is now a matter of concern. One of the factors for this is their depletion, primarily because of the non-realization of their short term and long term significance. Particularly, uneducated villagers and antisocial persons are not convinced on there importance. It is therefore imperative to make the common people aware on the significance of biodiversity around us. Although many realize the economic importance of plants and animals in the form of food, fodder, fuel and medicines, most do not realize their other aspects. Very often their non-use values are ignored. Present paper highlights the importance of biodiversity with respect to their leisure, cultural and aesthetic values. It is expected that the conservationists, ecologists, scientists and educationists will get some clues from this article to prepare innovative strategy for biodiversity management.

Key words: Biodiversity, importance, leisure value, cultural vale, aesthetic value

Introduction

Other than physical components including water, air and soil; earth also possesses biological components. Besides human being, the variety in plants, animals and other organisms on earth together can be called as biodiversity. The concept of biological diversity was used first by wildlife conservationist, Dasmann in 1968, but the term was actually adopted after 12 years (Soule, 1980). Biologists now define biodiversity as the "totality of genes, species, and ecosystems of a region" (Larsson, 2001). It is an essential component of nature and it ensures the survival of human species directly or indirectly by providing food, fuel, shelter, medicines and other resources to mankind. Biodiversity is not distributed evenly on the entire earth. In fact, the richness of biodiversity depends largely on the climatic conditions and other physical factors of the region. It is richest in the tropics. Irrespective of their distribution and size, all species have significance to mankind. Although this is being taught in all most all levels of education, man is yet busy in destroying the nature and the apparently looking useless plants and animals. Therefore, for their conservation, it is imperative to understand the real importance of biodiversity. Common people are aware of their economic, ecological & agricultural importance, but not on their non-use values (Chan *et al.*, 2012).The present article is an attempt to highlight the values of biodiversity with respect to their leisure, cultural and aesthetic properties. However, before we proceed, let us have an idea on their general significance.

General significance of biodiversity:

In true sense, all living beings on earth (including human) are interdependent and are closely linked for the balance and maintenance of the ecological systems. If one species is removed from the system, the cycle can break down, and the human life gets affected indirectly, at least in long run (Kar, 1992). Biodiversity also contributes significantly to both climate-change mitigation and adaptation. Tropical rainforests, in particular, provide many beneficial products in the form of natural food, medicines, fuel, fodder, pest control agents for agriculture and many vital needs of the man. Forests also provide natural vegetation cover which helps to create water catchments, regulates excessive water run off and protects us from the extreme drought and flood. Despite their great significance the population of biodiversity, particularly of wild lives is dwindling every year (status of different life forms have been mentioned in Table 1). One of the strategies to reverse the trend of destruction of biodiversity is to convince the common people on the importance of our biodiversity.

Table 1: Present status of different groups of biodiversity

Groups of biodiversity	Estimated no. of described species
Vertebrates	
Mammals	5,513
Birds	10,425
Reptiles	10,038
Amphibians	7,302
Fishes	32,900
Subtotal	*66,178*
Invertebrates	
Insects	1,000,000
Arachnids	102,248
Crustaceans	47,000
Molluscs	85,000

Groups of biodiversity	Estimated no. of described species
Corals	2,175
Velvet worms	165
Others	68,658
Subtotal	**1,305,246**
Plants	
Mosses	16,236
Gymnosperms	1,052
Fern and Alies	12,000
Green algae	4,242
Red algae	6,144
Flowering plants	268,000
Subtotal	**307,674**
Fungi and protists	
Brown algae	3,127
Lichen	17,000
Mushrooms	31,496
Subtotal	**51,623**
Total	**1,730,725**

Source: *www://cmsdocs.s3.amazonaws.com/summarystats/2014_3_Summary_Stats_Page_Documents/2014*

Leisure values of biodiversity:

An environment rich in biodiversity is usually beautiful and fascinating both to adults and to young individuals. Therefore, in the leisure time people love to visit garden / park / sanctuary / national park and zoo.

Biodiversity also inspires musicians, painters, sculptors, writers and other artists. Some cultures view themselves as an integral part of the natural world and so respect other living organisms. Popular activities such as gardening, fish keeping, and domesticating pet animals including dog, cat, guinea pig, white rat, rabbit, horse, duck, parrot, ornamental birds and specimen collection strongly depend on biodiversity. A recent paper examines and describes the welfare dimension of the recreational services of coastal ecosystems (Ghermandi, 2015). Gardening by most urban people is considered to be a relaxing activity for many people. Garden design is very often based on different themes including perennial, butterfly, wildlife, Japanese, water, tropical and shade gardens (Naydler, 2012). Other than plantation and gardening, animals are good source of leisure values, therefore need special mention.

For the entertainment value of animals, from ancient time, zoos have been regarded as suitable places in urban areas, where people go to enjoy the presence and behavior of the animals in their leisure time (Caroline, 1967; David, 2001). Some animals are used as source of entertainment. For example, dolphins, bear, dogs, elephants and monkeys doing tricks. Many people love circus, described as "one of the most brazen of entertainment forms" (Helen, 2000), where not only a special type of theatrical performance /acrobatics is exhibited by players, but also it involves performance of animals. People also often participate in many contests using animals, some times regarded as sports. For example, horse racing is regarded as a sport and also as an important source of entertainment. Interestingly, in Australia, the horse race contest is arranged on Melbourne Cup and the public regards the race as an important annual event. Another one is camel racing that also requires human riders. People find it entertaining to watch animals race competitively, whether they are trained, like horses, camels or dogs, or untrained, like cockroaches. Some people love fishing in river, pond and lake. Ecotourism has now become a major concept keeping in mind the leisure values of plants and animals. Private Protected Areas (PPAs) are often used for wildlife-based ecotourism as their primary means of generating business (Maciejewski and Kerley, 2014).

Fish keeping is a popular hobby concerned with keeping small fishes in home aquaria or garden ponds (Rani *et al*, 2014). In fact, ornamental fish production globally is a multibillion dollar industry. Although ornamental fish keeping was initially considered as one of the attractive hobbies, now it is also good source of revenue. Some common Cold-freshwater aquarium fishes are Gold Fish, Bloodfin Tetras & White Cloud; while some hot freshwater aquarium fishes are Danios, Black Molly, Black Skirt Tetra, Kuhli Loach, Platies, Swordtails, & Betta (please know more from: http:// home aquaria. com / best-freshwater-aquarium-fish/ # ixzz 3y4a6CTQ9).

Cultural values of biodiversity

Animals and cultural values

Since ages animals and plants have been given cultural importance. Sacred natural sites (SNS) are bright examples of bio-cultural landscapes protected for spiritual motives (Frascaroli *et al.*, 2015). These sites frequently host important biological values in areas of Asia and Africa and the local community takes care of the traditional resource management.

On animal worship, a good deal of literature is available in Wikipedia. To mention few, in Hinduism, the cow is regarded as a symbol of *ahimsa* (non-violence), mother goddess and bringer of good fortune and wealth (Claus, 2003). For this reason, cows are domesticated and worshiped.During Maker shankranti, some community also decorate the cattle and offer special pooja for them. Similarly in Thailand it is presumed that a white elephant may contain the soul of a dead person, perhaps a Buddha. It is normally not bought or sold and mourned like a human being at its death. In Surat, India unmarried Anāvil girls participate in a holiday referred to as Alunām (Naik, 1958) to honor the goddess Pārvatī, in which, a clay elephant is prepared (most likely to celebrate Pārvatī's creation of Ganesha from a paste of either turmeric or sandalwood) that is worshiped. Another animal, sheep, Amun, the god of Thebes, Egypt, was represented as ram-headed. His worshippers held the ram to be sacred. Egyptian ram-headed god was *Banebdjed*, a form of Osiris. Similarly in Greece, Italy and Egypt, the goat was worshipped in both goat form and phallic form possibly for developing increased child bearing capacity (Dorinda, 1988). Dogs too have a religious significance among the Hindus in Nepal and some parts of India. The dogs are worshipped as a part of a five-day Tihar festival that falls in/around November every year. In Hinduism, it is believed that the dog is a messenger of *Yama*, the god of death, and dogs guard the doors of Heaven. Socially, they are believed to be the protectors of our homes and lives. In order to please the dogs they expect to meet at Heaven's doors after death, so they would be allowed in Heaven. People mark the 14th day of the lunar cycle in November as Kukur-tihar, as known in Nepali language for the dog's day. In Hindu mythology, many animals are associated with God or Godesses. Few to mention, snake and bull with lord Shiva, lion/tiger with Durga, rat-Ganesha, peocock-Kartikeya, owl-Parvati and Swan-with Swaraswati.

Plants and cultural values

In addition to their ecological values, many plants have cultural significance. In fact, large old trees are part of a social realm and as such provide numerous social-cultural benefits to people (Blicharska and Mikusiński, 2014). In Indian Hindu culture, many plants including, Pipal, Baniyan, Goos berry, Mango and Banana are worshiped. Grasses and leaves as well as flowers of different plants are offered to God & Goddesses.for example, Tulsi (*Ocimum sanctum*) or Holy basil is a sacred plant in Hindu belief (Mondal, 2009). Hindus regard it as an earthly manifestation of the goddess Tulsi, a consort of the god Vishnu. The offering of its leaves is mandatory in ritualistic worship of Vishnu and his form, Krishna. The dried root of turmeric is ground into a powder which is used on several religious occasions in Hinduism. In fact, India is the largest producer of turmeric in the world. The Indian Lotus, also known as the Sacred Lotus, is a culturally significant plant in many Asian cultures in general and Indian culture in particular. It is a plant native to the Indian subcontinent, but now is found as an ornamental plant world wide. It is a symbol of purity and innocence. It is the seat of Goddess Lakshmi, Lord Vishnu and Lord Brahma, of course most significantly associated with the latter. Lotus flowers are also used as offerings in Indian temples. In fact, many plants are known to have cultural significance. Few have been mentioned in Table 2.

Aesthetic values of biodiversity

"A thing of beauty is the joy forever", as said by famous poet John Keats primarily refers to the beauty of biodiversity in nature. Everyone admires the prettiness and freshness of the plants, flowers, butterflies and animals including birds present in the nature. In fact, thousands of plant species are cultivated for their aesthetic purposes as well as to provide shade, modify temperatures, reduce wind, abate noise and provide privacy. Plants are considered as the basis of the tourism industry, which includes travel to historic gardens, national parks, forests /rain forests, sanctuaries and islands.

Table 2: Some common plants of India with religious significance

Species name	In Hindi	Religious significance
Mimusops elengi (Spanish cherry)	Bakul, Maulsari	Lord Krishna played flute underneath a Bakul tree in Vrindavan.Its flowers are offered in his worship..
Jasminum sambac (jasmine)	Bela, Mogra	Garlands made from Jasmine flowers are adorned by women during special occasions.
Nelumbo nucifera (Lotus,)	Kamal	It is the seat of Goddess Saraswati, Goddess Lakshmi, Lord Vishnu and Lord Brahma. Also offered in Indian temples.
Phyllanthus emblica (Indian Gooseberry)	Amla	It is associated with Amala Ekadasi, celebrated on 11th day of Phalguna sukla; used in the worship of Radha-Krsna.
Saraca indica (Ashoka)	Ashok	Worshiped as it has a symbolic importance in Buddhism. Buddha was born under Ashoka tree.
Santalum album (Indian Sandalwood)	Chandan	Chandan pastes are applied to the forehead of Gods and of the worshipers. Sandalwood oil is also used in incense sticks
Ocimum tenuiflorum (Holy Basil)	Tulsi	Many Hindus grow in /near their home, often in special Tulsi pots. It is also offered to Vishnu God.
Curcuma longa (Common Turmeric)	Haldi	Turmeric paste is applied to the body of the bride to cleanse the body and make it more radiant. Used by Hindus in pooja.
Magnifera indica (Mango)	Amra / Amm	On auspicious occasions, mango leaves are tied to a string and hung on doors
Ficus benghalensis (Indian banyan)	Bargad	In Savitri Vrata, married women worship by circumambulating, tying with the sacred protective thread.
Azadirachata indica (Neem tree)	Neem	Goddess Sitala who is said to reside in the Neem tree.
Aegle marmelos (Bael fruit)	Bael	Leaves are offered to Lord Siva.
Musa balbisiana (Banana tree)	Kela	It is offered to Lord Visnu and Laksmi and to the Sun God.

Sources:http://www.biodiversityofindia.org/index.php?title=Plants_of_religious_ significanc and http://www.esamskriti.com/essay-chapters/Sacred-Trees-of-the-Hindus-1. aspx

In private outdoor gardens, lawn grasses, shade trees, ornamental trees, shrubs, vines, herbs and bedding plants are used. In gardens plants are cultivated in the naturalistic state. In fact, gardening is the most popular leisure activity in India and abroad. Very often, plants are grown or kept indoors as house plants or in specialized structures such as green-houses that are designed for the care and cultivation of desired plants. Venus Fly trap, sensitive plant and resurrection plant are examples of plants sold as novelties. There are also art forms specializing in the arrangement of cut or living plant, such as bonsai, ikebana, and the arrangement of cut or dried flowers. Images of plants are often used in painting and photography, as well as on textiles, stamps, flags and coats of arms.

It is generally seen that people are more productive and less stressed in area with good plantation than those who work in environment with no plants. A study of Washington State University demonstrates that plant transpiration in office environment releases moisture, creating a humidity level exactly matching the recommended human comfort range of 30-60% (http://www.plantcultureinc.com/whygreen.html). It is now well understood that, indoor plants offer a guarantee of positively enhancing perception and contributing to overcome "Sick Building syndrome" that concentrates toxins inside sealed office buildings (http://news.softpedia.com/news).

For many years Dr. Billy C. Wolverton and his aids in the Environmental Research Laboratory of John C. Stennis Space center conducted innovative research employing natural biological processes for air purification. They found plants to suck these chemicals out of the air and postulated that plants can act as the lungs and kidneys of the buildings. They clean contaminated office air in two ways, by absorbing office pollutants into their leaves and by transmitting the toxins to their roots, where they are transformed into a source of food for the plant (Wolverton and Takenaka, *2010)*

Natural landscapes at undisturbed places are a delight to watch and also provide opportunities for recreational activities like picnic, animal /bird watching, photography etc. Enriching the zoological and botanical gardens promotes eco-tourism which further generates revenue. In addition national parks and sanctuaries attract visitors for the beauty of plants and animals they have (http://www.allsubjects4you.com/value-of-biodiversity.htm).

Ferns, the flowerless plants are also of great aesthetic value due to their grace and delicate beauty and are cultivated as ornamental plants (Mannan *et al.*, 2008). They can be grown on ground or in pots, on tree trunks or in hanging baskets. Several species of *Lycopodium Linn* are used in the decoration. Mostly these are used in Christmas wreaths and are popularly known as "Christmas green". The leaves of *Selaginella serpens*, which in the morning are bright green in color and during the day they gradually become paler and in the night they again resume their green color. In fact many plants can be grown as ornamentals in the gardens in ground or as epiphytes or as in baskets for indoor decoration.

Every body knows the magic of flowers, one of the most beautiful things in the nature that provide grace to a place and spread fragrance. They are pleasing to the eyes, refresh our body and mind and cheer up the emotions. A living room has special attraction and soothing effect that is beautified with fresh flowers. A bunch of untied flowers is commonly given as gift, bouquet in diff occasion such as meetings, celebrations, marriage, and other occasions. Even the use of single flower to cherish mood and feel affection of love can make the place a centre of attraction. A lonely daisy or some other flower may be used to reveal fondness.In valentine day, single rose plays a major role. *Valley of Flowers,* an Indian national park, located in West Himalaya, in the state of Uttarakhand attracts lots of tourists. Ggarlands of flowers are also offered to and worn by loved ones. A **gajra** is a flower garland which women in South Asia wear during festive occasions, weddings or as part of everyday traditional attire. They are made usually made of jasmine flowers, but rose, crossandra and barleria are also widely used in gajras (*Randhawa and Mukhopadhyay, 1986*).

Practitioners of Buddhism, Taoism, and Confucianism place flowers on their altars, a practice which dates back to 618-906 CE. They created paintings, carvings, and embroidered items with depictions of flowers. The paintings can be found on vases, plates, scrolls and silk, while carvings were done on wood, bronze, jade and ivory.

Egyptians used to decorate with flowers as early as 2,500 BCE. They regularly placed cut flowers during burials, for processions, and simply as table decorations. Illustrations of arranged flowers have been found on Egyptian carved stone relieves and painted wall decorations.

Interestingly, flowers were selected according to symbolic meaning, with emphasis on religious significance. The lotus flower or water lily, for example, was considered sacred to Isis and was often included in arrangements. Many other flowers have been found in the tombs of the ancient Egyptians, and garlands of flowers were worn by loved ones and left at the tombs.

Henna has been used as a cosmetic hair dye for many years, not only in India,but also in other parts of the world including North Africa, the Arabian Peninsula, the Near East and South Asia. Commercially packaged henna, intended for use as a cosmetic hair dye, is now available in many countries, such as Middle East, Europe, Australia, Canada and the United States. (https://en.wikipedia.org/wiki/Henna#History).

Aesthetic values of animals

Animals and animal parts have been considered as source of aesthetic values from ages as the beauty of organisms such as butterflies, birds, ornamental /reef fishes, and some animals has been widely recognized from time to time. In palaces very often parts of the wild animals (say tusk, antlers/horns, skin, feathers etc.) are kept for decoration. Now a days beautiful ornaments, usable articles such as wallet, bag, belt, bangles, ear rings etc. are made from skin, bones, scales & teeth of animals. This is the reason that some wild animals are also killed. However, we should realize that we can afford these items only when we have excess number of the animals.

In addition, beauty of the zoos & national parks largely depends on the animals they posses. In fact, presence of some beautiful/ colorful birds is the source of attraction of the tourists. People also keep them in their houses as pet birds. Even many less well-known fauna, when viewed objectively appear attractive. For example, some tiny wasps and flies, if seen under a microscope, appear to be beautiful. Their diversity of form and behavior is a potential source of infinite fascination in the world. The insects are probably the least appreciated as aesthetic resources. But, some are beautiful and can be kept as ornamental species.

Many marine animals are enjoyed just by watching them. People love to watch whale, dolphin, shark, jelly fishes and other sea animals in their natural environment. The natural beauty of coral reefs is known to attract millions of tourists worldwide resulting in substantial revenues (Haas *et al.,* 2015).Because of their beautiful body color, many small marine fishes are also kept in drawing room and in zoos, which are always appreciated by children, young and adults.

Conclusion

Biodiversity enriches leisure activities such as natural history study, hiking, bird watching, fishing and weekend travel /long drive. It also inspires musicians, painters, sculptors, writers and other artists. In fact, biodiversity is an extremely important part of life on Earth. It should not be considered only as the variety of living organisms on our planet, but also as the factors that bind all the organisms directly or indirectly including humans. Particularly their lesser known importance such as cultural, aesthetic & leisure values offer monetary values also for the leisure and recreation industry. Moreover, men preserve & conserve those, whom they understand as of great values. Therefore, not only biodiversity management people but also common people must be aware of their importance. We must have vision of a world in which humankind values, protects and preserves the biodiversity on earth.

References

1. Blicharska, M. and Mikusiński, G. (2014). Incorporating social and cultural significance of large old trees in conservation policy. *Conserv Biol.* 28(6): 1558-1567.

2. Caroline, J (1967). The Value of Zoos for Science and Conservation. *Oryx.* 9(2): 127-136.

3. Chan, K.M.A.http://www.sciencedirect.com/science/article/pii/ S0921800911004927 - cr0005, Satterfield, T. and Goldstein, J. (2012). Rethinking ecosystem services to better address and navigate cultural values. *Ecological Economics.* 74(2): 8-18.

4. Clause, P.J., Sarah, D. and Margaret, A.M. (2003). South Asian folklore. Taylor & Francis.

5. Dasmann, R.F. (1968). *A different kind of country.* New York, MacMillan.

6. David, H. (2001). A different nature: the paradoxical world of zoos and their uncertain future. Berkeley. University of California Press.

7. Dorinda, N. (1988). The Witch in Early 16th-Century German Art. *Woman's Art Journal.* 9(1): 3-9.

8. Frascaroli, F., Bhagwat, S., Guarino, R., Chiarucci, A. and Schmid, B. (2015). Shrines in Central Italy conserve plant diversity and large trees. Ambio. *[In press].*

9. Ghermandi, A. (2015). Benefits of coastal recreation in Europe: identifying trade-offs and priority regions for sustainable management. *J Environ Manage.* 152: 218-229.

10. Haas, A.F., Guibert, M., Foerschner, A., Co, T., Calhoun, S., George,E., Hatay, M., Dinsdale, E., Sandin, S.A., Smith, J.E., Vermeij, J., Felts,B., Dustan, P., Salamon, P. and Rohwer, F. (2015). Can we measure beauty? Computational evaluation of coral reef aesthetics. *PeerJ.* 3: e1390.

11. Helen, S. (2000). Rings of Desire: Circus History and Representation. Manchester & New York, Manchester University Press. 13.

12. Kar, A. (1992). Reasonable measures to correct and reverse the existing damage and to prevent the new destruction to wild life. In *Environment, Yesterday, Today & Tomorrow.* In P. Bhatt & S. Bhatt (Eds), N. Delhi: Galgotia publication. 258-262.

13. Larsson, T.B. (2001). *Biodiversity evaluation tools for European forests.* New York USA. Wiley-Blackwell. 178.

14. Maciejewski, K. and Kerley, G.I. (2014). Understanding tourists' preference for mammal species in private protected areas: is there a case for extralimital species for ecotourism? *PLoS One.* 9(2): e88192.

15. Mannan, M.M., Maridass, M. and Victor, B. (2008). A Review on the Potential Uses of Ferns. *Ethno-botanical Leaflets.* 12: 281-285.

16. Mondal, S., Mirdha, B.R. and Mahapatra, S.C. (2009). The science behind sacredness of Tulsi (*Ocimum sanctum* Linn.). *Indian J Physiol Pharmacol.* 53(4): 291-306.

17. Naik, T.B. (1958). Religion of the Anāvils of Surat. *The Journal of American Folklore,* 71(281).

18. Naydler, J. (2012). Gardening as a Sacred Art. *London: Floris Books.*

19. Randhawa, G.S. and Mukhopadhyay, A. (1986). Floriculture in India. Allied Publishers. 607.

20. Rani, P., Immanuel, S. and Kumar, N.R. (2014). Ornamental Fish Exports from India: Performance, Competitiveness and Determinants. *Internat J of Fishries and Aquatic Studies (IJFAS).* 1(4): 85-92.

21. Soulé, M.E. and Wilcox, B.A. (1980). Conservation biology: an evolutionary-ecological perspective. *Sunderland Massachusetts: Sinauer.*

22. Wolverton, B.C. and Takenaka, K. (2010). Plants: Why You Can't Live Without Them. *Roli Books (New Delhi).*

2

Taxonomic Review on the Potter Wasp Genus *Pseumenes* Giordani Soika (Hymenoptera: Vespidae: Eumeninae) from India

P. Girish Kumar and P.M. Sureshan

Western Ghat Regional Centre, Zoological Survey of India,
Kozhikode-673006, Kerala, India
E-mail: kpgiris@gmail.com

Abstract

Potter wasp genus *Pseumenes* Giordani Soika, 1935, is reviewed from India. A new synonymy is proposed, i. e., the subspecies *P. depressus annulatus* van der Vecht under, 1963, is synonymized under the nominate subspecies *P. depressus depressus* (de Saussure, 1855). *P. depressus depressus* is recorded for the first time from Kerala and Maharashtra. An updated checklist of Oriental species is provided.

Keywords: Hymenoptera, Vespidae, Eumeninae, *Pseumenes*, new synonymy, new record, checklist, India, Oriental Region.

Introduction

`Giordani Soika (1935) described *Pseumenes* as a subgenus of the genus *Pareumenes* de Saussure based on the type species *Eumenes eximius* Smith, 1861. Van der Vecht (1963) raised it to the generic rank. This genus is distributed in the Australian, Oriental and Palearctic Regions of the world. Five species with additional seven subspecies are recorded from the Oriental Region till date of which only a single species, namely, *Pseumenes depressus* (de Saussure, 1855) is recorded from the Indian subcontinent with two subspecies, *viz., P. depressus depressus* (de Saussure, 1855) and *P. depressus annulatus* van der Vecht, 1963. In this paper, we synonymize the subspecies *P. depressus annulatus* van der Vecht, 1963, under the nominate subspecies *P. depressus depressus* (de Saussure, 1855) after studying more specimens from different localities. This species is recording here for the first time from Kerala and Maharashtra. An updated checklist of Oriental species is also provided.

Materials and Methods

The specimens examined under LEICA M60 stereozoom microscope and images captured with the camera model LEICA DFC-450. The studied specimen is added to the 'National Zoological Collections' of the Western Ghat Regional Centre, Zoological Survey of India, Kozhikode (= Calicut), India (ZSIK).

Abbreviations used for the Museums: BMNH—British Museum (Natural History), London, England; MP—Muséum National d'Histoire Naturelle, Paris, France; ZSIK—Western Ghat Regional Centre, Zoological Survey of India, Kozhikode (= Calicut), India.

Abbreviations used for the terms: H = Head; M = Mesosoma; OOL = Ocellocular Length; POL = Posterior Ocellar Length; S = Metasomal sterna; T = Metasomal terga.

Results

Genus *Pseumenes* Giordani Soika, 1935

Pseumenes Giordani Soika, 1935: 145, subgenus of *Pareumenes* de Saussure. Type species: *Eumenes eximius* Smith, 1861, by original designation.

Diagnosis: Forewing with prestigma shorter than pterostigma; female without cephalic fovea; mesoscutum without prescutal grooves; mesepisternum without epicnemial carina; propodeum dorsally with elongate fovea from which carina runs to propodeal orifice, with dentiform projections above propodeal valvulae; propodeal orifice rounded dorsally; axillary fossa narrower than long, slit-like; tegula with narrow posterior lobe which about equal parategula posteriorly; second submarginal cell acute basally; midtibia with one spur.

Distribution: Australian, Oriental and Palearctic Regions.

Pseumenes depressus depressus (de Saussure, 1855) (Figure 1-12)

Eumenes depressus de Saussure, 1855: 135, female (in division *Pareumenes*)-"Les Indes Orientales" (MP).-Smith, 1857: 23 (*depressa*; cat.); 1871: 372 (*depressa*; cat.).-Dalla Torre, 1894: 22 (*depressa*; cat.).-Bingham, 1897: 334 (key), 337 (*depressa*; male; Tenasserim).-Tosawa; 1934: 4 (key), 6, pl. 1 fig. 3 (*depressa*; Taiwan).

Pareumenes depressus; Dalla Torre, 1904: 19 (cat.).-Dover, 1929: 44 (Sarawak, Borneo).-Liu, 1936: 102 (cat.).-van der Vecht, 1937: 273, fig. 3b (in subgenus *Pseumenes*; notes; distribution).-Sonan, 1938: 77 (*Pareumenes* [!]).-Liu, 1941: 284 (notes on cotypes unavailable).

Pseumenes depressus depressus; van der Vecht, 1963: 21, 25 (misidentified by Piel, 1935, and Liu, 1941; no localities in Palaearctic region; India: West Bengal, Sikkim; Thailand; Malaya; Indo-China; Taiwan).-van der Vecht and Fischer, 1972: 124 (cat.; ? Palaearctic Region).-Gusenleitner, 2006: 695 (India: Meghalaya).

Pseumenes depressus annulatus van der Vecht, 1963: 26, female-"Southern India: Bolampati Valley, Coimbatore District" (BMNH). **Syn. nov.**

Diagnosis: *Female*: Clypeus convex, broadly subpyriform and deeply emarginate at apex; front closely punctured; POL 1.31x OOL; prestigma 0.55x pterostigma; propodeum smooth except at lateral sides with scattered punctures; apex of T1 with a median longitudinal groove.

Colour: Female: Body black with the following yellow markings: a spot at base of mandible; clypeus; a vertical streak above it; scape below; ocular sinus; a line on the inner orbits not reaching the vertex; temple (sometimes extends towards vertex); vertex (sometimes back median area); pronotum (sometimes black posteriorly); two hook-shaped marks back to back on mesoscutum; two large marks on mesopleuron; tegula except a small brown spot at middle; parategula; square spot at each side of scutellum; spots at metanotum (sometimes absent); either side of propodeum with a black at the middle of yellow area (sometimes yellow spot absent); two elongate streak on T1 above, a medially incised band at apex which extends through lateral margin and sometimes fused with the dorsal spot; two large marks at basal side of T2, a medially thin band at apex; apical margins T3-T5; sometimes a median spot at T6; large mark on S2. Legs black variegated with yellow. Wings fuscohyaline with a fulvous tinge.

Size (H+M+T1+T2): *Female*, 17-20 mm.

Material examined: INDIA: **Kerala**, Palakkad district, Walayar forest, 1♀, 1957, Coll. P. S. Nathan, ZSI/WGRS/I.R-INV.5015; **Maharashtra**, Kolhapur district, Talewadi, 1♀, 3-10.x.1916, Coll. S. Kemp, ZSI/WGRS/I.R-INV.5014; **Sikkim**, exact collection locality and date of collection unknown, 1♀, Coll. Niceville, ZSI/WGRS/ I.R-INV.5013; **West Bengal**, exact collection locality, date of collection and name of collector unknown, 1♀, ZSI/WGRS/I.R-INV.5012.

Synonymy of Pseumenes depressus annulatus **van der Vecht, 1963, under the nominate subspecies** *P. depressus depressus* **(de Saussure, 1855)**: Van der Vecht (1963) described a new subspecies, namely, *Pseumenes depressus annulatus* van der Vecht, 1963, based on a single female specimen from Tamil Nadu (Southern India) and provided its colour differences from the nominate subspecies *P. depressus depressus* (de Saussure, 1855) (van der Vecht, 1963: page 26). This subspecies differs from the nominate subspecies in having more extensive yellow markings. In the present study, we examined female specimens from four different localities of India, such as, Sikkim, West Bengal, Maharashtra and Kerala. The colour pattern differs as follows:

Colour pattern	*P. depressus depressus*	Sikkim specimen	West Bengal specimen	Maharashtra specimen	Kerala specimen	*P. depressus annulatus*
Median area of vertex	Entirely black without yellow spots.	Entirely black without yellow spots (Fig. 7).	Black with two yellow spots (Fig. 10).	Almost entirely yellow with two narrow black stripes (Fig. 12).	Almost entirely yellow with two narrow black stripes (Fig. 3).	Almost entirely yellow
Dorsal area of pronotum	Not entirely yellow; posteriorly black.	Not entirely yellow; posteriorly black (Fig. 7).	Not entirely yellow; posteriorly black (Fig. 10).	Almost entirely yellow (Fig. 12).	Almost entirely yellow (Fig. 3).	Almost entirely yellow

Colour pattern	*P. depressus depressus*	Sikkim specimen	West Bengal specimen	Maharashtra specimen	Kerala specimen	*P. depressus annulatus*
The recurved yellow lines on the mesoscutum extending to the parategula	With a large gap.	With a large gap (Fig. 7).	With a large gap (Fig. 10).	With a small gap (Fig. 12).	Without any gap (Fig. 3).	Without any gap.
Dorsal area of propodeum yellow on each side of the median concavity,	With black spot.	With black spot (Fig. 9).	With black spot.	With black spot.	With black spot.	Without black spot.
Spots on petiole	Not connected with the apical band.	Not connected with the apical band.	Not connected with the apical band.	Not connected with the apical band.	Not connected with the apical band.	Connected with the apical band.

From this, it is very clear that these differences of colour pattern in different specimens are only colour variations in different population. So, we herewith synonymize the subspecies *P. depressus annulatus* van der Vecht, 1963, under the nominate subspecies *P. depressus depressus* (de Saussure, 1855).

Distribution: India: Kerala (**new record**), Maharashtra (**new record**), Meghalaya, Sikkim, Tamil Nadu, West Bengal; China; Myanmar; Thailand; Malaysia; Vietnam; Taiwan.

Checklist of Oriental species of the genus *Pseumenes* Giordani Soika, 1935

(1a). *P. depressus depressus (de Saussure, 1855)—India: Kerala (new record), Maharashtra (new record), Meghalaya, Sikkim, Tamil Nadu, West Bengal; China; Myanmar; Thailand; Malaysia; Vietnam; Taiwan.*

(1b). *P. depressus hamanni van der Vecht, 1963—Indonesia: Sulawesi.*

(1c). *P. depressus insignis van der Vecht, 1963—Indonesia: Sumba.*

(1d). *P. depressus palawanensis Giordani Soika, 1993—Philippines.*

(1e). *P. depressus pictifrons (Smith, 1861)—Indonesia: Sulawesi.*

(1f). *P. depressus thoracicus (van der Vecht, 1937)—Indonesia: Java, Bali.*

(2a). *P. eximius arcuatoides van der Vecht, 1963—Indonesia: Moluccas, Papua.*

(2b). *P. eximius eximius (Smith, 1861)—Indonesia: Moluccas.*

(3).   *P. imperatrix (Smith, 1857)—China.*

(4).   *P. laboriosus (Smith, 1861)—Indonesia: Sulawesi.*

(5).   *P. polillensis (Giordani Soika, 1941)—Philippines.*

Acknowledgement

The authors are grateful to Dr. Kailash Chandra, Director-in-Charge, Zoological Survey of India, Kolkata, for providing facilities and encouragements.

References

1. Bingham, C.T. (1897). *The Fauna of British India, including Ceylon and Burma, Hymenoptera, I. Wasps and Bees*.pp. 1-579+ i-xxix. London, Taylor & Francis.

2. Dalla Torre, K.W.Von. (1904). *Vespidae, Genera Insectorum*. 19: 1-108.

3. Dalla Torre, K.W.Von. (1894). *Catalogus Hymenopterorum* 9, Vespidae (Diploptera). 1-181. Leipzig.

4. Dover, C. (1929). Wasps and bees in the Raffles Museum, Singapore. *Bulletin of the Raffles Museum*. 2: 43-70.

5. Giordani Soika, A. (1935). Richerche sistematiche sugli *Eumenes y Pareumenes* dell'-Archipelago Malese e della Nova Guinea. *Annali del Museo civico di storia naturale di Genova*. 57: 114-151.

6. Gusenleitner, J. (2006). Uber Aufsammlungen von Faltenwespen in Indien (Hymenoptera, Vespidae). *Linzer Biologische Beitrage*. 38(1): 677-695.

7. Liu, C.L. (1936). A bibliographic and synonymic catalogue of the Vespoidea of China, with a cross-referring index for the genera and species (1). *Peking Natural History Bulletin*. 11: 91-114.

8. Liu, C.L. (1941). Revisional studies of the Vespidae of China. I. The genus *Pareumenes* Saussure, with description of six new species. *Notes d-Entomologie Chinoise*. 8: 245-289.

9. Piel, O. (1935). Biologie de *Pareumenes quadrispinosus* Saussure (Hymènoptéres Vespides) et de ses parasites, en particulier: *Calosota chinensis* Ferriere. *Notes d-Entomologie Chinoise*. 2: 105-139.

10. Saussure, H.De. (1854-1856). *Études sur la famille des vespides. Toisième partie comprenant la Monographie des Masariens et un supplement a la Monographie des Euméniens*. V. Masson, Paris & J. Kessmann& J. Cherbuliez, Genève, 352 pp. + 15 pls. (1854) 1-48 + pl. 1-5; (1855) 49-288 + pl. 6-14; (1856) 289-352 + pl. 15, 16. [Dates of publication after Griffin 1939].

11. Smith, F. (1857). Catalogue of Hymenopterous insects in the collection of the British Museum. 5: 1-147.

12. Smith, F. (1871). A catalogue of the Aculeate Hymenoptera and Ichneumonidae of India and eastern Archipelago, with introductary remarks by A.R. Wallace. *Journal of the Proceedings of the Linnean Society of London, Zoology*. 11: 285-415.

13. Sonan, J. (1938). Notes on the Vespoidea in Japan (Hymenoptera). *Transactions of the Natural History Society Formosa*. 28: 77-81.

14. Tosawa, N. (1934). On *Eumenes* of Japan Empire. *Transactions of the Kansai Entomological Society*. 5: 3-16.

15. Vecht, J.Van.Der. (1963). Studies on Indo-Australian and East Asiatic Eumenidae (Hymenoptera: Vespoidea). *Zoologische Verhandelingen Leiden*. 60: 1-116.

16. Vecht, J.Van.Der and Fischer, F.C.J. (1972). Palearctic Eumenidae. *Hymenopterum Catalogus (new edition)*, 8: i-v+199.

17. Vecht, J.Van.Der. (1937). Descriptions and records of Oriental and Papuan solitary Vespidae. *Treubia*. 16: 261-293.

Plate I

Figure 1-6: *Pseumenes depressus depressus*(de Saussure) ♀ Walayar specimen. 1. Body profile; 2. Head frontal view; 3. Head and mesosoma dorsal view; 4. Mesosoma lateral view; 5. Metasoma dorsal view; 6. Ventral side of petiole.

Plate II

7

8

9

10

11

12

Figure 7-12: *Pseumenes depressus depressus* (de Saussure) ♀. 7-9 Sikkim specimen. 7. Head and mesosoma dorsal view; 8. Apical half of forewing; 9. Propodeum. 10-11 West Bengal specimen. 10. Head and mesosoma dorsal view; 11. Metasoma dorsal view. 12 Talewadi specimen Head and mesosoma dorsal view.

3

Studies on Water Quality Status and Plankton Diversity of Krishnagiri Reservoir, Tamilnadu, India

R. Ramanibai*, K. Vikram and M. Gomathi Jeyam

Unit of Aquatic Biodiversity,
Department of Zoology, University of Madras,
Guindy Campus, Chennai-600025, Tamil Nadu, India
E-mail: rramani8@hotmail.com

Abstract

Reservoirs are the important freshwater sources for mankind. Zooplankton community constitutes an important component in the faunal composition of the water body. They are sensitive indicators of pollution in comparison with phytoplankton. The present investigation was carried out in Krishnagiri reservoir (KRP). The physico-chemical parameters and zooplankton diversity were studied for a period of January and May, 2009. The phytoplankton were identified and counted by using Heamocytometer where as counting of zooplankton community were studied by using sedgewick rafter counting chamber. Water quality parameters were also analyzed using standard methods. Four groups of phytoplankton, Cyanophycea, Cholorophycea, Bacilloriphyceae, Dinophyceae were dominant as compared to other groups of algae. Zooplankton was five groups were recorded; dominant rotifer species were *Keratella* sp. and *Branchionous* sp. The uppermost concentration of phosphate showed the eutrophication nature of KRP reservoir. Hence, it is necessary to focus on conservation, monitoring and management strategies in places where there is a synergy between richness and diversity of planktons.

Keywords: Water quality, Plankton, Reservoir

Introduction

Planktons respond quickly to environmental changes and are considered good indicators of water quality and trophic environment because of their short making time and fast population regeneration. Unlike chemical analyses, biological monitoring is not limited to urgent conditions within the environment or single contaminants, but integrates information about past disturbances and the effects of multiple factors. The occurrence of planktonic organisms under natural conditions is related to the patience range (biological most advantageous) dependent on abiotic environmental factors (temperature, oxygen concentration and pH) as well as on the biotic interactions among organisms. In the multidimensional space (ecological niche), the occurrence of organisms is affected by numerous environmental factors, both anthropogenic and nonanthropogenic (Lampert and Sommer, 2001).

In India, recently the diversity of phytoplankton in different freshwater wetlands along with their physicochemical characteristics was studied (Tiwari and Shukla, 2007, Senthil kumar and Das, 2008). Though there are more than 3513 inland freshwater bodies including a good number of temple ponds in North Eastern Region of India, a little work have so far, been done on algal diversity of those water bodies (Baruah and Kakati, 2009).

The benefit of freshwater innate reservoirs, which can be used to the probable benefit of mankind, cannot be overemphasized (Mahapatra *et al.*, 2012). A least amount of water is required for utilization on a daily basis for endurance and therefore access to some form of water is essential to life (WHO, 2004). Changes in the aquatic environment associated anthropogenic contamination are a cause of upward concern and require monitoring of the plane waters and the organisms inhabiting them (Vandysh, 2004). The trophic state of a lake is usually assessed by commonly monitoring a variety of physical and chemical indicators. If changes in species diversity and population abundance result from either direct or indirect environmental stressors, then changes in biota may be used to elucidate changes in the environment. In this context, indicator species are those which, by their presence or plenty, provide some indication of the widespread environmental conditions (Thakur *et al.*, 2013).

Krishnagiri reservoir is the important fresh water source for mankind. Indian reservoir by large, have wide array representive of biotic communities. The reservoir is being used for multipurpose utility such as irrigation, fishing and washing. The main sources for nutrients are from rivers, soil erosion etc and improvement of organic material from sediments were significant source for the lakes water eutrophication (Gold *et al.*, 2002). The aim of the present exploration was diversity and distribution of plankton may provide the basis for determining the Krishnagiri reservoir. Based on the presence, absence and the abundance of different organisms observed during the present investigation, bioindicators of the trophic status of reservoir have been indomitable. More prominence is given on indicator species than the physico-chemical parameters.

Materials and Methods

Study area

Krishnagiri Reservoir is located in Krishnagiri, Dharmapuri district of Tamil Nadu at the latitude of 12°28′ N on the longitude of 78°11′ E. The KRP was constructed across the Ponnaiyar River (also called as Ponniyar) near Periyamuttur village about 10 km from Krishnagiri town. The KRP has two canals one on the left side and the other on the right side of the reservoir, running parallel to Ponnaiyar river (Figure 1).

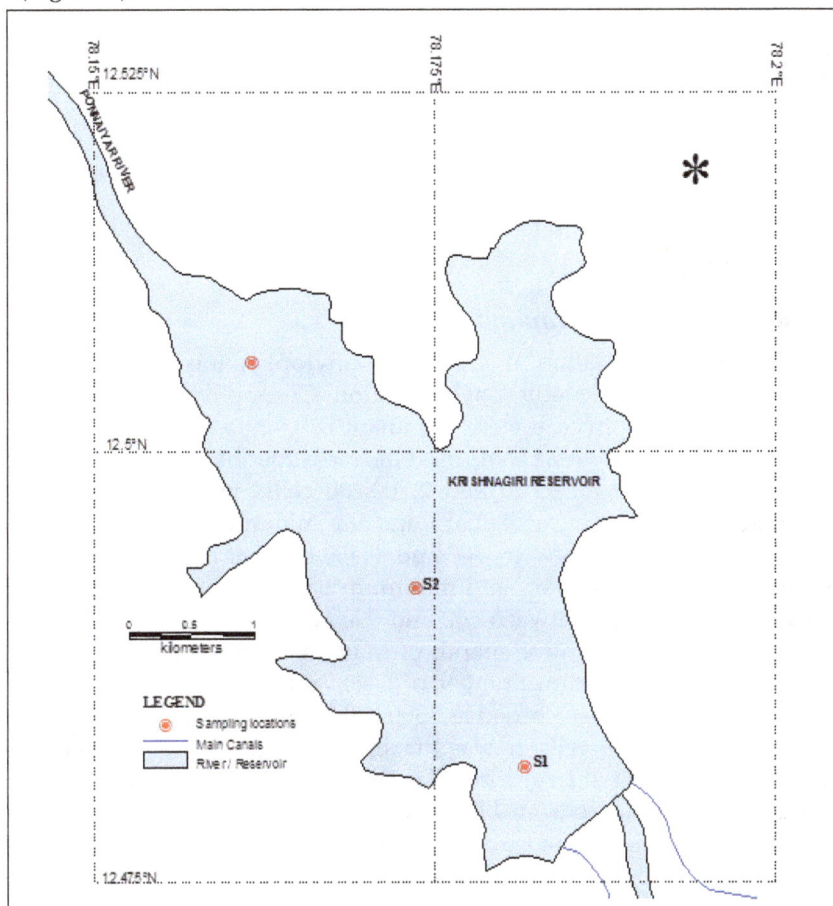

Figure 1: Study area

Sampling protocol

Plankton samples were collected between January and May, 2009 from inflow and outflow of KRP. All of them were narcotized using chloroform and preserved with 4% formalin. Plankton was enumerated using a Sedgwick-Rafter counting chamber under the compound microscope at 40-100X magnification. Identification was carried out up to species level where ever possible. Surface water samples were collected in clean polypropylene containers of 1 liter capacity and transported to the laboratory for the analysis of Physico-Chemical and biological parameters.

Plankton counting

The enumeration of plankton samples were done by Sedgewick Rafter Cell Counter. The values were expressed as cells /m^3 for phytoplankton and individuals /m^3 for zooplankton (Santhanam, 1989).

Formula:

$$N = \frac{n \, \acute{n} \, v \, \acute{n} \, 1000}{V}$$

Where,

N = Number of organisms or individuals or cells/m^3.

n = 1 ml of sample.

v = volume of sample.

V = volume of sample filtered.

Results and Discussion

Species diversity and phytoplankton

The present investigation, 4 groups of phytoplankton was verified from study areas. In present study phytoplankton Cyanophycea, Cholorophycea, Bacilloriphyceae, Dinophyceae were dominant as compared to other groups of algae (Figure 2). Cyanophyceae comprised the most abundant group of *Microcystis aeruginosa* ranged between 9,000 and 2, 00,200 cells/ m^3. It was found to be maximum at location 3 during January and the minimum at L2 during March. *Chrocoocus sp* was ranged between 300 and 88400 cells/m^3 and it was found to be maximum at L2 during January and minimum was recorded at L2 during May. The *Oscillatoria sp* ranged between 600 and 9900 cells/m^3 and it was found to be maximum in L2 of April. In the month of May, the minimum was found in L2. Whereas in L3, the maximum number of *Gomphosphaeria aponina* (34,400 cells / m^3) and the minimum was recorded as 4300 cells / m^3. *Merismopedia glauca* ranged between 1200 and 88400 cells/m^3 was maximum at L2 during January and the minimum was recorded at L2 during May 09. *Spirulina meneghiniana* was ranged between 2000 and 19,500 cells/ m^3. The maximum was found at L1 during March. The minimum was recorded at L4 during January. *Gleocapsa nigreseens* was ranged between 100 and 1600 cells / m^3. The maximum was recorded at L4 during March and the minimum was recorded at L3 during May. *Phormidium subpuscum* was ranged between 200 and 42,800 cells/ m^3. The maximum was recorded at L3 of January and the minimum was recorded at L1 during May 09.

The group of Cholorophycea, *Cylindrocapsa geminella* species was ranged between 700 and 24,600 cells/m^3. In this the maximum was observed at L4 during January and minimum was recorded at L2 during March. *Gloeotilopsis planktonick* was ranged between 400 and 23,800 cells / m^3. The maximum was recorded at L3 during the month of January and minimum was recorded at L1 during March. *Spirotaenia condensate* ranged between 1000 and 9800 cells/m^3. In this maximum was found at L2 during April and the minimum was recorded at L4 during the month of April. *Hormidium flaccidum* was ranged between 200 and 30200 cells/m^3.The maximum was recorded at L1 during January and the minimum was recorded at L1 during month of May. A*nkistrodesmus convolutus* was ranged between 1000 and 39100 cells/m^3. The maximum and the minimum were found at L1 during March and January respectively. *Micractinium pusillium* was ranged between 200 and

9100 cells / m³. The maximum was recorded at L3 during April and the minimum was recorded at L2 during the month of May. *Crucigenia tetrapedia* was ranged between 3500 and 18000 cells/m³. The maximum was found at L4 during January and the minimum was found at L2 during May. *Pachycladon umbrinus* was ranged between 200 and 1600 cells / m³. The maximum was found at L2 during May and the minimum was found at L3 during May. *Scenedesmus sp* was ranged between 1200 and 17700 cells/m³. The maximum was recorded at L1 during March and the minimum was recorded at L3 during the month of January. *Trebauria triappendicula* was ranged between 200 and 1400 cells/m³. The maximum and the minimum were found at L1 during the month of January and May respectively.

Bacilloriphyceae, *Cyclotella* sp., was ranged between 9400 and 39200 cells / m³.The maximum and minimium was found at L3 during April and January respectively. *Cymbella* sp., was ranged between 13800 and 74800 cells/m³. The maximum was recorded at L3 during January and the minimum was recorded at L2 during April. *Neidium sp.,* was ranged between 2700 and 17200 cells/m³. The maximum was found at L2 during January and the minimum was found at L4 during April.

Dinophyceae groups were *Navicula cuspidata* was ranged between 4100 and 14100 cells/m³. The maximum and minimum was found at L3 during April and May respectively. *Navicula rhyncocephala* was ranged between 5000 and 14200 cells/m³. The maximum was recorded at L3 during April and the minimum was recorded at L2 during the month of May.

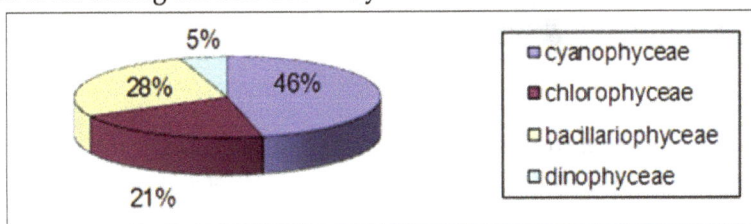

Figure 2: Abundasnce of phytoplankton in surface water sample of KRP reservoir.

Cynophyceae are well known to occur in diverse physicochemical conditions with varying degree of abundance and can tolerate wide fluctuations in chemical factors. Elayaraj and Selvaraju (2014) have been reported the microalgal distribution in Thachan pond, Chidambaram, Cyanophyceae (41%) were *Microcystis aeruginosa, Chroococcus minor, Oscillatoria curviceps, Oscillatoria tenuis, Anabaena spiroides, Nostoc pruniforme* and *Aphanocapsa grevillei* was the dominant species. The diversity found of phytoplankton belonging to Cyanophyceae, Chlorophyceae and Bacillariophyceae classes from GirijaKund and Maqubara pond, Faizabad, India during May 1999-June 2000. The seasonally allocation of algal diversity shows ascendancy scenery as Cyanophyceae > Bacillariophyceae > Chlorophyceae (Dwivedi and Pandey, 2002). Ravi Kumar et al., (2009) noted that the 26 different species/genera of phytoplankton and also large majority of theme belonged to diatom. The pragmatic biological assessment of river Mutha Pune, India during 2005 they recorded that the blue green and diatoms like, Oscillatoria and Anabena throughout the investigation occur abundantly and repeatedly

(Jafari and Gunale, 2006). In our study, Cyanophyceae (46 %), Bacillariophyceae (28 %), Chlorophyceae (21 %) and Dinophyceae (5 %) were recorded in KRP.

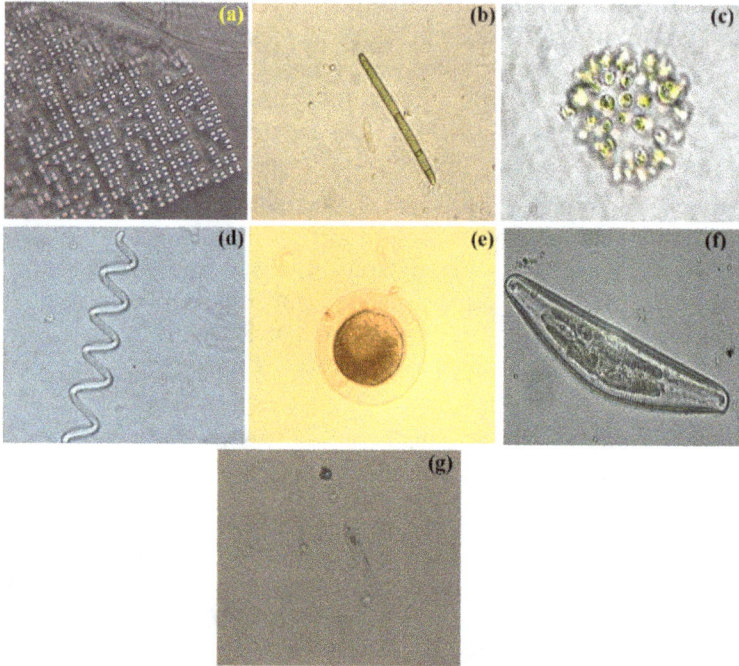

Figure 3: Diatom species from KRP reservoir. (a)*Merismopedia glauca,* (b) Oscillatoria sp, (c) *Microcystis aeruginosa* (d) *Spirulina* sp, (e) *Cyclotella* sp, (f) *Cymbells* sp and (g) *Neidium* sp.

Species diversity and Zooplankton

The diversity of the phytoplankton from KRP was 5 groups, Cyclopoid, Calanoid, Rotifers, Ostracods and cladocerans (Figure 4). The Cyclopoids were ranged between 6,800 and 41,800 ind/m^3 and the maximum was recorded at L3 during March 2009. The minimum was recorded at L1 during May 2009. Calanoid, the other group which was recorded as rare. The maximum number of Calanoid was recorded at L1 during January 2009 and the minimum was recorded at L3 of May 2009. In January, the maximum of ostracoda group found in L4 as 48,000 ind/m^3 and the minimum 2300 ind/m3 during the month of April, 09 in L3. The rotifer was ranged between 7,500 and 1, 59,200 ind/m^3 and the maximum number at L3 during March, 09. The minimum number was noticed in L1 during January, 09. Among rotifers, *Keratella himalis, Keratella vulga, Keratella quadrata* and *Branchionous* sp were identified. During the month of January, the abundance of cladocera group was found to be maximum in L4 and the minimum in L1 during the month of January 09.

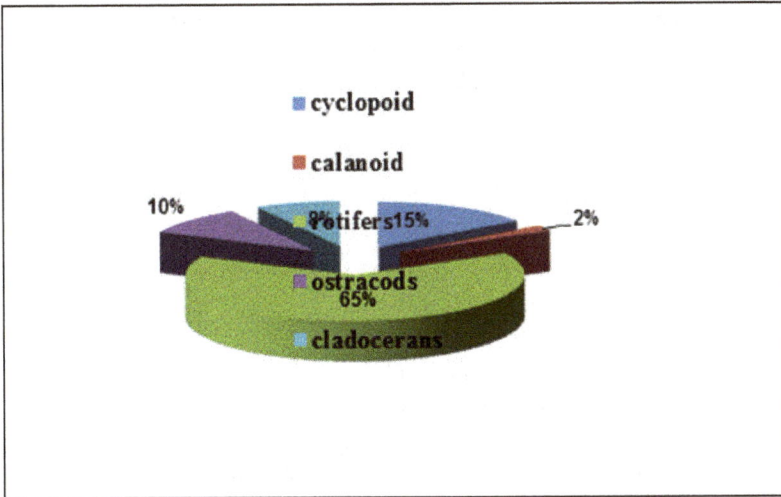

Figure 3: Abundance of zooplankton in surface water sample of KRP reservoir

In the present study the most dominant rotifer species were *Keratella sp* and *Branchionous* sp. The rotifer fauna of Tamil Nadu is characterized by greater higher diversity. This generalization is endorsed by the fact that out of 67 genera of the Phylum recorded so far from India, 56 genera (83.6%) and all (25) families of Eurotatoria known from this country are represented in Tamil Nadu. The salient features of the rich species, generic and family diversity attach special biodiversity importance to the studied fauna and, reflect greater environmental heterogeneity as well as microhabitat diversity of the aquatic environs of this state (Sharma, 2009). Nair and Nayar (1971) reported 18 species of rotifers from freshwater habitats in Irijalakuda, Kerala. Anitha (2003) recorded 44 species of rotifers belonging to 16genera under 12 families from two estuaries located in southern part of Kerala. Prabahar et al. (2012) reported 17 species in the nearby Adirampattinam mangrove area. Among the KRP reservoir habitats of Maximum numbers of genera was recorded during the present study.

Physico-chemical factors

The maximum temperature of 26.8ºC was recorded in the month of May 2009 at L4. The minimum of 23.1ºC was recorded during January at L2 and L4 (Table 1). The maximum pH was noted during January at L1 as 9.7 and in April, the minimum pH was recorded as 8 at L3. The maximum dissolved oxygen was noted as 1.8 mg/L at L3 during May and March, the minimum dissolved oxygen was recorded as 0.5mg/L at L1.During January, the maximum amount of phosphate was recorded as 7.56 mg/L at L1 and the minimum was noticed as 2.84 mg/L at L4 during the month of April, 2009 (Table 1). During the month of March, carbonate was observed as 12 mg/L at L2 as minimum and the maximum was observed as 63mg/L at L3 during May. Bicarbonate was observed as maximum as 146.4 mg/L at L2 during January and the minimum was observed as 6.1mg/L at L1 and L3 during the month of January and March respectively.

Table 1: Physico-chemical analysis of KRP reservoir

Parameters	Value
Temperature (°C)	25.04
pH	8.63
DO (mg/l)	1.11
Phosphate (mg/l)	4.74
Carbonate (mg/l)	39.25
Bicarbonate (mg/l)	62.31

The pH values of the surface water of KRP showed alkaline in nature, this may be due to the presence of mineral salts (Gupta *et al.,* 2001). Low input of nitrogen and phosphorus appears to limit the Phytoplankton productivity. Phosphorous is the main limiting nutrient in the red tide-area and the adjacent (Paul *et al.,* 2005). The data showed that phosphate concentration decreased quickly with an increase in phytoplankton abundance, and in most cases phosphate concentration was lower than the level where Phytoplankton is limited (Gao and Song, 2005). High plankton count and high phosphate content were noticed in KRP reservoir. The highest concentration of phosphate showed the eutrophication nature of KRP reservoir.

Conclusion

In Conclusion, Phytoplankton represented by 26 species and 5 groups of Zooplankton were identified. Species varied in numerical abundance both spatially and temporally between the locations. This data provides valuable information about the distribution and abundance of plankton in KRP reservoir. The regulation of the inflow and outflow of this reservoir influences the importance of the different variables shaping the planktonic community structure.

Acknowlegement

We acknowledged UGC New Delhi (Grant No: F. No 33-362 / 2007 (SR) through MRP for financial support to carry out this work.

References

1. Anitha, P.S. (2003). Studies on certain selected live feed organisms used in aquaculture with special reference to rotifers (Family: Brachionidae). Ph.D. Thesis, C.l.F.E., Mumbai.
2. Baruah, P.P. and Kakati, B. (2009). Studies on phytoplankton community in a highland temple pond of Assam, India. Indian *J. Environ. Ecoplan.* 16(1): 17-24.
3. Dwivedi, B.K. and Pandey, G.C. (2002). Physico-chemical factors and algal diversity of two ponds (GirijaKund and Maqubara pond) of Faizabad, India. *Poll. Res.* 21(3): 361-370.
4. Elayaraj, B. and Selvaraju, M. (2014). Water quality variation and screening of microalgal distribution in thachan pond Chidambaram taluk of Tamil nadu. Inter. *J. Biol. Res.* 2(2): 90-95.

5. Gao, X. and Song, J. (2005). Phytoplankton distributions and their relationship with the environment in the Changing Estuary, China. *Mar. Poll. Bull.* 50: 327-335.

6. Gold, C., Feurtet-Mazel, A., Coste, M. and Boudou, A. (2002). Field transfer of periphytic diatom communities to assess short term structural effects of metals (Cd, Zn) in rivers. *Water Res.* 3654-3664.

7. Gupta, S., Rose, C., Rathose, G.S. and Methas, C.D. (2001). Hydrochemistry of Udaipur lakes. *Indian J Environ. Health.* 1: 333.

8. Jafari, N.G. and Gunale, V.R. (2006). Hydro biological Study of Algae of an Urban Freshwater River. *J. Appl. Sci. Environ. Mgt.* 10(2): 153-158.

9. Lampert, W. and Sommer, U. (2001). Ekologia wód sródladowych [Ecology of inland waters]. PWN, Warszawa: Wyd. Nauk.

10. Mahapatra, S.S., Sahu, M., Patel, R.K. and Panda, B.N. (2012). Prediction of water quality using principal component analysis. *Water Quality Exposure and Health.* 4(2): 93-104.

11. Nair, K.K.N. and Nayar, C.K.O. (1971). A preliminary study on the rotifers of Irinjalakuda and neighbouring places. *J. Ker. A cad. Bioi.* 3(1): 31-43.

12. Paul, J., Elifermstad, L. and Tonumols, A. (2005). Impacts of increased nitrogen supply on Norwegian lichen rich Alpine. 10 years experiments. *J. Ecol.* 93: 472-481.

13. Prabhahar, C., Saleshrani, K.., Arsheed Hussain, D. and Tharmaraj, K.. (2012). Studies on the ecology and distribution of zooplankton composition in Adirampattinam mangrove region, Tamil nadu India. *Int. J. Rec. Sci. Res.* 3(5): 356-359.

14. Ravi Kumar, M.S., Ramaiah, N. and DanLing Tang, (2009). Morphometry and cell volume of diatoms from a tropical estuary of India. *Indian J. Mar. Sci.* 38(2): 160-165.

15. Santhanam, R., Velayutham, P. and Jegathesan, G., (1989). A manual of fresh water ecology. *Daya Publishing House. New Delhi.* 1–125.

16. Senthilkumar, P.K. and Das, A.K. (2008). Distribution of phytoplankton in some freshwater reservoirs of Karnataka. *J. Inland. Fish. Soc. India.* 33: 29-36.

17. Sharma, B.K. (2009). Biodiversity and distribution of freshwater Rotifers (Rotifera: Eurotatoria) of Tamil nadu. *Rec. zool. Surv. India.* l09(3): 41-60.

18. Thakur, A.K., Rath, S. and Mandal K.G. (2013). Differential responses of system of rice intensification (SRI) and conventional flooded-rice management methods to applications of nitrogen fertilizer. *Plant Soil.* 370: 59-71.

19. Tiwari, D. and Shukla, M. (2007). Algal biodiversity and trophic status of some temporary water bodies of Kanpur. *Nat. Environ. Pollut. Tech.* 6: 85-90.

20. Vandysh, O.I. (2004). Zooplankton as an indicator of the state of lake ecosystems polluted with mining wastewater in the Kola Peninsula. *Russian J. Ecol.* 35(2): 110-116.

21. WHO. (2004). Water, sanitation and hygiene links to health: facts and figures. Geneva, Switzerland: World Health Organization.

4

Exploring the Soil Microbial Diversity and Ecosystem Functions with respect to Climate Change

Kanchan Vishwakarma[1], Shivesh Sharma[1]*, Neha Upadhyay[1],
Nitin Kumar[1], Ashish Tiwari[1], Vivek Kumar[2]

[1]*Department of Biotechnology,*
Motilal Nehru National Institute of Technology,
Allahabad-211004, Uttar Pradesh, India
[2]*Amity Institute of Microbial Technology,*
Amity University, Noida-201313, Uttar Pradesh, India
E-mail: shiveshs@mnnit.ac.in; ssnvsharma@gmail.com

Abstract

Soil as one of the most important matter and has its direct and indirect impact on climatic conditions. On the other hand, climatic conditions can alter organic and mineral content of soil. Soil contains microorganisms ranging from bacteria, fungi, virus and algae to actinomycetes whose presence can define health of soil ecosystem. These microbial populations play an important role in decomposition, release of greenhouse gases, plant growth, carbon, nitrogen, phosphorus, nutrient and other mineral cycles. Soil microbe activities frequently depend upon environmental parameters like temperature, moisture and nutrient availability, all of which are affected by climate change. One of the major reasons of global climate changes include anthropogenic activities that lead to increase in greenhouse gases and thus cause decline in microbial diversity present in soil. Thus, there is increasing needs to overcome such changes for better sustainability of different ecosystems. Therefore, the present study comprises to review the influence of warming due to the changing global climate on microbial community and their activity.

Keywords: Climate change, Global warming, Soil microbial community, Temperature

Introduction

Soil is the incoherent matter on Earth's surface having organic and mineral content. It is subjected to environmental factors and hence shows effects of climate change, organisms over a period of time. Soil organic matter (SOM) is primarily composed of amides, stable material called humus, soil microbes and other organic molecules. Soil contains microorganisms which indicate soil microbial activity in soil. These microbes are the source of soil enzymes that have significant part in deposition of organic matter present in soil and nutrient cycling (Waldrop et.al. 2004). Soil is active pool of carbon, nitrogen, phosphorus and other minerals. A variable but significant pattern to this pool is contributed by microbial biomass C and N (Sicardi *et al.*, 2004).

Biodiversity is defined as "the variability among living organisms from different sources such as, terrestrial, marinal, aquatic and other ecosystems". Soil biodiversity refers to the number of heterotrophic species belowground (Hooper *et al.*, 2005). Different types of soil organisms include: microorganisms (eg. bacterium, fungus etc), microfauna (eg. protozoans, nematodes), mesofauna (eg. springtails) and macrofauna (eg. insects, earthworms). Among all above mentioned soil organisms, microorganisms are preferred for the maintenance of quality of soil and yield of plant crop.

Soil has abounding micro-scale creatures furnishing inside it which involve bacteria, fungi, virus, algae and actinomycetes. Among all these microscopic lives, bacteria are the most widespread and familiar i.e. accounting approx 95%. In general soil hosts a wide range of bacterial population having 10^8 to 10^9 cells per gram soil (Schoenborn *et al.*, 2004). On the other hand, stressed soils contains lesses number of bacteria i.e. 10^4 cells per gram soil (Timmusk *et al.*, 2011). The number and type of bacterial population is largely affected by condition of soil and climate (Glick *et al.*, 1999). Outcome of these stress factors is reduced growth of plant which further pours changes on soil microbial diversity.

Soil microbe activities frequently depend upon environmental parameters like temperature, moisture and nutrient accessibility, all of which are influenced by climatic changes (IPCC, 2007). The main uncertainty in prediction of climate change is soil microbe's reaction to warming temperature (Briones *et al.*, 2004). Several works showed that elevated temperature accelerates rate of microbial decomposition resulting in emission of increased CO_2 by soil respiration, thereby huge soil carbon losses and amplification of global warming occurs (Allison *et al.*, 2010). Enhanced concentration in the atmosphere is thought to be mitigated in part to sequester large amount of CO_2 by the ability of terrestrial forests (Schlesigner and Lichter, 2000). For better understanding in this context, the extent to which soil emits green house gas and processes that lead to such emissions must be reduced.

Microbes present in the soil are helpful in regulating carbon and nutrients in ecological-systems. Furthermore, activities of microbes are modulated by biotic as well as abiotic parameters like temperature, moisture etc. Any changes in atmosphere or climate will hamper both biotic and abiotic factors in ecosystem (Bardgett *et al.*, 2008).

Plant-microbe associations can be taken as a biotic conduit of carbon to the soil from environment. It will lead to the longest residing carbon pool and terrestrial ecosystem. For restoration of exhausted and barren lands, it would be beneficial to incorporate native plants which have close association with microbial communities (Nave, 2012). Carbon forms moving into soils to subsidize microbial activities often grow in symbiosis with roots. These microbes act as carbon source in the form of their detritus dead bacterial cells and fungal hyphae. Microbes produce enzymes and mucilage which binds to soil particles together and lead to decomposition of soil carbon.

One of the leading scientific and political challenges in our era is the progressing alterations in global climate which are majorly induced by anthropogenic activities thereby increasing GHGs (green house gases) (Magnani *et al.*, 2007). GHGs such as CO_2, CH_4 and NO are released and absorbed through climatic-feedbacks. To make this process feasible, terrestrial ecosystems have a significant role. Also, such ecosystems help in storage of carbon in bulk in vegetation, hence acting as noteworthy carbon sequester (Schimel *et al.*, 1994). But still, terrestrial ecosystems are being deteriorated by human activities, misuse of agriculture land (Smith *et al.*, 2008), excessive N storage (Beedlow *et al.*, 2004), sulfur deposition (Monteith *et al.*, 2007) and fluctuation in atmospheric O_3 concentration (Sitch *et al.*, 2007). An overview of impact of global warming on soil microbial diversity is depicted in Figure 1.

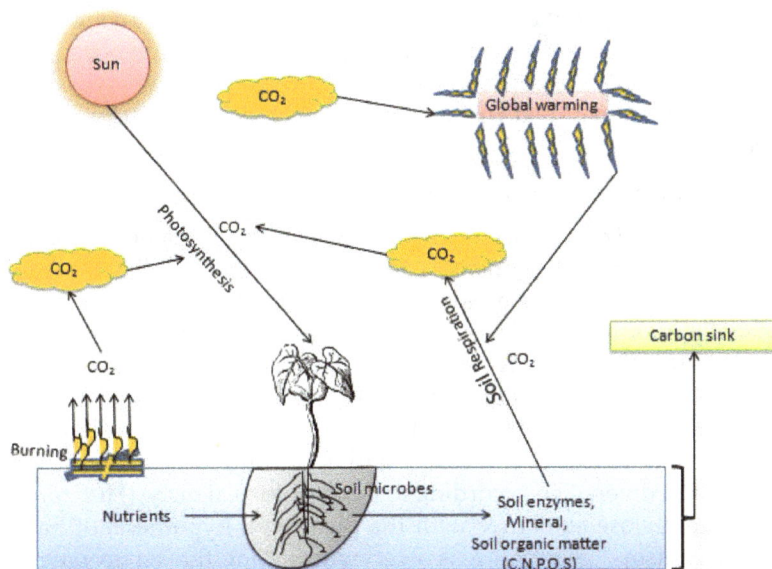

Figure 1: Effect of global warming on soil microbial community (Vishwakarma *et al.*, 2015)

Interactions between both climate & atmospheric changes drivers may cause different soil microbial diversities, and any change in these diversities may modulate our ecosystem (Castro *et al.,* 2010). As it is important to know that these climate variations have its effect on both direct & indirect effects of functionalization of soil microbial community which generate green house gases and maintain global warming: direct effects consists of influence on soil microbial diversity and temperature modulation by greenhouse gases, changes in precipitation & extreme weather. Indirect effects consists of climatic derived changes in crop yield and varieties which can alter physio-chemical nature of soils, carbon supply to soil & activities at microbial level involved with processes of decomposition and release of carbon from soil.

Heterotrophic soil microbial diversity modulates important processes which control nitrogen (N) & carbon (C) cycling. This represents the important connection between ecosystem functionalization and plant diversity. Also those microbial diversity present in soil helps in maintaining important nutrient & carbon cycles.

Due to human intervention we nowadays experience lot of climatic changes at high altitudes and latitudes, and there is lot of predictions that in future these changes are most probably going to be of lager magnitude and will have broader impact on climate. Whereas Arctic soils plays potential role in maintaining global carbon cycle because it contain an uneven large cockpit of carbon in soil (Post *et al.,* 1982). This amount of large carbon pool is susceptible to climate change; risk of mineralization of soil carbon via increases in microbial respiration because of higher temperatures can occur (McGuire, 1995). The next segments of the chapter take a closer look on how climatic changes affect the soil microbial diversity and the important ecological functions. The rapid and continuous changes occur in global climate and their consequences on microbial community and keystone functions has been explored further with the help of previously published and available literature.

Global climate change and soil microbial diversity

Soil is characterized as the most diverse habitat on this planet. A microorganism incorporates a greater component of soil diversity and soil biomass and as well as endowed with important ecosystem functions for example the breakdown of soil organic matter and their role in cycling of different nutrients. In spite of the fact that microbial species are responsible for maintaining a healthy and fertile soil, very less reports are available on the microbial responses towards the varying climatic conditions. The survival and diversity of microorganism are controlled by different environmental conditions and according to that the microorganism were sorted and diversified according to their ecological niche (Hutchinson, 1957). This has a very close similarity with the purported hypothesis of Baas-Becking for microorganism "everything is everywhere: but the environment selects" (BaasBecking, 1934) that shaped the basis of the presently formulated species-sorting paradigm (Leibold *et al.,* 2004; Holyoaketal, 2005) which was proved to be the main determinant for the assembly of microbial communities (Vander Guchtetal, 2007; Logueand Lindström, 2008).

Many ecosystems are undergoing speedy and predominant environmental transformations, obtaining a quantitative approach which figures out that the mechanisms that influence microbes and microbial communities is pivotal for predicting the responses of microbial communities to novel or changing selective forces and their implications on the regional and world scale. Soil microorganisms are an essential part of soil ecology and involve in maintaining soil health by participating in important soil functions for example carbon sequestration plant health and nutrient cycling. Nonetheless, the soil microbial diversity differs on a large scale and is greatly affected by anthropogenic activities, changing global climate and then ongoing environmental stress. The understanding of distribution and characteristic of microbial communities in soil is limited and with the help of novel molecular techniques it is feasible to understand the modern approaches of microbial ecology. It is an essential requirement to explore the ecological information on native microbial communities in soil, and their responses to global climate change in addition to hazardous soil conditions.

Climate changes are continuous variations in global environmental conditions due to human activities or as a result of natural approaches. The greenhouse gas pollution caused by various human activities has contributed considerably to the global warming. Several anthropogenic activities such as automobiles, agricultural practices and industrialization are the main source of greenhouse gases such as CO_2, CH_4, and N_2O into the surrounding environment at very high rate. As a result of increasing concentration of greenhouse gases in the atmosphere a rise in global temperature was observed. In the recent years around 0.6°C rise in temperature was observed around the earth surface and if the rate of greenhouse gas emission persist the temperature could rise up to 1.5 to 4.5 °C rapidly (Houghton *et al.,* 2001) which causes melting of snow, glaciers and the rise of sea levels. Global warming influences animal, plant or microorganism together by shifting the habitat and with the rising temperature.

Soil microorganism plays important role in the biochemical cycling of nutrients such as carbon (C), nitrogen (N) in ecosystem and the microbial activity is very much affected by the biotic and abiotic factors for example moisture, temperature, presence of contaminants. Both the climatic and atmospheric alterations affect the biotic as well as abiotic component in ecosystem plus the responses of ecosystem to these variations. Soil microbial communities might be able to regulate the response of ecosystem back to the atmosphere (Bardgett *et al.,* 2008). Even though the ecosystem functions in soils are quite regulated by the soil microbial communities but the phenomenon behind the correlation of the abundance and diversity of microbial communities with respect to climatic changes and their consequences on ecosystem processes is remain unclear. Till now most of the research based on the impact of climate change on ecosystem mainly focuses on the macroscale responses to climate change for example modification in plant growth (Norby *et al.,* 2005), effect in soil microbial processes such as nutrient cycling (Emmett *et al.,* 2004; Hungate *et al.,* 2006; Garten *et al.,* 2008).

However, the direction and magnitude of these responses is uncertain. For instance, rise in temperature may enhance the activity, processing and turnover of microorganisms causing the microbial community to shift in favor of legislative bodies adapted to higher temperatures and faster growth rates (Zogg *et al.*, 1997; Pettersson and Baath, 2003; Zhang *et al.*, 2005; Bradford *et al.*, 2008). Atmospheric and climatic changes are happening with respect to each other; as a result ecosystems experience enhanced levels of atmospheric CO_2, global warming along with fluctuations in precipitation in system.

Effect of climatic change on ecosystem functioning

The climatic changes are responsible for altering distributions of species and eventually the interactions within organism (Wookey *et al.*, 2009; van der Putten, 2012). Natural habitats are very complex and having a vast array of organism with different growing conditions. These interactions may be beneficial, pathogenic or neutral and vary with respect to changing environmental conditions (Vandenkoornhuyse *et al.*, 2015). Few studies were available on interaction of soil communities with the terrestrial ecosystem on varying climatic conditions (Schimel *et al.*, 2007; de Vries *et al.*, 2012).

Soil microorganism interacts with each other and with the host plant in numerous ways that form and maintain ecosystem characteristics. On the other hand the beneficial soil plant microbe interactions also responsible for maintaining landscape pattern of plants, composition and diversity of microbes as well as abundance of animals (Berg *et al.*, 2010; van der Putten *et al.*, 2013). Soil-plant-microbe interactions turned to be negative when the overall effect of soil microbes comprising pathogens, symbiotic microbe and decomposers reduces the plant growth and yield, where as a positive interaction referred to those when the soil microbial community are responsible in the enhancement of plant growth.

Consequently they play a significant role in maintaining ecosystem characteristics, and it an essential priority of research to unravel the soil microbe-microbe as well as soil plant-microbe interactions relating to climate change which will be helpful in understanding the important ecosystem functions such as storage of carbon in soil (Ostle *et al.*, 2009; Berg *et al.*, 2010; Fischer *et al.*, 2014). Approximately ~120 Gt flux of carbon produced per year into and out of terrestrial ecosystems which surpass the limit of the amount of carbon that is being generated by the combustion of fossil fuels (IPCC, 2013). Therefore, a minute change regarding carbon exchange within ecosystem and atmosphere may have significant impact on the upcoming concentrations of atmospheric carbon. Around more than 2.53 % of carbon is stored in soil in combination with plant biomass (Singh *et al.*, 2010). The ability of soil to hold large amount of carbon as well as their potential to sequester carbon may be aided to diminish rising atmospheric carbon dioxide.

The climatic change influences the soil microbial interactions and plant-microbe interactions either directly or indirectly. Several authors were reviewed the direct and indirect effects of climatic alteration on compositional and functional aspects of soil along with the associated microbes and plants (Blankinship *et al.*, 2011; Sistla *et al.*, 2013; Chen *et al.*, 2014; Delgado-Baquerizo *et al.*, 2014; Classen *et al.*, 2015).

Impacts of climate change on plant and associated microbial community

The quantity, diversity and functions of soil microbial communities alters greatly by changing climatic conditions because the survival of microorganism in soil varies in their growth rates, physiological conditions and temperature sensitivity (Castro *et al.,.* 2010; Gray *et al.,* 2011; Classen *et al.,* 2015). Microorganisms respond to global climate change such as warming and other effect due to resistance and its flexibility to returns to the initial state after the stress has passed (Allison and Martiny, 2008). Microbial shifts are responsible to changing the ecosystem functions when microorganism in soil varies in their functional characteristics (Schimel and Schaeffer, 2012). For example, a number of specific microbial species regulate ecosystem functions such as in biochemical cycling (e.g. nitrogen fixation) (Isobe *et al.,* 2011) and methanogenesis (Bodelier *et al.,* 2000).The climatic changes such as rising in temperature directly influence soil respiration rates because the microbial activities in soils are sensitive towards temperature. The continuous increase in atmospheric temperature causes the migration of plant species into higher latitudes (Walther *et al.,* 2002; Parmesan and Yohe, 2003). The previous studies suggest that, with the rising of community level, it was observed that in Arctic regions, woody shrubs have replaced grasses in numerous regions causes the alteration in ecosystem by modifying the function such as carbon feedbacks in these systems (Hinzman *et al.,* 2005; Lawrence and Swenson, 2011; Pearson *et al.,* 2013). The microbial community in soil, particularly that is closely associated with plants, may facilitate plant community transitions, such as, microorganism associated with plant root may strongly affect the growth and survival of plants, life cycle of plants and its associated functions (Van der Heijden *et al.,* 1998; Friesen *et al.,* 2011).

The distributions of plant and microbial community in soil changed with the varying temperature and climatic conditions and well studied previously by several authors (Grabherr *et al.,* 1994; Walther *et al.,* 2002; Parmesan and Yohe, 2003). The plant-microbe-soil interactions within soil rhizosphere are important to understand the regulator of soil nutrient cycles. If the soil microbial communities are affected by the changing climatic conditions, it eventually affects the development and progress of plant species and subsequently alters the ecosystem responses. Current studies on the reaction of soil microorganism make the plants survive in the presence of harsh environmental conditions such as drought (Lau and Lennon, 2012). This discussion leads to the conclusion that the soil ecosystem responses towards altering climatic conditions might be compensate if the plant community changed according to the climatic changes.

Mechanisms and Consequences of Warming Effects on Microbial Diversity

Several patterns emerge from the above warming experiments, the first of which is the lack of similarity between phylogenetic and functional diversity in the studies that measured both. The efficiency of microbes for conversion of carbon to biomass (carbon use efficiency), affected due to warming, emerges as second pattern. This is leading to shifts to a community characteristic of higher carbon use efficiency in some instances (Frey *et al.*, 2008), with no predictable change (Gray *et al.*, 2011) or a reduction in others (Schindlbacher *et al.*, 2011; Yergeau *et al.*, 2012). Inconsistent changes have been observed in microbial biomass, some are increasing at some sites (Sheik *et al.*, 2011; Zhou *et al.*, 2012; Yergeau *et al.*, 2012) and decreasing at others (Frey et al., 2008; Bradford et al., 2008; Gray et al., 2011; Allison and Treseder, 2008) while no change (Schindlbacher *et al.*, 2011) or a delayed response have been also observed. These irregularities may be methodological, site-specific, or due to plant communities, which are well known links to below ground communities (Wardle *et al.*, 2004).

The lack of apparent congruence between phylogeny and function may be due to a —priming effect. Here one stress (for example, drought or fire) favours growth of species or functions which in turn helps in providing resistance to a secondary stress (experimental warming), but not seen always. An initially diverse community raised a primed species (community diversity hypothesis), or a handful of organisms whose keystone role is their ability to weather the change in the environment. The augmented diversity under drought and warming at the Great Plains site implies the former (Sheik *et al.*, 2011), while Yergeau's Antarctic and sub-Antarctic study indicates the latter. Functional gene richness has been reduced without affecting microbial abundance or activity (Yergeau *et al.*, 2012). This may indicate that functional redundancy of the community was reduced by warming, and so the community may be more sensitive to further disturbances.

Conclusion and Future Prospects

The lack of uniformity in the classification of microbial community response towards global warming present in the available literature is due to the associated environmental factors. It has been observed that the influences of climatic changes on the soil microbial community and its associated functions changes broadly, however diverse effect has been obtained by different climatic drivers. To meet the intention of an improved expertise of methods influencing the microbial reaction to warming, there is a need to have advanced understanding of microbes, from the extent of the gene and the physiological condition of each single microbe on the way to microbial composition and interactions. To discover the application of microbes in ecosystem functioning, a wide variety of alternatives are available to assemble in terms of quantifying variety, and lastly, particularly few measures of the contributions of microbes to ecosystem function. The contributions of microbes in defining the ecosystem functioning is widely affected by the global climatic change. Lastly, to envisage and recover the outcomes of climate change for the functioning of soil ecosystem, it is necessary to analyze and quantify the role of microbial mechanism and its interaction in the soil microhabitat. This may be addressed with the aid of a experimental approaches having a precise observation with long-term field-based studies signifying the interacting climate changes on soil microbial community responses.

Acknowledgement

Authors are thankful to the Director, MNNIT, Allahabad for providing necessary facilities for conducting this study.

References

1. Allison, S.D. and Martiny, J.B.H. (2008). Resistance, resilience, and redundancy in microbial communities. *Proceedings of the National Academy of Sciences USA*. 105: 11512–11519.

2. Allison, S.D. and Treseder, K.K. Warming and drying suppress microbial activity and carbon cycling in boreal forest soils. *Global Change Biol*. 14: 2898–2909.

3. Baas Becking, L.G.M. (1934). Geobiologie of Inleiding Tot de Milieukunde. The Hague: W. P. Van Stockum and Zoon.

4. Bardgett, R.D., Freeman, C. and Ostle, N.J. (2008). Microbial contributions to climate change through carbon cycle feedbacks. *The ISME Journal*. 2: 805-814.

5. Beedlow, P.A., Tingey, D.T., Phillips, D.L., Hogsett, W. and Olszyk, D.M. (2004). Rising atmospheric CO_2 and carbon sequestration in forests. *Front Ecol Env*. 2: 315-322.

6. Berg, M.P., Kiers, E.T., Driessen, G., Van Der Heijden, M., Kooi, B.W., Kuenen, F., Liefting, M., Verhoef, H.A. and Ellers, J. (2010). Adapt or disperse: understanding species persistence in a changing world. *Global Change Biology*. 16: 587–598.

7. Blankinship, J.C., Niklaus, P.A. and Hungate, B.A. (2011). A meta-analysis of responses of soil biota to global change. *Oecologia*. 165: 553-565.

8. Bodelier, P.L.E., Roslev, P., Henckel, T. and Frenzel, P. (2000). Stimulation by ammonium-based fertilizers of methane oxidation in soil around rice roots. *Nature*. 403: 421-424.

9. Bradford, M.A., Davies, C.A., Frey, S.D., Maddox, T.R., Melillo, J.M., Mohan, J.E., Reynolds, J.F., Treseder, K.K. and Wallenstein, M.D. (2008). Thermal adaptation of soil microbial respiration to elevated temperature. *Ecol. Lett*. 11: 1316-1327.

10. Castro, H.F., Classen, A.T., Austin, E.E., Norby, R.J. and Schadt, C.W. (2010). Soil Microbial Community Responses to Multiple Experimental Climate Change Drivers. *Applied and Environmental Microbiology*. 76(4): 999-1007.

11. Chen, S., Zou, J., Hu, Z., Chen, H. and Lu, Y. (2014). Global annual soil respiration in relation to climate, soil properties and vegetation characteristics: summary of available data. *Agricultural and Forest Meteorology*. 198: 335-346.

12. Classen, A.T., Sundqvist, M.K., Henning, J.A., Newman, G.S., Moore, J.A., Cregger, M.A., Moorhead, L.C. and Patterson, C.M. (2015). Direct and indirect effects of climate change on soil microbial and soil microbial-plant interactions: What lies ahead? *Ecosphere*. 6(8): 1-21.

13. de Vries, F.T., Liiri, M.E., Bjornlund, L., Bowker, M.A., Christensen, S., Setala, H.M. and Bardgett, R.D. (2012). Land use alters the resistance and resilience of soil food webs to drought. *Nature Climate Change*. 2: 276-280.

14. Delgado-Baquerizo, M., Maestre, F.T., Escolar, C., Gallardo, A., Ochoa, V., Gozalo, B. and Prado-Comesana, A. (2014). Direct and indirect impacts of climate change on microbial and biocrust communities alter the resistance of the N cycle in a semiarid grassland. *Journal of Ecology*. 102: 1592-1605.

15. Emmett, B.A., Beier, C., Estiarte, M., Tietema, A., Kristensen, H.L., Williams, D., Penuelas, J., Schmidt, I. and Sowerby, A. (2004). The response of soil processes to climate change: results from manipulation studies of shrublands across an environmental gradient. *Ecosystems.* 7: 625-637.

16. Fischer, D.G., Chapman, K., Classen, A.T., Gehring, C.A., Grady, K.C., Schweitzer, J.A. and Whitham, T.G. (2014). Marschner Review: Plant genetic effects on soils under climate change. *Plant and Soil*. 37: 91-19.

17. Frey, S.D., Drijber, R., Smith, H. and Melillo, J. (2008). Microbial biomass, functional capacity, and community structure after 12 years of soil warming. *Soil Biol Biochem*. 40: 2904–2907.

18. Friesen, M.L., Porter, S.S., Stark, S.C., von Wettberg, E.J., Sachs, J.L. and Martinez-Romero, E. (2011). Microbially mediated plant functional traits. *Annual Review of Ecology, Evolution, and Systematics*. 42: 23-46.

19. Garten, C.T., Classen, A.T., Norby, R.J., Brice, D.J., Weltzin, J.F. and Souza, L. (2008). Role of N_2-fixation in constructed old-field communities under different regimes of $[CO_2]$, temperature, and water availability. *Ecosystems*. 11: 125-137.

20. Glick, B.R., Patten, C.L., Holguin, G. and Penrose, D.M. (1999). Biochemical and Genetic Mechanisms Used by Plant Growth Promoting Bacteria, Imperial College Press, London, UK.

21. Glick, B.R. (2012). Plant Growth-Promoting Bacteria: Mechanisms and Applications. Hindawi Publishing Corporation. *Scientifica..* Editors: T. Ano, G. Comi, and M. Shoda. 1-15.

22. Grabherr, G., Gottfried, M. and Pauli, H. (1994). Climate effects on mountain plants. *Nature*. 369: 448-448.

23. Gray, S.B., Classen, A.T., Kardol, P., Yermakov, Z. and Miller, R.M. (2011). Multiple climate change factors interact to alter soil microbial community structure in an old-field ecosystem. *Soil Science Society of America Journal*. 75: 2217-2226.

24. Hinzman, L.D. (2005). Evidence and implications of recent climate change in northern Alaska and other arctic regions. *Climatic Change*. 72: 251-298.

25. Holyoak, M., Leibold, M.A. and Holt, R.D. (2005). Meta communities: Spatial Dynamics and Ecological Communities. Chicago, IL; London: The University of Chicago Press.

26. Houghton, J.T., Ding, Y.D.J.G., Griggs, D.J., Noguer, M., van der Linden, P.J., Dai, X., Maskell, K. and Johnson, C.A. (2001). Climate change 2001: the scientific basis.

27. Hungate, B.A., Johnson, D.W., Dijkstra, P., Hymus, G., Stiling, P., Megonigal, J.P., Pagel, A.L., Moan, J.L., Day, F., Li, J.H., Hinkle, C.R. and Drake, B.G. (2006). Nitrogen cycling during seven years of atmospheric CO_2 enrichment in a scrub oak woodland. *Ecology*. 87: 26-40.

28. Hutchinson, G.E. (1957). Population studies-animal ecology and demography-concluding remarks. *Cold Spring Harb. Symp.* 22: 415-427.

29. IPCC. (2013). The Physical Science Basis. Contribution of Working Group I to the Fifth Assessment Report of the Intergovernmental Panel on Climate Change. Cambridge University Press, Cambridge, UK.

30. Isobe, K., Koba, K., Otsuka, S. and Senoo, K. (2011). Nitrification and nitrifying microbial communities in forest soils. *Journal of Forest Research.* 16: 351-362.

31. Lau, J.A. and Lennon, J.T. (2012). Rapid responses of soil microorganisms improve plant fitness in novel environments. *Proceedings of the National Academy of Sciences USA.* 109: 14058-14062.

32. Lawrence, D.M. and Swenson, S.C. (2011). Permafrost response to increasing Arctic shrub abundance depends on the relative influence of shrubs on local soil cooling versus large-scale climate warming. *Environmental Research Letters.* 6: 8.

33. Leibold, M.A., Holyoak, M., Mouquet, N., Amarasekare, P., Chase, J.M. and Hoopes, M.F. (2004). The metacommunity concept: a frame work for multi-scale community ecology. *Ecol. Lett.* 7: 601-613.

34. Logue, J.B. and Lindström, E.S. (2008). Biogeography of bacterio planktonin in land waters. *Freshw. Rev.* 1: 99-114.

35. Magnani, F., Mencuccini, M., Borghetti, M., Berbigier, P., Berninger, F. and Delzon, S. (2007). The human footprint in the carbon cycle of temperate and boreal forests. *Nature.* 447: 848-850.

36. McGuire, A.D. (1995). The responses of net primary production (NPP) and total carbon storage for the continental United States to changes in atmospheric CO_2, climate, and vegetation. *Bull. Ecol. Soc. Am.* 76: 177.

37. Monteith, D.T., Stoddard, J.L., Evans, C.D., de Wit, H.A., Forsius, M. and Hogasen, T. (2007). Dissolved organic carbon trends resulting from changes in atmospheric deposition chemistry. *Nature.* 450: 537-539.

38. Norby, R.J., Ledford, J., Reilly, C.D., Miller, N.E. and O'Neill, E.G. (2004). Fine-root production dominates response of a deciduous forest to atmospheric CO_2 enrichment. *Proc. Natl. Acad. Sci. U.S.A.* 101: 9689-9693

39. Ostle, N.J. (2009). Integrating plant–soil interactions into global carbon cycle models. *Journal of Ecology.* 97: 851-863.

40. Parmesan, C. and Yohe, G. (2003). A globally coherent fingerprint of climate change impacts across natural systems. *Nature.* 421: 37-42.

41. Pearson, R.G., Phillips, S.J., Loranty, M.M., Beck, P.S.A., Damoulas, T., Knight, S.J. and Goetz, S.J. (2013). Shifts in Arctic vegetation and associated feedbacks under climate change. *Nature Climate Change.* 3: 673-677.

42. Pettersson, M. and Baath, E. (2003). Temperature-dependent changes in the soil bacterial community in limed and unlimed soil. *FEMS Microbiol. Ecol.* 45: 13-21.

43. Post, W.M., Emanuel, W.R., Zinke, P.J. and Stangenberger, A.G. (1982). Soil carbon pools and world life zones. *Nature.* 298: 156-159.

44. Schimel, D.S., Braswell, B.H., Holland, E.A., McKeown, R., Ojima, D.S. and Painter, T.H. (1994). Climatic, edaphic, and biotic controls over storage and turnover of carbon in soils. *Global Biogeochem. Cycles.* 8: 279-293.

45. Schimel, J., Balser, T.C. and Wallenstein, M. (2007). Microbial stress-response physiology and its implications for ecosystem function. *Ecology.* 88: 1386-1394.

46. Schimel, J.P. and Schaeffer, S.M. (2012). Microbial control over carbon cycling in soil. *Frontiers in Microbiology 3. fmicb.* 00348.

47. Schindlbacher, A., Rodler, A., Kuffner, M., Kitzler, B., Sessitsch, A. and Zechmeister-Boltenstern, S. (2011). Experimental warming effects on the microbial community of a temperate mountain forest soil. *Soil Biol Biochem.* 43: 1417-1425.

48. Schoenborn, L., Yates, P.S., Grinton, B.E., Hugenholtz, P. and Janssen, P.H. (2004). Liquid serial dilution is inferior to solid media for isolation of cultures representative of the phylum-level diversity of soil bacteria. *Applied and Environmental Microbiology.* 70(7): 4363-4366.

49. Sheik, C.S., Beasley, W.H., Elshahed, M.S., Zhou, X., Luo, Y. and Krumholz, L.R. (2011). Effect of warming and drought on grassland microbial communities. *Isme J.* 5: 1692-1700.

50. Singh, H.B. (2010). Pollution influences on atmospheric composition and chemistry at high northern latitudes: boreal and California forest fire emissions. *Atmospheric Environment.* 44: 4553-4564.

51. Sistla, S.A. and Schimel, J.P. (2013). Seasonal patterns of microbial extracellular enzyme activities in an arctic tundra soil: identifying direct and indirect effects of long-term summer warming. *Soil Biology and Biochemistry.* 66: 119-129.

52. Sitch, S., Cox, P.M., Collins, W.J. and Huntingford, C. (2007). Indirect radiative forcing of climate change through ozone effects on the land-carbon sink. *Nature.* 448: 791-794.

53. Smith, P., Martino, D., Cai, Z., Gwary, D., Janzen, H.H., Kumar, P. (2008). Greenhouse gas mitigation in agriculture. *Phil Trans Royal Soc B.* 363: 789-813.

54. Timmusk, S., Paalme, V. and Pavlicek, T. (2011). Bacterial distribution in the rhizosphere of wild barley under contrasting microclimates. *PLoS One.* 6(3): e17968.

55. van der Heijden, M.G.A., Klironomos, J.N., Ursic, M., Moutoglis, P., Streitwolf-Engel, R., Boller, T., Wiemken, A. and Sanders, I.R. (1998). Mycorrhizal fungal diversity determines plant biodiversity, ecosystem variability and productivity. *Nature.* 396: 69-72.

56. van der Putten, W.H. (2013). Plant–soil feedbacks: the past, the present and future challenges. *Journal of Ecology.* 101: 265-276.

57. Van der Putten, W.H. (2012). Climate change, aboveground–belowground interactions, and species' range shifts. *Annual Review of Ecology, Evolution, and Systematics.* 43: 365-383.

58. Vandenkoornhuyse, P., Quaiser, A., Duhamel, M., Le Van, A. and Dufresne, A. (2015). The importance of the microbiome of the plant holobiont. *New Phytologist*. 206(4): 1996-1206.

59. Vander Gucht, K., Cottenie, K., Muylaert, K., Vloemans, N., Cousin, S. and Declerck, S. (2007). The power of species sorting: Local factors drive bacterial community composition over a wide range of spatial scales. *Proc. Natl. Acad. Sci. U.S.A.* 104: 20404-20409.

60. Vishwakarma, K., Sharma, S., Kumar, N., Upadhyay, N., Devi, S. and Tiwari, A. (2015). Contribution of Microbial Inoculants to Soil Carbon Sequestration and Sustainable Agriculture. Springer Book Series.

61. Walther, G.R., Post, E., Convey, P., Menzel, A., Parmesan, C., Beebee, T.J.C., Fromentin, J.M., Hoegh-Guldberg, O. and Bairlein, F. (2002). Ecological responses to recent climate change. *Nature*. 416: 389-395.

62. Wardle, D.A., Bardgett, R.D., Klironomos, J.N., Setälä, H., van der Putten, W.H., Wall, D.H. (2004). Ecological linkages between aboveground and belowground biota. *Science*. 304: 1629.

63. Wookey, P.A., Aerts, R., Bardgett, R.D., Baptist, F., Brathen, K.A., Cornelissen, J.H.C., Gough, L., Hartley, I.P., Hopkins, D.W., Lavorel, S. and Shaver, G.R. (2009). Ecosystem feedbacks and cascade processes: understanding their role in the responses of Arctic and alpine ecosystems to environmental change. *Global Change Biology*. 15: 1153-1172.

64. Yergeau, E., Bokhorst, S., Kang, S., Zhou, J., Greer, C.W., Aerts, R. and Kowalchuk, G.A. (2012). Shifts in soil microorganisms in response to warming are consistent across a range of Antarctic environments. *Isme J*. 6: 692-702.

65. Zhang, W., Parker, K.M., Luo, Y., Wan, S., Wallace, L.L. and Hu, S. (2005). Soil microbial responses to experimental warming and clipping in a tallgrass prairie. *Global Change Biol*. 11: 266-277.

66. Zhou, J., Xue, K., Xie, J., Deng, Y., Wu, L., Cheng, X., Fei, S., Deng, S., He, Z. and Nostrand, J.D.V. (2012). Microbial mediation of carbon-cycle feedbacks to climate warming. *Nat Clim Change*. 2: 106-110.

67. Zogg, G.P., Zak, D.R., Ringelberg, D.B., MacDonald, N.W., Pregitzer, K.S. and White, D.C. (1997). Compositional and functional shifts in microbial communities due to soil warming. *Soil Sci. Soc. Am. J*. 61: 475-481.

5

Understanding Cyanobacterial Diversity in Relation to Different Ecological Niches

N.K. Singh[1], D.B. Patel[1], B.G. Morad[1] and D.W. Dhar[2*]

[1]*Department of Microbiology, C.P. College of Agriculture,*
S.D. Agricultural University, Sardarkrushinagar-385506, Gujrat, India
[2]*Centre for Conservation and Utilization of Blue Green Algae,*
Division of Microbiology, Indian Agricultural Research Institute,
New Delhi-110012, India
E-mail: dollywattaldhar@yahoo.com, dwdhar@iari.res.in

Abstract

Ecological niche play an important role in deciding the ecotypes and the diversity of an organism in a particular selection pressure. The organisms undergo niche construction, when exposed to a particular environment and facilitate the availability of nutrient to the other species constituting the food chain and food web. Cyanobacteria are gram negative, microscopic, prokaryotic organisms, show oxygenic photosynthesis, and exhibit huge diversity in relation to different ecological niches. They play an important role in agroecosystems due to their ability for nitrogen fixation, phosphate solubilization, production of phytohormones (auxins, gibberellins, cytokinins), and antimicrobial substances. Besides, they act as ameliorant for soil physical and chemical properties.

Some cyanobacterial strains like *Nostoc, Phormidium* and *Plectonema* are abundant in the alkaline water samples. *Nostoc muscorum, Anabaena variabilis, Oscillatoria princeps,* and *Plectonema* have excellent scavenging ability for available phosphorous and ammonical and nitrate-nitrogen from sewage effluent. These are potent candidates for bioremediation of a large number of organic and inorganic pollutants. Cyanobacteria are one of the dominant phototrophs in Antarctic terrestrial and freshwater ecosystems. Their excellent ability to survive under repeated cycles of drying and rehydration and to withstand high UV levels makes them the first colonizers in the succession of microbial populations on stone.

The cyanobacterial species such as *Loriella osteophila* and *Scytonema julianum* are troglophiles (species that grow and reproduce in caves) whereas *Geitleria calcarea*, *G. floridana*, *Herpyzonema pulverulentum* and *Symphyonema cavernicolum* are obligate cavernicolous (cannot survive outside the cave). The marine microbiota is dominated mainly by the unicellular cyanobacteria *Synechococcus* and *Prochlorococcus*, which are photoautotrophic picoplankton whereas; the predominant cyanobacteria present in the coastal wetland saline ecosystems are *Aphanothece halophytica*, *Dunaliella viridis*, *D. salina*, *Microcoleus chthonoplastes*, *Pleurocapsa entophysaloides*, and *Tetraselmis contracta*. Alternation of generations in some cyanobacteria between vegetative planktic and the resting benthic phases offers them a strategy for successful colonization in the saline and hypersaline lagoons.

The success of cyanobacteria in adaptation to various environmental conditions is largely attributed to the presence of PSI and PSII, chlorophyll, carotenoids, phycobiliproteins (phycoerythrin, phycocyanin, allophycocyanin), zeaxanthin, canthaxanthin, β-carotene, myxoxanthophylls together with the adoption of defence mechanisms like migration, mat formation, DNA repair, heat dissipation, synthesis of UVabsorbing/screening compounds, and several antioxidant systems. They produce mycosporin-like amino acids and indol alkaloid scytonemin that play role as UV-absorbing compounds. The heat-shock proteins and extracellular polysaccharides protect them against extremes of heat and desiccation. Their environmental dominance is also due to production of a large number of cyanotoxins, which may be cyclic peptides, alkaloids or lipopolysaccharides and exhibit neurotoxicity, hepatotoxicity, and/ or dermatotoxicity. Moreover, the exopolysaccharides produced by cyanobacteria provide protection to the cells against desiccation, extreme temperature, high light intensity, salinity, and metallic elements, and thus, encourage the cyanobacterial cells to colonize special ecological niches.

Key words: Cyanobacteria, Diversity, Ecological Niches

Introduction

In the process of changing the selection pressure to which organisms are exposed, one organism modulates the availability of resources to the other species present in the vicinity what is referred to as niche construction. It is an important episode in the evolutionary process as it builds connecting link between the biotic components of ecosystems. Organisms in a niche also play significant non-trophic impacts like ecosystem structure, function, and biodiversity. Ecological niches define ecotypes and these taxonomic entities experience periodic severe selection, which acts as a cohesive force similar to that experienced in eukaryotes due to sexual reproduction (Johansen and Casamatta, 2005). It has now become a well known fact that only morphology based taxonomy is not adequate to describe the biodiversity in cyanobacteria. The microalgae, in particular, cyanobacteria are morphology-poor, and the genetic variation within many morphologically defined species is so great that it is sometimes concluded that these "species" are actually complexes of cryptic species (Boyer *et al.*, 2002). However, a set of morphological character in addition to the biochemical, molecular, environmental, and evolutionary studies are required to understand the cyanobacterial diversity in relation to different ecological niches.

Cyanobacteria are microscopic gram negative prokaryotic organisms capable of showing oxygenic photosynthesis and are known to occupy diverse ecological niches and exhibit enormous diversity in terms of their habitat, physiology, morphology and metabolic activities (Singh and Dhar, 2011a; Singh and Dhar, 2014). The numerical abundance of cyanobacteria in various niches empowers them to play a profound impact on almost all biochemical cycles that shape life on Earth. They are regarded as one of the most important players in global oxygen supply, carbon dioxide (CO_2) sequestration, nitrogen fixation, as well as the primary phototrophic production of biomass (Beck *et al.*, 2012). Cyanobacteria are ubiquitous in distribution and are often found in extreme environmental conditions such as hot springs, frozen lakes of Antarctica, freshwater bodies, brackish waters, salt ponds, oceans and deserts (Whitton and Potts, 2000) and various agroecosystems with different soil type and crops combinations.

Thus, cyanobacteria can be said to occur in virtually all the ecosystem habitats on Earth, a success that surely reflects their long evolutionary history of around 3.5 billion years. The ability of cyanobacteria to adapt to extremes of environmental stresses like tolerance of freezing, desiccation, freeze–thaw cycles, high light intensities, oligotrophic low-nutrient conditions, and ability to tolerate relatively high levels of heavy metals. Such an array of physiological adaptation clearly reflects their ability to colonize harsh environments and major ecological roles these organisms play in the aquatic and terrestrial ecosystems.

Factors affecting success of cyanobacteria

Microalgae including cyanobacteria comprises a vast group of microscopic, simple, single-celled organisms that occur as discrete individuals alone, in pairs, in clusters, or in sheets of individuals all looking alike; but cannot form roots, stems, or leaves. These are considered photosynthetic cell factories, have high surface area-to-volume ratio, primary synthesizers of organic matter, capable of rapid uptake of nutrients and CO_2, possess faster cell growth and have much higher photosynthetic efficiency (Singh and Dhar, 2011b). Although, photosynthetic apparatus of cyanobacteria is more or less similar to that of higher plants, but it has some specialties. They are known to possess PSI and PSII, chlorophyll, and accessory pigments like carotenoids and phycobiliproteins (phycoerythrin, phycocyanin and allophycocyanin). The presence of accessory pigments in addition to chlorophyll in the photosystems extend the spectrum of visible light usable for photosynthesis. Terrestrial cyanobacteria also possess other pigments such as zeaxanthin, canthaxanthin, b-carotene and the myxoxanthophylls in addition to the above pigments and may play an important role in heat dissipation and photoprotection (Albrecht *et al.*, 2001; Ibelings *et al.*, 1994; Lakatos *et al.*, 2001). Moreover, in-spite of several detrimental effects of UV-induced reactive oxygen species (ROS) and oxidative stress at cellular and biochemical levels, cyanobacteria still survive and grow well in their natural habitats with high solar insolation.

In the process of natural selection and evolution, cyanobacteria have evolved several defence mechanisms such as avoidance (eg. migration and mat formation), DNA repair and heat dissipation mechanisms, synthesis of UVabsorbing/ screening compounds and several antioxidant systems to counteract the damaging effects of UV-induced oxidative stress (Rastogi *et al.*, 2014; Banerjee *et al.*, 2013). cyanobacteria are able to minimize the UV-induced oxidative damage caused by ROS through the antioxidant defence mechanisms operated due to activities of a large number of enzymes (catalase, superoxide dismutase, ascorbate peroxidise, glutathione peroxidase, and glutathione reductase) and metabolites (carotenoids, ascorbic acid, α-tocopherols and reduced glutathione) (Rastogi and Madamwar, 2015). The presence of antioxidant systems may however, exclusively regulate the homeostasis of ROS formation in cells. Some other group of secondary compounds such as polyamines with free radical scavenging activity have also been reported in cyanobacteria (Incharoensakdi *et al.*, 2010).

Moreover, they are also able to produce highly effective UV-absorbing compounds, such as mycosporin-like amino acids (MAA) and the indol alkaloid scytonemin, which is deposited extracellularly (Garcia-Pichel and Castenholz, 1991; Karsten and Garcia-Pichel, 1996) and protect them from its harmful effects. In addition, synthesis of extracellular polysaccharides (EPS) also plays an important role in mitigation strategy against desiccation and harmful effects of UV radiation (Chen *et al.*, 2009). Many of the cyanobacterial strains (25 to 75%) are known to form blooms that produce secondary metabolites named cyanotoxins which may harm life forms like zooplankton, shellfish, fish, birds, and mammals. These cyanotoxins may belong to a diverse group of chemical substances like cyclic peptides, alkaloids or lipopolysaccharides which may be neurotoxic, hepatotoxic, and dermatotoxic (Singh *et al.*, 2008; Singh and Dhar, 2013) and hence, improves the competitive ability of these organisms.

Upon temperature upshift, cyanobacteria induce a set of proteins called the heat-shock proteins or Hsps, by transcriptional activation (Rajaram and Apte, 2010). The magnitude of induction of these proteins is determined by the growth temperature and the extent of temperature upshift. The most prominent Hsps that accumulate in the cyanobacterial cells are GroEL, small Hsps and GroES (Bhagwat and Apte, 1989). Thermotolerance of both unicellular and filamentous cyanobacterial species is enhanced upon pre-treatment at sublethal temperatures through involvement of heatshock genes/proteins (Blondin *et al.*, 1993). A brief description of the functions of Hsps in cyanobacteria is provided in Table 1.

Table 1: Heat-shock proteins of cyanobacteria (modified after Rajaram *et al.*, 2014)

Family	Gene	Function	Reference(s)
Hsp60 Hsp10	groEL, cpn60 (Synechocystis PCC6803, Anabaena L-31)	GroEL provide thermotolerance under nitrogen fixing conditions whereas Cpn60 under nitrogen replete conditions	Chaurasia and Apte (2009), Rajaram and Apte (2008)

Family	Gene	Function	Reference(s)
Hsp70 Hsp40 GrpE	dnaK1/2/3, seven dnaJ grpE (Synechocystis PCC6803, Synechococcus PCC7942)	DnaJ2 provide thermotolerance and DnaK2 produce RNA chaperone thermosensor	Varvasovszki *et al.*, (2003), Watanabe *et al.*, (2007), Barthel *et al.*, (2011)
Hsp90	htpG (Synechococcus PCC7942)	Innate and acquired thermotolerance, protection of photosynthetic apparatus	Tanaka and Nakamoto (1999), Sato *et al.*, (2010)
Hsp100	clpBI (Synechococcus PCC7942)	Help in acquired thermotolerance	Eriksson and Clarke (1996)
	clpBII Synechococcus PCC7942	Provide cold tolerance	Porankiewicz and Clarke (1997)
sHsp	hsp16.6/hsp17/ hspA (Synechocystis PCC6803, Synechococcus PCC7942)	Protects membrane fluidity and provide thermotolerance	Nakamoto *et al.*, (2000), Horvath *et al.*, (1998), Lee *et al.*, (2000)

Agroecosystems

A great diversity of cyanobacteria occur naturally in the agricultural soil and are also being used increasingly as biofertilizers in agriculture due to their role as diazotrophs, ameliorant for soil physical and chemical properties, establishing proficiency in diverse soil ecologies, and their ability to compete with native flora and fauna (Dhar *et al.*, 2009; Singh *et al.*, 2014). They are able to tolerate various stresses in salt-affected soils such as nutrient deficiency, salinity, drought and temperature rise. Their adaptation to salt stress consists of at least three phenomena: accumulation of internal osmoticum in the form of inorganic ions or organic solutes, contribution of ion transport processes, and metabolic adjustments (Singh and Dhar, 2010a). Various mechanisms adopted by cyanobacteria for salt tolerance include among others production of stress responsive proteins, restricted entry of Na^+, Na^+ efflux, Na^+-dependent K^+ uptake in prokaryotes, compatible solutes and lipids in salt tolerance, and enhancement of cyanobacterial salt tolerance by combined nitrogen.

A large number of heterocystous and non-heterocystous cyanobacteria are able to do nitrogen fixation, phosphate solubilization, production of phytohormones (auxins, gibberellins, cytokinins), and provide protection from other microorganisms by antagonism or by production antimicrobial substances. Thus these microorganisms on the one side promote plant growth and on the other hand show significant biocidal activity against important agricultural insect pests (Singh, 2013). Bioactive compounds like, hapalindoles, calothrixins, cyanobacterins LU-1 and LU-2 and γ–lactone are proposed to possess allelopathic effect. The antimicrobial substances produced include nostocyclyne A, nosto fungicidin, nostocin A, Ambigol A and B, hapalindoles, tjipanazoles and scytophycins and exhibit fungicidal activity against important plant pathogens. These biocontrol agents provide multiple benefits and act as useful pointers for improving cultivation practices and establishment of plants in diverse inhospitable/barren habitats (Singh *et al.*, 2014). Therefore, they depict a promising multifaceted agent in integrated nutrient management, integrated pest and disease management, and organic farming practices popular in the present day agriculture.

Extracellular polymeric substances (EPS) in Soil crusts/Microbial mats

Soil crusts are a common feature of all dry land soils that form at the soil surface by an intimate association between mineral grains and organic matter with varying proportions of cyanobacteria, algae, lichens and mosses. These soils contain very small amounts of organic C, particularly humic substances; however, EPS is often a major source of C (Mager and Thomas, 2011). Microbial mats often support abundant populations of phototrophic cyanobacteria and microalgae and the chemo heterotrophs like archaea, bacteria, fungi and protozoa. The cyanobacterial diversity in microbial mats includes *Microcoleus, Oscillatoria, Lyngbya, Pseudanabaena, Phormidium* and *Spirulina* as the most abundant among the filamentous type and *Chroococcus, Gloeopcasa, Synechocystis* and *Mixosarcina* among the unicellular type. These mats like structures are often embedded in a thick mucilaginous matrix of EPS, which may consist of 90% or more of polysaccharides and implicated in the adhesion of microorganisms to sediment particles, thereby influencing the stability of the sediment bed (Stal, 2012). However, under suitable meteorological and hydrological conditions, the EPS-producing cyanobacterial colonies in lakes aggregate and float up to form the mucilaginous cyanobacterial blooms.

Under unfavourable climatic condition, the cyanobacteria are adapted to secrete high molecular mass polymers, which released into the surrounding environment as extracellular polymeic substances (EPS) (Mager and Thomas, 2011). EPS are formed of polysaccharides, proteins, humic substances, nucleic acids, and lipids (Flemming and Wingender, 2010) and play a major role in the protection of cells against adverse environmental conditions like desiccation, extreme temperature, light intensity, salinity, metallic trace elements etc thus encourage the cyanobacteria cells to colonize special ecological niches. Biological EPS secreted by cyanobacteria are recognized as agents of soil particle aggregation, improve the fertility of the soils, immobilize nutrients in the surface that would otherwise be lost by leaching, and are associated with colony development (Gan *et al.*, 2012). However, the production and biochemical composition of EPS varies with the soil type, environmental conditions and nutritional levels.

Cyanobacteria can migrate to the soil surface when wet and move back down the subsoil upon subsequent drying in order to avoid desiccation and photo-damage (Garcia-Pichel and Pringault, 2001). Cyanobacteria can even remain indefinitely in a desiccated state partly due to the protection afforded to the cells by the EPS. The physical structure of the EPS facilitate the cyanobacteria to absorb water many times their dry weight (Satoh *et al.*, 2002). Water and light influence the ability of cyanobacterial filaments to create surface aggregates and production of EPS. If the light is low, red color develops in the biofilm. Most of the Microbial mats showed dominance of cyanobacteria (*Oscillatoria, Lyngbya, Phormidium*and *Microcoleus*) and diatoms (De Philippis *et al.*, 2011; Yang *et al.*, 2011). They reported that EPS in biofilms protects the cells by maintaining a highly hydrated layer surrounding the biofilm, and thus, prevent lethal desiccation in some natural biofilms and protect against diurnal variations in high light intensity, high and low temperatures, high salinity, UV radiation, heavy metals, changes in pH of the suspending fluid, conservation of extracellular enzyme activities, and shelter from predation.

Inland aquatic system

Cyanobacteria in inland aquatic systems has wide adaption qualities, show spatial and temporal variation in their genetic community structure and respond differently to factors such as nutrient deprivation, light intensity, and predation (Singh *et al.*, 2011). They are considered highly effective in uptake of nutrients like N and P and accumulation and degradation of different kinds of environmental pollutants, including xenobiotic compounds. They increase the O_2 content of waters *via* photosynthesis while on the other they remediate some heavy metals by biosorption, alkaline precipitation and production of metallothioneins, chelatins and polysaccharides. Moreover, these communities affect the biochemical oxygen demand, chemical oxygen demand, turbidity, inorganic nutrients and pathogens (coliform bacteria); and at certain densities keep the aerobic phase of the facultative ponds functioning (Singh and Dhar, 2010b).

Borase *et al.*, (2013) observed twenty three cyanobacterial strains in the alkaline water samples of three water bodies (Anasagar, Pushkar lake and Sambhar lake) of Rajasthan. They noted one colonial, nine heterocystous and the remaining non-heterocystous filamentous forms. There was abundance of *Nostoc, Phormidium* and *Plectonema* followed by *Anabaena* and *Westiellopsis* however, *Microcystis* and *Oscillatoria* were least abundant. *Nostoc* and *Phormidium* showed highest abundance in terms of CFU/mL followed by *Anabaena* and *Plectonema*. Among the nitrogen fixing blue green algae isolated from soils of J and K state Mishra *et al.*, (2004) concluded that *Gloeotrichia* sp. exhibited a maximum dry weight accumulation while *Calothrix membranacea* the least.

Sediment-water interface of the shallow water ecosystems represents a distinct macromolecular matrix in which microorganism assemblages coexist, typically attached to a surface, in which complex food webs occur (Larson and Passy, 2005; De Vicente *et al.*, 2006). In such an ecosystem, the photosynthetic activity by benthic microalgae is the primary source of fixed carbon, and they may contribute up to 50% of the total autotrophic production in some ecosystems (Underwood and Kromkamp, 1999), and sometime, the photosynthetic activity of benthic attached algae in wetlands may rival or even surpass primary productivity rates of aquatic macrophytes (Poulickova *et al.*, 2008).

Singh and Dhar (2007) observed significant correlation between N and P removal and dry weight and pigments of cyanobacteria which indicate the utility of sewage effluent for cultivation of microalgae with the efficient N and P scavenging ability. *Chlorella vulgaris* was most efficient in scavenging ammonical nitrogen while nitrate-scavenging ability was highest in *Oscillatoria*. However, *Nostoc muscorum, Anabaena variabilis, Oscillatoria princeps, Plectonema* sp and *Chlorella* were all able to remove available phosphorous efficiently.Microalgae are valuable sources of food and feed products, high-value oils, biofuels, chemicals, medicinal products and pigments. These are potent candidates for bioremediation of a large number of pollutants including the distillery effluents which is also referred to as spentwash/stillage/slop/vinasses (Singh and Patel, 2012), which are one of the most environmentally aggressive industrial effluents. However, the distribution of cyanobacteria in inland aquatic systems may get affected due to anthropogenic activities like habitat loss and fragmentation, disturbance and pollution, construction of dams, chemical pollution and global climatic changes.

Frozen lakes of Antarctic

Cyanobacteria have often been recorded as the dominant phototrophs in Antarctic terrestrial and freshwater ecosystems. On the basis of geographical distribution the cyanobacteria *Oscillatoria* cf. *subproboscidea* was considered endemic to Antarctica (Komarek, 1999). Most of the species found in the field microbial mat sample from Lake Fryxell of Antarctica and in an artificial cold-adapted sample have been found in various biotopes outside Antarctica and appear to have cosmopolitan distributions on the basis of morphological data. Therefore, the morphological results support the idea that endemism is rare among Antarctic cyanobacteria (Taton *et al.*, 2003). However, the molecular tools revealed an entirely different pattern and a higher number of phylotypes than morphotypes was detected. The phylogenetic analysis showed that sequences determined in this study along with the polar sequences available from the databases were distributed in 22 lineages, 2 of which were novel and 7 of which were exclusively Antarctic.

Cyanobacterial diversity and ecology on monuments

The microbiota on building stones represents a complex ecosystem that develops depending upon the environmental conditions and physicochemical properties of the material. Epilithic biofilms associated with the Mayan monuments at Uxmal (Yucatan, Mexico) revealed dominance of the cyanobacterial genera *Xenococcus, Gloecapsa, Gloethece, Synechocystis* and *Synechococcus* (Ortega-Morales *et al.*, 1999). De Miguel *et al.*, (1995) also found cyanobacterial genera *Phormidium, Plectonema, Scytonema, Chlorogloepsis* and *Gloecapsa* as the most representative one. They also noted that, except for *Chlorella*, eucaryotic algae were absent from most samples.

Cyanobacteria are photolithoautotrophic organisms, can survive repeated cycles of drying and rehydration, and are able to withstand high UV levels (Schwieger and Tebbe, 1998). These physiological attributes make them particularly important on exposed surfaces and they are considered the first colonizers in the succession of microbial populations on stone, although under certain circumstances oligotrophic heterotrophic microbes (bacteria and fungi) can develop without the need for nutrients from excreted metabolites or cyanobacterial biomass (Albertano and Urzi, 1999; Crispim and Gaylarde, 2005). Phototrophic microorganisms may grow on the stone surface (epilithic phototrophs) or may penetrate some millimetres into the rock pore system (endolithic phototrophs) (Golubic *et al.*, 1981). These epilithic and endolithic organisms can potentially con tribute to the breakdown of rock crystalline structures (Crispim and Gaylarde, 2005) through the release of their metabolic products, such as inorganic and organic acids. In addition, extracellular polymeric materials, principally polysaccharides, act as glues, trapping dirt and other particulate materials, increasing the damaging effects of the biofilm. Such hygroscopic biomolecules may cause mechanical stresses to the mineral structure due to shrinking and swelling cycles of these colloidal biogenic slimes inside the pore system, leading to the alteration of pore size and distribution, together with changes in moisture circulation patterns and temperature response.

Limestone caves

Limestone caves are highly specific environments scattered all over the world and karst caves are considered a specific case of extreme environment (Mulek and Kosi, 2008). Caves generally represent stable environments characterized by uniform temperatures throughout the year, high humidity and low natural light. However, cave characteristics such as dimensions, morphology, location, orientation and lithic substrate can play important role in deciding the biocommunity structure. Hoffmann (2002) has categorized the taxa recorded in caves into: 'troglobitic' species (obligatory cavernicoles) that cannot survive outside the cave, 'troglophilic' species that grow and reproduce in caves, and 'trogloxenic' species those accidentally reach the cave environment. The cyanobacterial species such as *Loriella osteophila* and *Scytonema julianum* can be considered as troglophiles. Very few are considered as obligatory cavernicolous such as *Geitleria calcarea*, *G. floridana*, *Herpyzonema pulverulentum* and *Symphyonema cavernicolum* (Asencio and Aboal, 2000; Hoffmann, 2002).

However, in a typical cave where light shows a clear gradient from the entrance inwards, the cyanobacterial community is organized into mosaics following the expected typical pattern of cyanobacterial distribution. The entrance community zone shows biofilms with the highest number of species mostly of the order Chroococcales (Roldan and Hernandez-Marine, 2009). The mucilaginous, temporary biofilms protected against dessication where direct light prevails at the entrance zone is dominated by *Aphanocapsa.*, *Chroococcus*, *Eucapsis minor*, *Leptolyngbya gracillima*, *L. perelegans* and *Pseudophormidium spelaeoides*. On the contrary, the dim light community zone is characterized by a lower number of species with Oscillatoriales prevailing over Chroococcales (Roldan & Hernandez-Marine, 2009). In such location, biofilms become thinner and less mucilaginous and cyanobacteria with calcified filaments that are able to survive at dim light are *Scytonema julianum* and *Iphinoe spelaeobio*. Lamprinou *et al.*, (2012) further noted that *Scytonema julianum* together with the species *Leptolyngbya palikiana*, *L. gracillima*, *Pseudophormidium spelaeoides* almost always present from the entrance to the deeper part of cave.

Marine cyanobacteria

Oceans cover more than two-third of the Earth's surface and represent vast sources and sinks for the biogeochemical cycling of various elements. Around 40% of global primary production, the fixation of carbon dioxide into biomass occurs in marine systems and around three-quarters of this take place in open ocean environments (Scanlan, 2001). Earlier large areas of oligotrophic oceans were regarded as biological deserts that support few marine organisms and represent only a small fraction of global oceanic productivity. However, the discovery of tiny, single-celled cyanobacteria as ubiquitous and abundant components of the marine microbiota entirely changed the composition and concept of the functioning of the marine ecosystems. These unicellular cyanobacteria are dominated mainly by two genera, *Synechococcus* and *Prochlorococcus*, which are also regarded as photoautotrophic picoplankton.

Prochlorococcus (0.6 μm diameter) was originally considered to be a member of the Prochlorophytes, the closest known relatives of higher plant chloroplasts. It represents oxyphotobacteria and is essentially cyanobacteria which have evolved chl *b* independently in the evolutionary process (Whitton and Potts, 2000). The presence of divinyl derivatives of chl *a* and *b* make this organism unique amongst this group, a feature which appears to be a direct adaptation to harvesting the longer wavelengths of blue light that penetrate deepest down the water column in oceanic environments (West and Scanlan, 1999). It also contains phycoerythrin, a phycobiliprotein normally required for light-harvesting. The success of this organism is due to several factors, but its small size, and hence high surface area/volume ratio, allows it to acquire nutrients even in ultra-oligotrophic waters, whilst an ability to fix carbon even at extremely low light levels suggests a tremendous photoacclimation or photoadaptation capacity. However, the sister genus, *Synechococcus* possess normal chl *a* and phycobilins as their pigment complement (Bryant, 1994). *Prochlorococcus* is although largely confined to 40°N-40°S latitudinal band with a density of 10^5–10^6 cells per ml and occupy a 100-200m deep layer in surface irradiance (Partensky *et al.*, 1999). However, presence of *Synechococcus* group is a ubiquitous feature of phytoplankton populations throughout the world's oceans and so can be found even in Polar Regions, albeit at lower cell concentrations than equatorial areas.

Saline and hypersaline lagoons

Hypersaline environments in the shallow and sheltered lagoon show temporal and spatial differences in salinity up to desiccation due to large fluctuations of temperature related to high rates of evaporation and low precipitation in the coastal desert. Benthic cyanobacteria are abundant in mangrove environments and 20% of species that occur in saline conditions are truly marine. Cyanobacteria can be divided into euryhaline (living in brine are hyper-saline) and stenohaline (living in brackish water) forms (Desikachary, 1959). These lagoons/mangrove forests are among the world's most productive ecosystems. They provide a huge amount of carbon and nitrogen supply to the coastal waters and maintaining a rich coastal biodiversity (Kathiresan and Bingham, 2001) and are known to occur in great diversity in mangrove communities throughout the world.

Most euryhaline algae and cyanobacteria found in the lagoon have been reported for other saline environments as well (Montoya, 2009). Hypersaline waters are usually dominated by coccoid colonies and filamentous cyanobacteria, unicellular diatoms and unicellular or filamentous chlorophytes (Borowitzka, 1981; Garcia-Pichel *et al.*, 1998). The predominant cyanobacteria that are reported to be present in the coastal wetland saline ecosystems are *Aphanothece halophytica, Dunaliella viridis, D. salina, Microcoleus chthonoplastes, Pleurocapsa entophysaloides,* and *Tetraselmis contracta* (Montoya and Golubic, 1991; Montoya and Olivera, 1993; Aguilar, 1998). However, Montoya (2009) reported presence of a large number of cyanobacterial species that dominate the coastal saline wetlands were *Aphanothece stagnina, Chroococcus turgidus, Gomphosphaeria aponina, Lyngbya aestuarii, Microcoleus chthonoplastes, Nodularia spumigena, Oscillatoria limnetica, O. tenuis, Phormidium hypolimneticum, Pleurocapsa entophysaloides,* and *Spirulina subsalsa.* Most of which also occur in other saline and hypersaline ecosystems.

The benthic biofilm colonization and species dynamics provide a laminated or flocculent colloidal texture that contributes to the sediment stabilization in the lagoon. The structural integrity of sediment assemblages is maintained by the cyanobacterial extracellular matrix that contains the extracellular polymeric substance (EPS). The highly hydrated EPS layer besides providing protection against desiccation and biogenic sediment stabilization also plays a functional role in motility within the biofilm (Cibic *et al.*, 2007; Otero and Vincenzini, 2004). Some species of *Oscillatoria* and *Lyngbya* are motile and they may migrate through the biofilm mucilage of *Chroococcus.* Moreover, microalgal communities of saline lagoons, solar salterns and harsh habitats of the desert crusts display a morphological diversity of cyanobacteria with different community structures (Garcia-Pichel *et al.*, 2001; Oren, 2002). The distribution of *Microcoleus chthonoplastes* in cyanobacterial mat is considered cosmopolitan in distribution (Garcia – Pichel *et al.*, 1996). *Microcoleus* species are known to produce polyhydroxyl carbohydrates which replace the water shell around cellular macromolecules, and thus, encourage the cells to tolerate desiccation (Potts, 1999).

Cyanobacteria like, *Aphanizomenon flos-aquae, Nodularia spumigena, Oedogonium pringsheimii, Prorocentrum cassubicum, Tetraselmis contracta* are known to evolve alternation of generations between vegetative planktic and resting benthic phases, which offers them a strategy for successful colonization. The resting stages are tolerant to extreme conditions of desiccation. They may survive buried in the sediment for many years, however, the level of tolerance can vary with the species and the amount of sediment disturbance. The resting stages in the microphytobenthos can serve as refuge populations (seed bank) for recolonization following harsh times of unfavorable conditions, as was reported for other algal groups (Gao and Ye, 2007). Therefore, the shallow sediments of the Grande coastal lagoon are important for the recruitment of species from sediments, to serve as inoculum enabling development of future blooms.

Some cyanobacterial strains growing slowly at high salinities also show phenotypic plasticity. The tendency to form compact mucilage that slows down cell dispersal results in the formation of large colonies (Dor and Hornoff, 1985; Garcia-

Pichel *et al.*, 1998). Significant size differences with smaller cells at higher salinities, as well as changes in shape were reported for *Cyanothece* or *Aphanothece* strain from a solar evaporation pond in Eilat, Israel (Yopp *et al.*, 1978). These findings support the idea that morphological divergence is not always reflected in genetic diversity (Palinska *et al.*, 1996). However, some cyanobacteria show congruence in morphological and phylogenetic traits (Casamatta *et al.*, 2005). The dynamic benthic biofilm ecosystem has been considered a depositional environment that entraps particles and precipitates, in which the extracellular matrix controls preservation and viability of the species. The growth and reproductive phases of phytoplankton (cyanobacteria and microalgae) during the turbid state of the lagoon showed the heterogeneity and complexity of their adaptive strategies.

Cyanobacteria have the capacity to grow under very low water potential, and hence, such species can resist desiccation and grow in the arid environments (deserts) or can tolerate high salinity to grow in hyper-saline ponds. They have evolved a broad variety of adaptations, including the accumulation of osmolytes, to cope with osmotic and ionic stress in such hostile and extreme environments (Oren, 2000). Extreme environments such as saline and hypersaline lagoons are hostile to most forms of life; however, they harbor significant populations of microorganisms. Their colonization by primary producers (algae and cyanobacteria) demonstrates that such organisms can adapt to extreme ecological niches.

References

1. Aguilar, S.C. (1998). Crecimiento e historia de vida de *Dunaliella salina* de las salinas de Los Chimus, Ancash y de Chilca, Lima, Perú. *Anales del IV Congreso Latino Americano de Ficología, Brasil*. 996: 309-324.

2. Albertano, P. and Urzi, C. (1999). Structural interactions among epilithic cyanobacteria & heterotrophic microorganisms in Roman Hypogea. *Microbial Ecol.* 38: 244-252.

3. Albrecht, M., Steiger, S. and Sandmann, G. (2001). Expression of a ketolase gene mediates the synthesis of canthaxanthin in *Synechococcus* leading to tolerance against photoinhibition, pigment degradation and UV-B sensitivity of photosynthesis. *Photochem Photobiol.* 73: 551-555.

4. Asencio, A. and Aboal, M. (2000). Algae from Seretta cave (Murcia, SE Spain) and their environmental conditions. *Algol. Stud.* 69: 59-78.

5. Banerjee, M., Raghavan, P.S., Ballal, A., Rajaram, H. and Apte, S.K. (2013). Oxidative stress management in the filamentous, heterocystous, diazotrophic cyanobacterium, Anabaena PCC7120. *Photosynth Res.* 118(1-2).

6. Barthel, S., Rupprecht, E. and Schneider, D. (2011). Thermostability of two cyanobacterial GrpE thermosensors. *Plant Cell Physiol.* 52: 1776-1785.

7. Beck, C., Knoop, H., Axmann, I.M. and Steuer, R. (2012). The diversity of cyanobacterial metabolism: genome analysis of multiple phototrophic microorganisms. *BMC Genomics* 13:56. http://www.biomedcentral.com/1471-2164/13/56.

8. Bhagwat, A.A. and Apte, S.K. (1989). Comparative analysis of proteins induced by heat shock, salinity, and osmotic stress in the nitrogen fixing cyanobacterium *Anabaena* sp. strain L-31. *J Bacteriol.* 171: 5187-5189.

9. Blondin, P.A., Kirby, R.J. and Barnum, S.R. (1993). The heat shock response and acquired thermotolerance in three strains of cyanobacteria. *Curr Microbiol.* 26: 79-84.

10. Borase, D., Dhar, D.W. and Singh, N.K. (2013). Diversity indices and growth parameters of cyanobacteria from three lakes of Rajasthan. *VEGETOS.* 26(2): 377-383.

11. Borowitzka, L.J. (1981). The microflora. Adaptations to life in extremely saline lakes, *Hydrobiologia.* 81: 33-46.

12. Boyer, S.L., Johansen, J.R. and Flechtner, V.R. (2002). Phylogeny and genetic variance in terrestrial *Microcoleus* (Cyanophyceae) species based on sequence analysis of the 16S rRNA gene and associated 16S-23S ITS region. *J Phycol.* 38: 1222-1235.

13. Bryant, D.A. (1994). The Molecular Biology of Cyanobacteria. Dordrecht: Kluwer Academic Publishers.

14. Casamatta, D.A., Johansen, J.R., Vis, M.L. and Broadwater, S.T. (2005). Molecular and morphological characterization of ten polar and near-polar strains within the Oscillatoriales (Cyanobacteria). *Journal of Phycology.* 41: 421-438.

15. Chaurasia, A.K. and Apte, S.K. (2009). Over expression of the groESL operon enhances the heat and salinity stress tolerance of the nitrogen fixing cyanobacterium *Anabaena* sp. strain PCC7120. *Appl Environ Microbiol.* 75: 6008-6012.

16. Chen, L.Z., Wang, G.H., Hong, S., Liu, A., Li, C., Liu, Y.D. (2009). UV-B-induced oxidative damage and protective role of exopolysaccharides in desert cyanobacterium *Microcoleus vaginatus. J Integr Plant Biol.* 51: 194-200.

17. Cibic, T., Blasutto, O., Hancke, K. and Johnsen, G. (2007). Microphytobenthic species composition, pigment concentration, and primary production in sublittoral sediments of the Trondheimsfjord (Norway). *Journal of Phycology.* 43: 1126-1137.

18. Crispim, C.A. and Gaylarde, C.C. (2005). Cyanobacteria and biodeterioration of cultural heritage: a review. *Microbial Ecol.* 49: 1-9.

19. De-Miguel, J.M.G., Sanchez-Castillo, L., Ortega-Calvo, J.J., Gil, A. and Saiz-Jimenez, C. (1995). Deterioration of building materials from the Great Jaguar pyramid at Tikal, Guatemala. *Build Environ.* 30: 591-598.

20. De-Philippis, R., Colica, G. and Micheletti, E. (2011). Exopolysaccharide-producing cyanobacteria in heavy metal removal from water: molecular basis and practical applicability of the biosorption process. *Appl Microbiol Biotechnol.* 92: 4697-4708.

21. De-Vicente, I., Amores, V. and Cruz-Pizarro, L. (2006). Instability of shallow lakes: A matter of the complexity of factors involved in sediment and water interaction? *Limnetica.* 5: 253-270.

22. Desikachary, T.V. (1959). Cyanophyta. Indian Council of Agricultural Research, New Delhi. 686.

23. Dhar, D.W., Saxena, S. and Singh, N.K. (2009). BGA biofertilizer: production, constraints and future perspectives. In: Mallik CP, Wadhwani C, Kaur B (eds) Crop breeding and biotechnology. *Pointer Publishers, Jaipur.* 209-226.

24. Dor, I. and Hornoff, M. (1985). Salinity-temperature relations and morphotypes of a mixed population of coccoid cyanobacteria from a hot, hypersaline pond in Israel. *Marine Ecology.* 6: 13-25.

25. Eriksson, M.J. and Clarke, A.K. (1996). The heat shock protein ClpB mediates the development of thermotolerance in the cyanobacterium *Synechococcus* sp. strain PCC 7942. *J. Bacteriol.* 178: 4839-4846.

26. Flemming, H.C. and Wingender, J. (2010). The biofilm matrix. *Nat Rev Microbiol.* 8: 623-633.

27. Gan, N., Xiao, Y., Zhu, L., Wu, Z., Liu, J., Hu, C., Song, L. (2012). The role of microcystins in maintaining colonies of bloom-forming *Microcystis* spp. *Environ Microbiol.* 14: 730-742.

28. Gao, K. and Ye, C. (2007). Photosynthetic insensitivity of the terrestrial cyanobacterium *Nostoc flagelliforme* to solar UV radiation while rehydrated or desiccated. *J Phycol.* 43: 628-635.

29. Garcia-Pichel, F. and Castenholz, R.W. (1991). Characterization and biological implications of scytonemin, a cyanobacterial sheath pigment. *J Phycol.* 27: 395-409.

30. García-Pichel, F., Lopez-Cortes, A. and Nubel, U. (2001). Phylogenetic an morphological diversity of cyanobacteria in soil desert crusts from the Colorado Plateau. *Applied and Environmental Microbiology.* 67: 1902-1910.

31. Garcia-Pichel, F., Nübel, U. and Muyzer, G. (1998). The phylogeny of unicellular, extremely halotolerant cyanobacteria. *Archives Microbiol.* 169: 469-482.

32. Garcia-Pichel, F. and Pringault, O. (2001). Cyanobacteria track water in desert soils. *Nature.* 413: 380-381.

33. Garcia-Pichel, F., Prufert-Bebout, L. and Muyzer, G. (1996). Phenotypic and phylogenetic analyses show *Microcoleus chthonoplastes* to be a cosmopolitan cyanobacterium. *Applied and Environ Microbiol.* 62: 3284-3291.

34. Golubic, S., Friedmann, I. and Schneider, J. (1981). The lithobiontic ecological niche, with special reference to microrganisms. *J. Sedimentary Petrology.* 51: 475-478.

35. Hoffmann, L. (2002). Caves and other low-light environments: Aerophytic photoautotrophic microorganisms. In: Bitton G (ed) Encyclopedia of Environmental Microbiology. New York: John Wiley & Sons. 835-843.

36. Horvath, I., Glatz, A., Varvasovszki, V., Torok, Z., Pali, T., Balogh, G., Kovacs, E., Nadasdi, L., Benko, S., Joo, F. and Vigh, L. (1998). Membrane physical state controls the signaling mechanism of the heat shock response in *Synechocystis* PCC 6803: identification of hsp17 as a "fluidity gene". Proc Natl Acad Sci USA 95:3513-3518.

37. Ibelings, B.W., Kroon, B.M.A., Mur, L.R. (1994). Acclimation of photosystem II in a cyanobacterium and a eukaryotic green alga to high fluctuating photosynthetic photon flux densities, simulating light regimes induced by mixing in lakes. *New Phytol.* 128: 407-424.

38. Incharoensakdi, A., Jantaro, S., Raksajit, W. and Maenpaa, P. (2010). In: Mendez-Vilas A (ed) Polyamines in cyanobacteria: biosynthesis, transport and abiotic stress response. In: Current Research, Technology and Education Topics in Applied Microbiology and Microbial Biotechnology, Microbiology Book Series No. Formatex, Spain. 23-32.

39. Johansen, J.R., Casamatta, D.A. (2005). Recognizing cyanobacterial diversity through adoption of a new species paradigm. *Algol. Stud.* 117: 71-93.

40. Karsten, U. and Garcia-Pichel, F. (1996). Carotenoids and mycosporine-like amino acid compounds in members of the genus Microcoleus (Cyanobacteria): a chemosystematic study. *System Appl Microbiol.* 19: 285-294.

41. Kathiresan, K. and Bingham, B.L. (2001). Biology of mangroves and mangrove ecosystems. *Advances in Marine Biology.* 40: 81-251.

42. Komarek, J. (1999). Diversity of cyanoprokaryotes (cyanobacteria) of King George Island, maritime Antarctica-a survey. *Arch Hydrobiol.* 94: 181-193.

43. Lakatos, M., Bilger, W. and Budel, B. (2001). Carotenoid composition of terrestrial cyanobacteria: response to natural light conditions in open rock habitats in Venezuela. *Eur. J. Phycol.* 36: 367-375.

44. Lamprinou, V., Danielidis, D.B., Economou-Amilli, A. and Pantazidou, A. (2012). Distribution survey of Cyanobacteria in three Greek caves of Peloponnese. *Int J Speleology.* 41(2): 267-272.

45. Larson, C. and Passy, S.I. (2005). Spectral fingerprinting of algal communities: a novel approach to biofilm analysis and biomonitoring. *Journal of Phycology.* 41: 439-446.

46. Lee, S., Owen, H.A., Prochaska, D.J. and Barnum, S.R. (2000). HSP16.6 is involved in the development of thermotolerance and thylakoid stability in the unicellular cyanobacterium, *Synechocystis* sp. PCC 6803. *Curr Microbiol.* 40: 283-287.

47. Mager, D.M. and Thomas, A.D. (2011). Extracellular polysaccharides from cyanobacterial soil crusts in dry land surface processes. *Journal of Arid Environments.* 75: 91-97.

48. Mishra, U., Pabbi, S., Dhar, D.W. and Singh, P.K. (2004). Floristic abundance and comparative studies on some specific nitrogen fixing blue green algae isolated from soils of J and K state. *Ad Plant Sci.* 2: 635-640.

49. Montoya, H. (2009). Algal and cyanobacterial saline biofilms of the Grande Coastal Lagoon, Lima, Peru. Natural Resources and Environmental Issues. 15: 127-134.

50. Montoya, H.T. and Golubic, S. (1991). Morphological variability in natural populations of mat forming cyanobacteria in the salines of Huacho, Lima, Perú. Algological Studies. 64: 423-441.

51. Montoya, H.T. and Olivera, A. (1993). Dunaliella salina Teodoresco (Chlorophyta) from saline environments at the Central Coast of Perú. *Hydrobiologia*. 267: 155-161.

52. Mulek, J. and Kosi, G. (2008). Algae in the aerophytic habitat of Racise ponikve cave (Slovenia). Natura Sloveniae. 10: 39-49.

53. Nakamoto, H., Suzuki, N. and Roy, S.K. (2000). Constitutive expression of a small heat-shock protein confers cellular thermotolerance and thermal protection to the photosynthetic apparatus in cyanobacteria. FEBS Lett. 483: 169-174.

54. Oren, A. (2000). Salts and brines. In: Whitton BA, Potts M (eds) The Ecology of Cyanobacteria. Their Diversity in Time and Space. Kluwer Academic Publishers, Dordrecht. 281-306.

55. Oren, A. (2002). Halophilic Microorganisms and their Environments. Kluwer Academic Publishers, Dordrecht.

56. Ortega-Morales, O., Hernández–Duque, G., Borges–Gómez, L. and Guezennec, J. (1999). Characterization of epilithic microbial communities associated with Mayan stone monuments in Yucatan, Mexico. *J Geomicrobiol*. 16: 221-232.

57. Otero, A. and Vincenzini, M. (2004). Nostoc (Cyanophyceae) goes nude: extracellular polysaccharides serve as a sink for reducing power under unbalanced C/N metabolism. *Journal of Phycology*. 40: 74-81.

58. Palinska, K.A., Liesack, W., Rhiel, E. and Krumbein, W.E. (1996). Phenotype variability of identical genotypes: the need for a combined approach in cyanobacterial taxonomy demonstrated on Merismopedia-like isolates. Archives of Microbiology. 166: 224-233.

59. Partensky, F., Hess, W.R. and Vaulot, D. (1999). Prochlorococcus, a marine photosynthetic prokaryote of global significance. Microbiol Mol Biol Rev. 63: 106-127.

60. Porankiewicz, J. and Clarke, A.K. (1997). Induction of the heat shock protein ClpB affects cold acclimation in the cyanobacterium Synechococcus sp. strain PCC 7942. *J Bacteriol*. 179: 5111-5117.

61. Potts, M. (1999). Mechanisms of desiccation tolerance in cyanobacteria. *European Journal of Phycology*. 34: 319-328.

62. Poulickova, A., Hasler, P., Lysakova, M. and Spears, B. (2008). The ecology of freshwater epipelic algae: an update. *Phycologia*. 47: 437-450.

63. Rajaram, H. and Apte, S.K. (2008). Nitrogen status and heat-stress dependent differential expression of the cpn60 chaperonin gene influences thermotolerance in the cyanobacterium Anabaena. *Microbiology*. 154: 317-325.

64. Rajaram, H. and Apte, S.K. (2010). Differential regulation of groESL operon expression in response to heat and light in Anabaena. *Arch Microbiol*. 192: 729-738.

65. Rajaram, H., Chaurasia, A.K. and Apte, S.K. (2014). Cyanobacterial heat-shock response: role and regulation of molecular chaperones. *Microbiology*. 160: 647-658.

66. Rastogi, P.R. and Madamwar, D. (2015). UV-Induced Oxidative Stress in Cyanobacteria: How Life is able to Survive? *Biochem Anal Biochem.* 4: 173.

67. Rastogi, R.P., Sinha, R.P., Moh, S.H., Lee, T.K., Kottuparambil, S., Kim, Y.J., Rhee, J.S., Choi, E.M., Brown, M.T., Hader, D.P. and Han, T. (2014). Ultraviolet radiation and cyanobacteria. *J Photochem Photobiol* B. 141: 154-169.

68. Roldan, M. and Hernandez-Marine, M. (2009). Exploring the secrets of the three-dimensional architecture of phototrophic biofilms in caves. Int J Speleology. 38: 41-53.

69. Sato, T., Minagawa, S., Kojima, E., Okamoto, N. and Nakamoto, H. (2010). HtpG, the prokaryotic homologue of Hsp90, stabilizes a phycobilisome protein in the *cyanobacterium Synechococcus* elongates PCC 7942. *Mol Microbiol.* 76: 576-589.

70. Satoh, K., Hirai, M., Nishio, J., Yamaji, T., Kashino, Y. and Koike, H. (2002). Recovery of photosynthetic systems during rewetting is quite rapid in a terrestrial cyanobacterium, *Nostoc commune. Plant and Cell Physiology.* 43: 170-176.

71. Scanlan, D. (2001). Cyanobacteria: ecology, niche adaptation and genomics. *Microbiol Today.* 28: 128-130.

72. Schwieger, F. and Tebbe, C. (1998). A new approach to utilize PCR-single–strand–conformation polymorphisms for 16S rRNA gene-bases microbial community analysis. *Appl Environ Microbiol.* 64: 4870-4876.

73. Singh, N.K. (2013). Plant Growth Promoting Rhizobacteria (PGPR) in Agricultural Sustainability. In: Rodriguez HG, Ramanja-neyulu AV, Sarkar NC, Maiti R (eds) Advances in Agro-technology A text book. *Puspa Publishing House, Kolkata.* 21-42.

74. Singh, N.K., Borase, D. and Dhar, D.W. (2011). Diversity analysis of cyanobacteria from aquatic bodies. *Proc Indian Natn Sci Acad.* 77(4): 351-371.

75. Singh, N.K. and Dhar, D.W. (2007). Nitrogen and phosphorous scavenging potential in microalgae. *Indian J Biotechnol.* 6: 52-56.

76. Singh, N.K. and Dhar, D.W. (2010a). Cyanobacterial reclamation of salt affected soil. In: Lichtfouse E (ed) Genetic engineering, biofertilisation, soil quality and organic farming. *Springer: Dordrecht, Heidelberg, London, New York, Sustainable Agriculture Reviews.* 4: 243-275.

77. Singh, N.K. and Dhar, D.W. (2010b). Microalgal Remediation of sewage effluent. *Proc Indian Natn Sci Acad.* 76(4): 209-221.

78. Singh, N.K. and Dhar, D.W. (2011a). Phylogenetic relatedness among *Spirulina* and related cyanobacterial genera. *World J Microbiol Biotechnol.* 27: 941-951.

79. Singh, N.K. and Dhar, D.W. (2011b). Microalgae as second generation biofuel. *A review. Agronomy Sust. Developm.* 31: 605-629.

80. Singh, N.K. and Dhar, D.W. (2013). Cyanotoxins, related health hazards on animals and their management: A Review. *Indian J Ani Sci.* 83(11): 1111-1127.

81. Singh, N.K. and Dhar, D.W. (2014). Diversity analysis of the geographically isolated strains of *Spirulina* and related genera. *Journal of Environmental Biology.* 35(1): 197-203.

82. Singh, N.K., Dhar, D.W. and Tabassum, R. (2014). Role of cyanobacteria in crop protection. *Proc Natl Acad Sci, India, Sect B Biol Sci.*

83. Singh, N.K. and Patel, D.B. (2012). Microalgal remediation of distillery effluent: a review. In: Lichtfouse E (ed) Farming for Food and Water Security. *Springer: Dordrecht, Heidelberg, London, New York.* 83-109.

84. Singh, N.K., Saxena, S., Tiwari, O.N. and Dhar, D.W. (2008). Cyanobacterial toxins and public health issues. In: Khattar JIS, Singh DP, Kaur G (eds) Algal biology and biotechnology. I.K. *International Publishing House Pvt. Ltd., New Delhi.* 179-203.

85. Stal, L.J. (2012). Cyanobacterial mats and Stromatolites. *Ecology of Cyanobacteria II.* 561-591.

86. Tanaka, N. and Nakamoto, H. (1999). HtpG is essential for the thermal stress management in cyanobacteria. *FEBS Lett.* 458: 117-123.

87. Taton, A., Grubisic, S., Brambilla, E., De-Wit, R. and Wilmotte, A. (2003). Cyanobacterial Diversity in Natural and Artificial Microbial Mats of Lake Fryxell (McMurdo Dry Valleys, Antarctica): a Morphological and Molecular Approach. *Appl Environ Mirobiol.* 69(9): 5157-5169.

88. Underwood, G.J.C. and Kromkamp, J. (1999). Primary production by phytoplankton and microphytobenthos in estuaries. *Advances in Ecological Research.* 29: 93-153.

89. Varvasovszki, V., Glatz, A., Shigapova, N., Josvay, K., Vigh, L. and Horvath, I. (2003). Only one dnaK homolog, dnaK2, is active transcriptionally and is essential in *Synechocystis. Biochem Biophys Res Commun.* 305: 641-648.

90. Watanabe, S., Sato, M., Nimura-Matsune, K., Chibazakura, T. and Yoshikawa, H. (2007). Protection of psbAII transcript from ribonuclease degradation in vitro by DnaK2 and DnaJ2 chaperones of the cyanobacterium *Synechococcus elongatus* PCC 7942. *Biosci Biotechnol Biochem.* 71: 279-282.

91. West, N.J. and Scanlan, D.J. (1999). Niche-partitioning of *Prochlorococcus* populations in a stratified water column in the eastern North Atlantic Ocean. *Appl Environ Microbiol.* 65: 2585-2591.

92. Whitton, B.A. and Potts, M. (2000). Introduction to the cyanobacteria. In: Whitton BA, Potts M (eds) The ecology of cyanobacteria: their diversity in time and space. Kluwer Academic Publishers, Dordrecht. 1-11.

93. Yang, L., Liu, Y., Wu, H., Ho, N., Molin, S. and Song, Z.J. (2011). Current understanding of multispecies biofilms. *Int J Oral Sci.* 3: 74-81.

94. Yopp, J.H., Tindall, D.R., Miller, D.M. and Schmidt, W.E. (1978). Isolation, purification and evidence for the halophilic nature of the blue-green alga *Aphanothece halophytica. Phycologia.* 17: 172-178.

6

Biodiversity and Characteristics of Endolithic Cyanobacteria from the Antarctic

Meenakshi Bhattacharjee

Department of BioSciences,
Rice University, Houston, Texas-77005, United States
E-mail: meenakshi.b.bhattacharjee@rice.edu

Abstract

Cyanobacteria evolved under the harsh conditions of the Precambrian, and their modern representatives retain remarkable abilities to grow and survive in extreme environments and conditions existing on the border line of life. They dominate polar environments especially the endolithic systems of Antarctic, even though they do not seem to be specifically adapted to optimal growth at low temperatures. Because of their rich biodiversity they play a major ecological role as they often are primary colonizers of substrates and major primary producers in these ecosystems. The application of molecular tools in combination with classic morphological techniques has begun to provide new insights into the real biodiversity of cyanobacteria and their biogeographical distribution in cold environments. Recent studies suggest complex distributional patterns of cyanobacteria, with cosmopolitan, endemic, and habitat-specific genotypes. Ongoing research will help to identify specific geographical areas that have unique microbial communities. However, many more studies are needed to unravel the enormous diversity of cyanobacteria and to better define their biogeographical patterns in cold environments. In Antarctic within the endoliths, cyanobacterial phototrophs exist as permanently immobilized state forming a green or brown layer few millimeters below the surface and are associated with lichens and heterotrophic bacteria that is an exciting aspect of these extreme ecosystems. These naturally immobilized systems are very similar to immobilized laboratory cultures. It is suggested that study of natural immobilized systems is important for the development of procedures to use immobilized phototrophs for practical purposes.

Introduction

Perennially cold environments in which temperatures remain below 5°C are common throughout the biosphere (Margesin and Haggblom, 2007; Davila *et al.*, 2015). In these habitats, the persistent cold temperatures are often accompanied by freeze–thaw cycles, extreme fluctuations in irradiance (including ultraviolet radiation), and large variations in nutrient supply and salinity. As a result of these constraints, such cold ecosystems contain a reduced biodiversity, with prokaryotes contributing a major component of the total ecosystem biomass as well as species richness. Cyanobacteria are of particular interest because they often represent the predominant phototrophs in such ecosystems. Current research shows that a diverse range of cyanobacteria can be found in Antarctic habitats, and that they show a remarkable ability to tolerate the abiotic stresses that prevail in these cold environments. Their presence was already observed during the early explorations of the Polar Regions at the end of the nineteenth century (Vincent, 2007).

The widespread distribution of cyanobacteria in Antarctic communities, particularly in ice-based environments, makes them of great interest for the reconstruction of microbial life and diversification on early Earth (Vincent *et al.*, 2004; Dance, 2015). These ice-based habitats with their sustainable microbial communities are potential analogues for biotopes present during the major glaciation events of the Precambrian. The fossil record suggests that cyanobacteria would have been present throughout these Proterozoic events, and perhaps during earlier periods of global cooling (Schopf, 2000).

Antarctic Endolithic Habitat

Antarctic microbes, including cyanobacteria, are also of interest to astrobiologists studying the prospect of life beyond our planet. Antarctica has been proposed as an analogue to an early stage of Mars where liquid water occurred and where life could have evolved at a similar time to the development of cyanobacteria on early Earth (Friedmann, 1986; Davila *et al.*, 2010; Billi *et al.*, 2013). Endolithic microbial communities play an important role in global processes, such as the weathering of rocks and the cycling of nutrients and elements. The endolithic environment is thought to buffer microbial communities from intense solar radiation, temperature fluctuations, wind and desiccation in environments where such environmental factors inhibit epilithic growth (Friedmann, 1982; Friedmann and Ocampo-Friedmann, 1984; Bell, 1993). Other authors suggest the ability to colonize endolithic environments is an adaptive strategy to escape competition for space and nutrients on the surface (Matthes-Sears *et al.*, 1997). This hypothesis, however, does not explain the common observation that conspicuous, photosynthetic endolithic growth occurs mostly in the absence of substantial epilithic growth (Bell, 1993). Epilithic growth probably supports endolithic communities, although little is known about such communities. Finally, it has been suggested that mineral deposits that often form over endolithic communities can create a relatively isolated environment and closed ecosystem, which must efficiently recycle nutrients (Friedmann and Ocampo-Friedmann, 1984). Nevertheless, little is known about the compositions and dynamics of endolithic communities. To date, the best-studied examples of endolithic communities are those of the polar deserts of the McMurdo Dry Valleys in Antarctica, where such communities constitute the major forms of life. The McMurdo Dry Valleys region of South Victoria Land, Antarctica, is one of the largest ice-free regions of the Antarctic continent.

The climate of the Dry Valleys is characterized by extreme cold and dryness (**McKay et al.,** 1993, Banerjee *et al.,* 2000a). In winter, a period of continuous darkness, temperatures reach as low as −60°C, and winds of up to 100 km/h sweep through the valleys. During summer, average daily air temperatures range from −35°C to +3°C, with oscillations of up to 15°C due to cloud cover occurring on the order of minutes throughout the day. Continuous summertime sunshine can raise the internal temperature of rocks above the freezing point, up to 10°C above the ambient temperature. In addition to the cold, the region is also one of the driest on the planet. In 1986, only 16 snowfall events (the only form of precipitation available) were recorded, most during the summer (**McKay et al.,** 1993). However, rocks wetted by snowmelt retain liquid water internally for several days. The combination of such rapid oscillations in temperature and wetting-drying cycles contributes to the creation of an extremely hostile living environment that has been proposed to be a potential analogue for environmental conditions on Mars (**McKay,** 1993; Robinson *et al.,* 2015).

One way to cope with extreme environmental conditions is to retreat inside rocks, either within macroscopic cracks as 'chasmoendoliths', or within the subsurface pore spaces of the rock as 'cryptoendoliths'. The micro-environmental conditions inside the rock are very different from the macroclimatic regimen and allow life to maintain itself and grow under macro-climatic conditions that are apparently hostile to life, albeit at slow rates of growth (Johnston and Vestal 1991; Sun and Friedmann, 1999). The use of the clement micro-environment associated with the inside of rocks as a refugia is probably an ancient innovation and may even date back to the latter part of the Precambrian (Campbell, 1982). As well as inhabiting sedimentary rocks that have sufficient porosity to allow subsurface invasion and lateral growth, particularly sandstones and limestones (Friedmann and Ocampo, 1976; Weber *et al.,* 1996), microorganisms can also inhabit crystalline rocks that have been altered and made more porous.

Subsurface endolithic life also extends at least 1 km beneath the surface, which has an average porosity of ~3% (Whitman *et al.,* 1998; Gold, 1999). Such porosity is sufficient volume to contain at least 105 cells cm-3, a cell density commonly observed in subsurface endolithic ecosystems. Most photosynthetic endolithic ecosystems previously studied occur in sedimentary rocks with typical porosities of 10-20%, which is sufficient space to support relatively lush and complex communities compared to those of the subsurface. Estimates of the global microbial population are on the order of 1030 cells (Whitman *et al.,* 1998). This represents a microbial biomass of ~500 Pg of carbon (1 Pg = 1015 g), which rivals that of macro-organisms. An estimated 90% of the Earth's microbial population inhabits the vast subsurface environment, of which ~80% is endolithic. Very little is known about subsurface life beyond its ubiquitous presence. Nonetheless, subsurface endolithic life may represent the largest reservoir of biomass on Earth and, possibly, the largest reservoir of genetic diversity.

Early studies

Interest in endolithic life was sparked when extensive endolithic communities were discovered in polar deserts of Antarctica, where tests were underway for the Voyager mission to Mars (Friedmann and Ocampo, 1976; Friedmann, 1982). The 9 primary goal of the Viking mission to Mars was life-detection, and the Antarctic desert was considered the closest analog to Martian environments (Friedmann, 1982; Friedmann and Ocampo-Friedmann, 1984b). As with hot desert endolithic communities, the Antarctic desert endolithic communities are most of the extant life. The first molecular phylogenetic analysis of endolithic communities from Antarctica supported the hypothesis that endolithic communities are among the simplest of microbial ecosystems known (Torre *et al.*, 2003). This comprehensive molecular survey also found that traditional microbiological methods had misidentified cyanobacterial species. This study also highlighted an abundant, presumably non-photosynthetic microbiota that is considered critical to the formation of endolithic ecosystems (Friedmann and Ocampo-Friedmann, 1984a), but has been largely ignored in previous studies.

Colonization factors for endolith formation.

The "bioreceptivity" of rock types is thought to be influenced by the physical and chemical properties of rock substrates, such as mineral composition, pore structure and permeability, as well as environmental factors, such as water availability, pH, climatic exposure and nutrient sources (reviewed in Warscheid and Braams, 2000). While chemical and mineralogical factors are known to contribute to where and how rocks are colonized, few specific details are known. Environmental factors such as water availability and light are understood better.

Water

Water is among the most important factors in endolithic ecosystems (Friedmann and Galun, 1974; Potts and Friedmann, 1981). Endolithic organisms are highly adapted to withstand frequent and sudden desiccation and rewetting. Endolithic communities resume high metabolic activity within minutes of rehydration. Water retention molecules (compatible solutes), such as trehalose and sucrose, are common among organisms cultivated from endolithic communities (Friedmann *et al.*, 1993) and are detected in desiccated Antarctic communities by Fourier transform infrared (FTIR) spectroscopy (Wynn-Williams and Edwards, 2000). Antarctic "lichen-dominated" endolithic communities can utilize water vapor ≥70% relative humidity for growth (Palmer and Friedmann, 1990). Long-term measurements of Antarctic endolithic microclimates have indicated that such suitable growth conditions exist for ~400-800 hours per year, despite exterior relative humidity's that rarely reach 70% and are often 80% and never 90% for photosynthesis. Physical characteristics, such as pore size and structure, influence moisture properties of rock substrates (Franzen and Mirwald, 2004). Biogenic mineral crusts and community biofilm matrixes also influence moisture properties of the endolithic environment. Some studies find crusts and biofilms help absorb water from the surface and reduce outward diffusion of water vapor (Kemmling *et al.*, 2004).

Temperature

Studies of thermal regimes in Antarctic endolithic communities find that temperature mostly influences colonization and growth due to its effect on liquid water (Friedmann, 1982). Temperatures rarely rise above-15º C and most precipitation falls as snow. Antarctic communities rely on solar radiation to raise temperatures to ~0º C for photosynthesis and to melt snow (Friedmann, 1978). At higher elevations, temperatures decrease and only rocks angled directly toward the summer sun are colonized (McKay and Friedmann, 1985). Many isolates fromAntarctic communities are psychrophiles, organisms with optimal growth temperatures from ~0-20º C (Siebert *et al.*, 1996).

Light

The location of the highly pigmented photosynthetic zone of endolithic communities varies, and typically occupies a vertical profile of a few millimeters that starts within ~1 cm beneath the exposed surface. Rock color and mineralogy determine the depth of light penetration. Light penetration is suggested to determine the extent and location of the photosynthetic zone (Bell, 1993). The photosynthetic zone in Antarctic communities extends to ~0.01% of incident solar radiation in dry samples, which typically corresponds to a few millimeters in depth (Nienow *et al.*, 1988). Radiation increases ~10 times at this depth in dampened samples as a result of increased light transmission due to reflectance. Maximum intensity of solar radiation has also been shown to determine the upper extent of the photosynthetic zone.

Nutrients

Most rock substrates are considered low nutrient (oligotrophic environments). This is consistent with cultivation studies that find isolates grow preferentially in oligotrophic media (Siebert *et al.*, 1996; Warscheid and Braams, 2000). Studies also find that addition of host rock extracts to defined media improves growth (Siebert *et al.*, 1996). This suggests endolithic organisms obtain nutrients from their host rock, as suggested by Bell (1993). Studies of the nitrogen economy of endolithic communities find that atmospheric deposition from precipitation and dust far exceeds biological demand (Friedmann and Kibler, 1980). These authors find evidence for *in situ* nitrogen fixation is exceptional, as measured by acetylene reduction, and conclude Antarctic communities are limited more by environmental conditions than nutrient availability.

How then do endoliths meet their Phosphate and Nitrogen requirements?

A first study on the phosphatase activity of endolithic communities of McMurdo Dry Valley in Southern Victoria Land, Antarctica which is one of the most extreme microbial habitats on earth, with low biodiversity and exceedingly low productivity was conducted by Banerjee *et al.*, 2002 a b. Under the prevailing arid conditions there are extensive areas of rock and soil without snow or ice cover. Apart from infrequent epilithic lichens (Nienow and Friedmann, 1993), the microbial communities forming the only biota are confined to a narrow zone beneath the surface of the rock (Friedmann, 1982; Friedmann *et al.*, 1988). Such endolithic communities always include one or more phototrophs. The crust of the rock, although permeable to water and gases, is a barrier to penetration of organisms and the cryptoendolithic microenvironment is largely separated from the outside environment. Phosphorus is scarce in Beacon Sandstone of the McMurdo Dry Valleys, Antarctica, and any input from precipitation is minimal. In endolithic microbial communities recycling of P by the action of phosphatases may therefore be important. The phosphatase activities of three different types of endolithic communities in the McMurdo Dry Valley, Antarctica, were studied in the laboratory. The dominant phototrophs were *Chroococcidiopsis*, mixed *Gloeocapsa* and *Trebouxia*, and *Trebouxia*.Bacteria were also visually conspicuous in the latter two communities, and the *Trebouxia* in both cases formed a lichenized association with fungal hyphae. In each case marked phosphomonoesterase (PMEase) activity was found in assays with 4-methylumbelliferyl phosphate (MUP) or *p*-nitrophenyl phosphate as substrate, and phosphodiesterase activity with bis-*p*-nitrophenyl phosphate as substrate. In view of the greater dependence of these communities on the rock for their sole supply of P than for C and probably N, it is suggested that the cycling of P within the communities is a key factor influencing their overall metabolic activity when moisture permits their activation. Long-established communities are presumably almost entirely dependent on P turnover, whereas they may receive further inputs of C from CO_2 and perhaps also N from N_2 fixation, as discussed later. Unless molecules released during breakdown of one organism are incorporated without degradation into other organisms, P availability for growth depends on there being a range of phosphatases and other enzymes capable of hydrolyzing organic phosphates (McComb *et al.*, 1979). Such enzymes may be released during breakdown of cells, present on the surface of live cells or released extracellularly by live cells. An overview of the features influencing phosphatase activities of phototrophs is given in (Hern'andez *et al.*, 2000).

Despite their existence in xeric habitats near the limits of life, phosphatase activity associated with endolithic cyanobacteria and algae was demonstrated on rehydration. All the samples showed phosphatase activities, with maxima close to the ambient pH. This and the staining of cyanobacterial sheaths suggest that "surface" phosphatases of the organisms make an important contribution to the activity detected (Banerjee and Sharma, 2004-2005; Wynn-William, 2000; Wynn-William *et al.*, 1997). Apart from any inorganic phosphate released from the rock, almost all the phosphate available for growing organisms must be that released from decaying organisms and thus initially organic phosphates, together with an extremely limited occasional input from snow meltwater and aeolian dust. It is likely that the phosphate released in the present experiments was enhanced considerably from that in nature due to the disturbance of the community, the optimal hydration state, experimental mixing, and high concentrations of substrate.

Nitrogen fixation by endolithic communities

Analysis of total nitrogen and nitrate reveals that there is negligible total nitrogen and no nitrate present in the Antarctic rocks. Therefore it is evident that some other processes must be operating to provide nitrogen for the phototrophs present. Although there are no reports of nitrogen fixation operating in the endoliths of Antarctic, reports of Banerjee and Verma (2008) prove otherwise. The cyanobacterial strain *Chroococcidiopsis* used in this study is present in the phototrophic zone of the Antarctic endolith. It is a non-heterocystous unicellular strain. A major factor that has attracted scientists to the study of nitrogen fixation in non-heterocystous cyanobacteria has been the apparent paradox of these organisms that simultaneously fix nitrogen and photo evolve O_2. In the present study, distinctly higher activity in the dark and under anaerobic conditions suggests that the spatial separation of N_2 fixation and photosynthesis may be very important for naturally occurring *Chroococcidiopsis* under endolithic conditions. Anaerobic or microaerobic conditions may be prevailing at the interphase of the photoautotrophic zone, where this cyanobacterium appears to be permanently immobilized, and the sandstone layer above it. The phototrophic zone is few millimeters below the surface of the rocks, and therefore the O_2 concentration available to the cyanobacterium may be less than at the surface, and the actively respiring bacteria may remove some of it, thus creating a micro anaerobic condition preventing the oxygen sensitive nitrogenase from being inactivated. There is probably a gradient of tolerance to oxygen among non-heterocystous N_2 fixing cyanobacteria and some strains are more sensitive to O_2 than others. Mechanisms the *Chroococcidiopsis* inside the rocks could use to minimize the deleterious effects of O_2 on N_2 fixation may include behavioral strategies, physical barriers, and metabolic strategies. Behavioral strategies include avoidance of O_2, a strategy that is practiced by those cyanobacteria fixing N_2 under anoxic or microaerobic conditions such as the cryptoendolithic niche where the prevailing concentrations of O_2 is low. Another explanation of higher nitrogenase activity in the dark compared to light in both aerobic and anaerobic conditions is that *Chroococcidiopsis* could be driving a heterotrophic N_2 fixation process using the osmolytes that are generally sugars and are associated with these organisms to protect the cells from desiccation and osmotic shocks in a chemoautotrophic pathway for reductant and energy supply to drive nitrogenase activity.

Cyanobacterial Taxonomy and Diversity

Cyanobacteria are Gram-negative oxygenic photosynthetic bacteria that, according to the fossil record, achieved most of their present morphological diversity by two billion years ago (Schopf, 2000). Cyanobacteria were initially described as algae in the eighteenth century and the first classification system was based on the International Code of Botanical Nomenclature as described by Oren (2004). In the botanical taxonomy, two major works can be noted. Firstly, Geitler (1932) produced a flora that compiled all European taxa, which already encompassed 150 genera and 1,500 species based on the morphology. Secondly, the recent revisions by Anagnostidis and Komarek (e.g., Komarek and Anagnostidis, 2005) aimed to define more homogeneous genera, still based on the morphology. Current taxonomical studies on cyanobacteria are now adopting a polyphasic approach, which combines genotypic studies with morphological and phenotypic analyses.

Early studies on the diversity and biogeographical distribution of cyanobacteria were based on the identification of the organisms entirely on the basis of morphological criteria. Cyanobacteria often have quite simple morphologies and some of these characters exhibit plasticity with environmental parameters, so that their taxonomic usefulness can be limited. Moreover, a number of botanical taxa have been delimited based on minute morphological differences (e.g., sheath characteristics, slight deviations in cell dimensions or form), and many authors have shown that the genetic diversity does not always coincide with that based on morphology (Taton *et al.*, 2006). To address these problems, studies on environmental samples (natural mixed assemblages of microorganisms) are typically based at present on clone libraries or DGGE (Denaturating Gradient Gel Electrophoresis) using molecular taxonomic markers, most often the 16S rRNA gene. The obtained 16S rRNA sequences are compared and generally grouped into OTUs (Operational Taxonomic Units) or phylo-types on the basis of their similarity. Studies on the molecular cyanobacterial diversity in Antarctica, using a culture independent approach, have focused on the following regions to date: the Prydz Bay region (Bowman *et al.*, 2000; Taton *et al.*, 2006), the McMurdo Dry Valleys (Priscu *et al.*, 1998; Gordon *et al.*, 2000; Banerjee *et al.*, 2000 a b; de la Torre *et al.*, 2003; Smith *et al.*, 2006), the McMurdo Ice Shelf (Jungblut *et al.*, 2005) and the Antarctic Peninsula region (Hughes *et al.*, 2004).

In the Dry Valleys of Antarctica, the soils are old, weathered and have low carbon and nutrient concentrations (Vincent, 1988). Thus, the colonization by cyanobacteria increases soil stability and nutrient concentrations through, for example, nitrogen fixation. Terrestrial dark crusts are found throughout Antarctica and are commonly dominated by cyanobacteria (e.g., Broady, 1996; Mataloni and Tell, 2002; Adams *et al.*, 2006). Cyanobacteria are also often identified in biofilms below and within the rocks where the microclimate gives protection against environmental stresses such as high UV radiation, temperature extremes, desiccation and physical removal by wind. They can be found in depth below the rock surface depending on the optical characteristics of the rocks and the level of available photosynthetically active radiation (PAR). Depending on the spatial location of the communities, they are hypolithic (beneath rocks), endolithic (in pore spaces of rocks), chasmoendolithic (in cracks and fissures of rocks), or cryptoendolithic (in the pore space between mineral grains forming sedimentary rocks) (Hughes and Lawley, 2003). Molecular analysis of such communities revealed a few cryptoendolithic cyanobacterial sequences in beacon sandstone of the Dry Valleys (De-la-Torre *et al.*, 2003), and in granite boulders of Discovery Bluff (de los Rios *et al.*, 2007). This group of sequences also has affinities (93.5% similarity) to the chlorophyll *d* containing *Acaryochloris marina* (Miyashita *et al.*, 1996) and De-Los-Rios *et al.*, (2007) hypothesized that some cryptoendoliths could possess this pigment and that its particular absorption spectrum would be beneficial in environments with little light. Another well-known cryptoendolithic cyanobacterium belongs to the genus *Chroococcidiopsis* and was found in sandstones of the Dry Valleys (Friedmann, 1986; Banerjee, 2000a b). It is remarkably resistant to desiccation and has close relatives in hot deserts (Banerjee *et al.*, 2009).

Endolithic microbial communities occur universally, nevertheless, little is known about the biodiversity, compositions and dynamics of these communities. To date, the best-studied examples of endolithic communities are those of the polar deserts of the McMurdo Dry Valleys in Antarctica, where such communities constitute the major forms of life. These communities and their environmental settings have been studied by microscopic observations to characterize morphologically distinct primary producers (**Friedmann *et al.*, 1988**) and by laboratory cultivation to study heterotrophic constituents of the communities (**Hirsch *et al.*, 1988; Siebert and Hirsch,** 1988; **Siebert *et al.*, 1996**). This study is the first rRNA-based molecular survey of endolithic communities made independent of laboratory cultivation. The results indicate that these communities are indeed diverse, as predicted from morphological analyses. Based on the frequencies of rRNA genes in clone libraries, the communities seem to be dominated by relatively small numbers of particularly abundant organisms. These abundant organisms do not necessarily correspond to those identified by morphological analysis or cultivation. Because of their abundance, the organisms represented by the abundant rRNA sequences are expected to be significant primary producers in these ecosystems.

That the frequencies with which different rRNA genes are obtained in PCR-based libraries must be interpreted cautiously. Organisms and their relative abundances as detected by PCR need to be confirmed with other methods. The relative ratios of rRNA genes in clone libraries do not necessarily correspond to the abundances of the represented organisms in the original ecosystem. It is believed that factors such as differential efficiencies of PCR amplification or cloning (Suzuki and Giovannoni, 1996) or different numbers of rRNA operons in different organisms an influence the proportions of rRNA phylotypes in clone libraries nonetheless, it is likely that the most abundant organisms in these ecosystems.

For both types of communities examined, rRNA-based molecular surveys suggest a lower diversity of the primary producers, as reflected by the abundances of rRNA genes in clone libraries, than has been proposed from morphological studies of these communities (**Friedmann et al 1988; Nienow and Friedmann,** 1993). Two principal classes of endolithic communities in the McMurdo Dry Valleys were defined by previous studies on the basis of the morphologically identifiable primary producers: lichen-and cyanobacterium-dominated communities.

Lichen-dominated communities are the most prevalent in the Dry Valleys. Because these lichens colonize the interstitial spaces of the rock substrate, have no epilithic morphology, and do not form mating structures, morphology-based identification of these organisms was not possible. Several species of filamentous fungi have been cultivated from these communities, but no rDNA sequences from these cultivars are yet available in public sequence databases. Researchers identified only one principal fungal phylotype from two different lichen-dominated samples. This endolithic fungal sequence was closely related (>97% nucleotide sequence identity) to rDNA sequences of epilithic lichens (*Buellia* spp.) similar to those found in the Dry Valleys (**Friedmann,** 1982). These results support the previously suggested hypothesis that the endolithic lichens are the same as the epilithic lichens, but with different cellular morphologies (**Nienow and Friedmann,** 1993).

Morphological and physiological characterizations of the endolithic lichen-dominated community suggested the presence of many species of chlorophycean algae. These photobionts were morphologically distinctive and were identified as *Trebouxia* sp. and *Pseudotrebouxia* sp. (**Archibald et al., 1983**). As in the case of the lichen mycobionts, molecular analysis revealed only one dominant green alga rDNA sequence. This sequence was nearly identical (>99% nucleotide sequence identity) to the published rDNA sequence of the lichen photobiont *T. jamesii*. Only one class of chloroplast rDNA sequence, presumably that of the *T. jamesii*-related photobiont was identified. The various photobionts identified by previous microscopic studies may be morphological variants of a single organism or may be present in the samples at levels represented by less than a few percent of the total rDNA sampled. Local small-scale heterogeneity in sampling sites may also account for variability in different studies.

Previous morphological studies of the Dry Valleys cyanobacterium-dominated endolithic communities suggested the presence of several different species of cyanobacteria, including *Hormathonema* sp., *Gloeocapsa* sp., *Anabaena* sp., *Aphanocapsa* sp. and *Lyngbya* sp. These organisms were often present in different combinations but were not always found in all samples of the cyanobacterium-dominated communities. Only one principal cyanobacterial rDNA sequence, most closely related to *Plectonema* sp. strain F3 and to other Antarctic environmental cyanobacterial sequences identified in soil and in sediments in the Dry Valleys ice covers (**Priscu et al., 1998; Smith et al., 2000**). Unfortunately, a comparison of these molecular results with observations from previous studies of Antarctic cryptoendolithic communities is complicated by the absence of rDNA sequence data for most of the organisms identified previously by morphology. For example, there are no rDNA sequences for relatives of *Hormathonema* sp., only one rDNA sequence for *Gloeocapsa* sp. and a few sequences closely related to *Chroococcidiopsis* sp. in public sequence databases. No close relatives of these organisms were detected in our molecular surveys.

Because of their conspicuous nature, it has been presumed that photosynthesis by lichens and cyanobacteria provides the main sustenance of Antarctic endolithic communities. However reports of Banerjee *et al.*, 2002 a b; Banerjee and Sharma, 2004-2005; Banerjee and Verma, 2009, clearly reveal that pohosphatase and nitrogenase activities may be principal way endolithic organisms supplement their P and N requirements. The high abundances of lichen and cyanobacterial rDNA sequences that were observed are consistent with leading roles for these organisms in these ecosystems. In addition to the sequences of known photosynthetic organisms identified in the cyanobacterium-dominated community other kinds of organisms that, because of the abundances of their rDNA sequences, were likely to contribute substantially to primary productivity in this community. Members of the family *Deinococcaceae* are best known for their ability to withstand extremely large amounts of ionizing radiation-induced damage to their DNA (Battista *et al.*, 1999). However, studies have suggested that this ability may be incidental and a consequence of an adaptation to repair DNA damage induced by desiccation. Such properties would be advantageous to organisms living in the desiccating environment of the Dry Valleys and have indeed been shown to exist in one cyanobacterium isolated from Antarctic cryptoendolithic communities (Billi *et al.*, 2000). Nevertheless, the high abundance of *Deinococcus*-like rDNA sequences seen in the cyanobacterium-dominated community was completely unexpected. This high abundance strongly suggests that the representative organisms are engaged in primary productivity. In the Antarctic cryptoendolithic communities, photosynthesis is probably the main or only source of primary productivity. The results of analyses of the clone libraries might therefore suggest that the organisms represented by clone FBP266 possess phototrophic properties. However, no phototrophic examples of the *Thermus-Deinococcus* phylogenetic group are known. Various investigators were able to culture *Deinococcus*-like organisms from samples of the lichen-dominated community (**Siebert and Hirsch,** 1988). Preliminary molecular characterization of these organisms (*Deinococcus* sp. strain AA692; (Hirsch and Stackebrandt, personal communication) places them well within the *Deinococcus* clade, most closely related to *Deinococcus radiopugnans* (>96% rDNA sequence identity). Despite being termed *Deinococcus*-like, the organisms represented by FBP266 are only distantly related (ca. 90% rDNA sequence identity) to cultivated examples of *Deinococcus* spp. and so are likely to have many physiological differences from the cultured organisms.

The molecular surveys revealed the unexpected presence of other potential phototrophs that were not identified in previous morphological or cultivation-based studies of the cyanobacterium-dominated community. One of the most abundant rDNA sequences in clone libraries prepared from the cyanobacterium-dominated community belongs to a member of the α-Proteobacteria most closely related to *B. ursincola* (94% rDNA sequence identity; approximately genus-level variation (Goebel and Stackebrandt, 1994) and other aerobic, anoxygenic phototrophs (Yurkov and Beatty, 1998). clones; (Smith *et al.*, 2000). Cultivated representatives of these organisms contain bacteriochlorophyll *a* incorporated into a photo chemically active reaction center and light-harvesting complexes. Laboratory cultures of these organisms seem unable to use light as their sole source of energy (Yurkov and Beatty, 1998). However, the considerable abundance of α-proteobacterial sequences related to clone FBP255 in clone libraries prepared from the endolithic cyanobacterium-dominated community suggests that the corresponding organisms engage in primary productivity and may be phototrophic. The occurrence in our the libraries of approximately equal numbers of rRNA genes of one each of a specific cyanobacterium, a specific α-proteobacterium, and a specific *Deinococcus*-like representative indicates approximately equivalent biomasses of the organisms and suggests that the representative organisms may be involved in a tightly regulated, syntrophic relationship. Further studies of additional examples of the cyanobacterium-dominated community will be needed to confirm the ubiquity of these organisms and to determine the character of their relationships.

Significance of diverse cyanobacterial phototrophs in naturally immobilized state

Endolithic communities dominated by cyanobacterial phototrophs forming a green or brown layer few millimeters below the surface have been described for many desert and semi desert sandstones like the Antarctic. Such microorganisms share several features with the immobilized cell in those ab systems where limited metabolic activity is required. For instance their location is fixed in relation to the substratum and in most circumstances it would seem advantageous for the organism to grow only very slowly. Laboratory studies immobilized cyanobacteria and eukaryotic algae have been carried out with several aims including long term preservation, removal of materials from the environment, production of specific extracellular products, and development of biological systems as sensitive environmental probes. In all but the first case some metabolic activity is required even if a very low levels. The two cyanobacteria Chroococcus *minutus* and *Nostoc insulare* required only very low light flux (0.5-1.5umol photon m^{-2}s^{-1}) when maintained in immobilized chamber to produce exopolysaccharides. It was impossible to grow the organisms at higher flux values to increase biomass and exopolysaccharide formation. So a study of naturally immobilized systems such as that found in the Antarctic endoliths might provide information useful to develop such laboratory systems. A comparison of the properties of a naturally immobilized cyanobacterial endolithic sample with the properties of the same cyanobacterium growing freely in culture revealed that there was a vast difference in growth and phosphatase activities between the two (Banerjee *et al.*, 2000b). The natural samples from the Antarctic were very sensitive to elevated temperature and light flux suggesting very low requirement by the cyanobacterium of these parameters. It is suggested that study of natural immobilized systems is important for the development of procedures to use immobilized phototrophs for practical purposes (Banerjee and Verma, 2008).

References

1. Akademischer, V.H., Archibald, P.A., Friedmann, E.I. and Ocampo-Friedmann, R. (1983). Representatives of the cryptoendolithic flora of Antarctica. *J. Phycol.* 19:7.

2. BA, Potts M (eds) The ecology of cyanobacteria. *Kluwer, Dordrecht.* 13-35.

3. Banerjee, M. and Sharma, D. (2004). Regulation of phosphatase activity of Chroococcidiopsis isolates from two diverse habitats: Effect of light, pH and temperature. *Applied Ecology and Environmental Research.* 2(1): 71-82.

4. Banerjee, M. and Sharma, D. (2005). Comparative studies on growth and phosphatase activity of endolithic cyanobacterial isolate Chroococcidiopsis from hot and cold deserts. *J Microbiology and Biotechnology.* 15 (1): 125-130.

5. Banerjee, M. and Verma, V. (2008). Nitrogen fixation in endolithic cyanobacterial communities of the Mc Murdo Dry Valley Antarctica. *Science Asia.* 35 (3): 35-38.

6. Banerjee, M., Whitton, B. and Wynn-Williams, D.D. (2000b). Surface phosphomonoesterase activity of a natural immobilized system: Chroococcidiopsis in an Antarctic desert rock. *J Applied Phycology.* 12: 549-552.

7. Banerjee, M., Whitton, B. and Wynn-Williams, D.D. (2000a). Phosphatase activities of endolithic communities in rocks of the Antarctic Dry Valley. *Microbial Ecology.* 39: 80-91.

8. Battista, J.R., Earl, A.M. and Park, M.J. (1999). Why is *Deinococcus radiodurans* so resistant to ionizing radiation? *Trends Microbiol.* 7: 362-365.

9. Bell, R.A. (1993). Cryptoendolithic algae of hot semiarid lands and deserts. *Journal of Phycology.* 29(2): 133-139.

10. Billi, D., Baque, M., Smith, H. and McKay, C. (2013). Cyanobacteria from Extreme Deserts to Space. *Advances in Microbiology.* 3(6): 80-86.

11. Billi, D., Friedmann, E.I., Hofer, K.G. Caiola, M.G. and Ocampo-Friedmann, R. (2000) Ionizing-radiation resistance in the desiccation-tolerant cyanobacterium*Chroococcidiopsis. Appl Environ Microbiol.* 66: 1489-1492.

12. Bowman, J.P., Rea, S.M., McCammon, S.A. and McMeekin, T.A. (2000). Diversity and community structure within anoxic sediment from marine salinity meromictic lakes and a coastal meromictic Code. *Int J Syst Evol Microbiol.* 54: 1895-1902.

13. Campbell, S.E. (1982). Precambrian endoliths discovered. *Nature.* 299: 429-431.

14. Dance, A. (2015). Inner Workings: Endoliths hunker down and survive in extreme environments. *Proc Natl Acad Sci. USA.* 112(8): 2296.

15. Davila, A.F., Duport, L.G., Melchiorri, R., Janchen, J., Valea, S. and De-Los-Rios, A. (2010). Hygroscopic salts and the potential for life on Mars. *Astrobiology.* 10: 617-628.

16. Davila, A.F., Hawes, I., Araya, J.G., Gelsinger, D.R., DiRuggiero, J.C. Ascaso, A., Osano and Wierzchos, J. (2015). In situ metabolism in halite endolithic microbial communities of the hyperarid Atacama Desert. *Front Microbiol.* 6: 1035.

17. De-la-Torre, J.R., Goebel, B.M., Friedmann, E.I. and Pace, N.R. (2003). Microbial diversity of cryptoendolithic communities from the McMurdo Dry Valleys, Antarctica. *Applied and Environmental Microbiology.* 69(7): 3858-3867.

18. De-la-Rios, A., Grube, M., Sancho, L.G. and Ascaso, C. (2007). Ultrastructural and genetic characteristics of endolithic cyanobacterial biofilms colonizing Antarctic granite rocks. *FEMS Microbiol Ecol.* 59: 386-395.

19. Franzen, C. and Mirwald, P.W. (2004). Moisture content of natural stone: static and dynamic equilibrium with atmospheric humidity. *Environmental Geology.* 46(3-4): 391-401.

20. Friedmann, E.I. (1978). Melting snow in the Dry Valleys is a source of water for endolithic microorganisms. *Antarctic Journal of the United States.* 13(4): 162-163.

21. Friedmann, E.I. (1986) The Antarctic cold desert and the search for traces of life on Mars. Adv.

22. Friedmann, E.I. and Galun, M. (1974). Desert algae, lichens and fungi. Desert Biology. *G. W. Brown. New York, Academic Press.* 2: 165-212.

23. Friedmann, E.I. and Kibler, A.P. (1980). Nitrogen economy of endolithic microbial communities in hot and cold deserts. *Microbial Ecology*. 6(2): 95-108.

24. Friedmann, E.I. and Ocampo, R. (1976). Endolithic blue–green algae in the dry valleys: primary producers in the Antarctic desert ecosystem. *Science*. 193: 1247-1249.

25. Friedmann, E.I. and Ocampo-Friedmann, R. (1984). Endolithic microorganisms in extreme dry environments: analysis of a lithobiontic microbial habitat. *Current Perspectives Microbial Ecol XII*. 177-185.

26. Friedmann, E.I. (1982). Endolithic microorganisms in the Antarctic cold desert. *Science*. 215: 68-74.

27. Friedmann, E.I., Kappen, L., Meyer, M.A. and Nienow, J.A. (1993). Long-term productivity in the cryptoendolithic microbial community of the Ross desert Antarctica. *Microbial Ecology*. 25(1): 51.

28. Friedmann, E.I., Hua, M. and Ocampo-Friedmann, R. (1988). Cryptoendolithic lichen and cyanobacterial communities of the Ross Desert Antarctica. *Polarforschung*. 58: 251-259.

29. Goebel, B.M. and Stackebrandt, E. (1994). Cultural and phylogenetic analysis of mixed microbial populations found in natural and commercial bioleaching environments. *Appl. Environ Microbiol*. 60: 1614-1621.

30. Gold, T. (1999). The Deep Hot Biosphere. New York, Copernicus, Springer-Verlag.

31. Gordon, D.A., Priscu, J. and Giovannoni, S. (2000). Origin and phylogeny of microbes living in permanent Antarctic lake ice. *Microb Ecol*. 39: 197-202.

32. Hern'andez, I., Niell, F.X. and Whitton, B.A. (2000). Phosphatase activity of benthic marine algae: An overview. In: Whitton BA, Hernandez I, (eds) Phosphatases in the Environment. Kluwer, Dordrecht.

33. Hirsch, P., Hoffmann, B., Gallikowski, C.C., Mevs, U., Siebert, J. and Sittig, M. (1988). Diversity and identification of heterotrophs from Antarctic rocks of the McMurdo Dry Valleys (Ross Desert). *Polarforschung*. 58: 261-269.

34. Hughes, K.A. and Lawley, B. (2003). A novel Antarctic microbial endolithic community within gypsum crusts. *Environ Microbiol*. 5: 555-565.

35. Hughes, K.A., McCartney, H.A. Lachlan-Cope, T.A. and Pearce, D.A. (2004). A preliminary study of airborne microbial biodiversity over Peninsular Antarctica. *Cell Mol Biol*. 50: 537-542.

36. Johnston, C.G. and Vestal, J.R. (1991). Photosynthetic carbon incorporation and turnover in Antarctic cryptoendolithic microbial communities: are they the slowest growing communities on Earth? *Appl Environ Microbiol*. 57: 2308-2311.

37. Jungblut, A.D., Hawes, I., Mountfort, D., Hitzfeld, B., Dietrich, D.R., Burns, B.P. and Neilan, B.A. (2005). Diversity within cyanobacterial mat communities in variable salinity meltwater ponds of marine basin, Vestfold Hilds Eastern Antarctica. *Environ Microbiol*. 2: 227-237.

38. Kemmling, A., Kamper, M., Flies, C. Schieweck, O. and Hoppert, M. (2004). Biofilms and extracellular matrices on geomaterials. *Environmental Geology.* 46(3-4): 429-435.

39. Komarek, J. and Anagnostidis, K. (2005). Cyanoprokaryota 2. Teil Oscillatoriales, Spektrum.

40. Margesin, R. and Haggblom, M. (2007). Thematic issue: Microorganisms in cold environments. *FEMS Microbiol Ecol.* 59:215-216.

41. Matthes-Sears, U., Gerrath, J.A. and Larson, D.W. (1997). Abundance, biomass, and productivity of endolithic and epilithic lower plants on the temperate-zone cliffs of the Niagara escarpment, Canada. *International Journal of Plant Sciences.* 158(4): 451-460.

42. McComb, R.B., Bower, G.N. and Posen, S. (1979). Alkaline Phosphatase. Plenum Press, New York.

43. McKay, C.P. (1993). Relevance of Antarctic microbial ecosystems to exobiology, p. 593-601. In E. I. Friedmann (ed.), Antarctic microbiology. Wiley-Liss, New York, N.Y.

44. McKay, C.P. and Friedmann, E.I. (1985). The cryptoendolithic microbial environment in the Antarctic cold desert: Temperature-variations in nature. *Polar Biology.* 4(1): 19-25.

45. McKay, C.P., Nienow, J.A., Meyer, M.A. and Friedmann, E.I. (1993). Continuous nanoclimate data (1985-1988) from the Ross Desert (McMurdo Dry Valleys) cryptoendolithic microbial ecosystem. *Antarct Res Ser.* 61: 201-207.

46. McMurdo, I.S. Antarctica. *Environ Microbiol.* 7: 519-529.

47. Meenakshi, B. and Verma, V. (2008). Extremophilic cyanobacteria and their novel biotechnological applications. *J Applied Bioscience.* 34(2): 7-14.

48. Meenakshi, B. Everwood, C. and Castenholz, R.W. (2009). Characterization of a novel clade of thermo-hallophillic clade of cyanobacteria from saline thermal springs and endolithic habitats by molecular techniques. *Extremophiles.* 13: 707-716.

49. Miyashita, H., Ikemoto, H., Kurano, N., Adachi, C.K., Chihara, M. and Miyashi, S. (1996). Chlorophyll *d* as a major pigment. *Nature.* 383-402.

50. Nienow, J.A. and Friedmann, E.l. (1993). Terrestrial lithophytic (rock) communities. In: Friedmann El (ed) Antarctic Microbiology. Wiley-Liss, New York. 353-412.

51. Nienow, J.A., McKay, C.P. and Friedmann, E.I. (1988). The cryptoendolithic microbial environment in the Ross Desert of Antarctica: Light in the photosynthetically active region. *Microbial Ecology.* 16(3): 271-289.

52. Oren, A. (2004). A proposal for further integration of the cyanobacteria under the Bacteriological

53. Palmer, R.J. and Friedmann, E.I. (1990). Water relations and photosynthesis in the cryptoendolithic microbial habitat of hot and cold deserts. *Microbial Ecology.* 19(1): 111.

54. Potts, M. and Friedmann, E.I. (1981). Effects of water-stress on cryptoendolithic cyanobacteria from hot desert rocks. *Archives of Microbiology.* 130(4): 267-271.

55. Priscu, J.C., Fritsen, C.H., Adams, E.E., Giovannoni, S.J., Paerl, H.W., McKay, C.P., Doran, P.T., Gordon, D.A., Lanoil, B.D. and Pinckney, J.L. (1998). Perennial Antarctic lake ice: an oasis for life in a polar desert. *Science.* 280: 2095-2098.

56. Robinson. C. K., Wierzchos, J., Black, C., Crits-Christoph, A., Ma, B. and Ravel, J. (2015). Microbial diversity and the presence of algae in halite endolithic communities are correlated to atmospheric moisture in the hyper-arid zone of the Atacama Desert. Environ Microbiol. 17: 299-315.

57. Schopf, J.W. (2000). The fossil record: tracing the roots of the cyanobacterial lineage. In: Whitton. *Space. Res.* 6: 265-268.

58. Siebert, J. and Hirsch, P. (1988). Characterization of 15 selected coccal bacteria isolated from Antarctic rock and soil samples from the McMurdo-Dry Valleys (South-Victoria Land). *Polar Biol.* 9: 37-44.

59. Siebert, J.P. Hirsch, B. Hoffmann, C.G. Gliesche, K.P. (1996). Cryptoendolithic microorganisms from Antarctic sandstone of linnaeus terrace (Asgard range): Diversity, properties and interactions. *Biodiversity and Conservation.* 5(11): 1337-1363.

60. Smith, J.J., Tow, L.A., Stafford, W., Cary, C. and Cowan, D.A. (2006). Bacterial diversity in three different Antarctic cold desert mineral soils. *Microbial Ecology.* 51: 413-421.

61. Smith, M.C., Bowman, J.P., Scott, F.J. and Line, M.A. (2000). Sublithic bacteria associated with Antarctic quartz stones. *Antarct Sci.* 12: 177-184.

62. Sun, H.J. and Friedmann, E.I. (1999). Growth on geological time scales in the Antarctic cryptoendolithic microbial community. *Geomicrobiol J.* 16: 193-202.

63. Suzuki, M.T. and Giovannoni, S.J. (1996). Bias caused by template annealing in the amplification of mixtures of 16S rRNA genes by PCR. *Appl Environ Microbiol.* 62:625-630.

64. Taton, A., Grubisic, S., Ertz, D., Hodgson, D.A., Piccardi, R., Biondi, N., Tredici, M.R., Mainini, M., Losi, D., Marinelli, F. and Wilmotte, A. (2006). Polyphasic study of antarctic cyanobacterial strains. *J Phycol.* 42: 1257-1270.

65. Vincent, W.F. (2007). Cold tolerance in cyanobacteria and life in the cryosphere. In: Seckbach J (ed) Algae and cyanobacteria in extreme environments. *Springer.*

66. Vincent, W.F., Mueller, D., Van Hove, P. and Howard-Williams, C. (2004). Glacial periods on early Earth and implications for the evolution of life. In: Seckbach J (eds) Origins: Genesis, evolution and diversity of life. *Kluwer, Dordrecht.* 481-501.

67. Warscheid, T. and Braams, J. (2000). Biodeterioration of stone: a review. *International Biodeterioration & Biodegradation.* 46(4): 343-368.

68. Weber B Wessels DCJ and Bu¨ del B (1996) Biology and ecology of cryptoendolithic cyanobacteria of a sandstone outcrop in the Northern Province, South Africa. *Algological Studies.* 83: 565-579.

69. Whitman, W.B., Coleman, D.C. and Wiebe, W.J. (1998). Prokaryotes: The unseen majority. *Proceedings of the National Academy of Sciences of the United States of America.* 95(12): 6578-6583.

70. Wynn-Williams, D.D. (2000). Cyanobacteria in deserts-Life at the limits. In: Whitton BA, Potts M (eds) Ecology of Cyanobacteria: Their Diversity in Time and Space. *Kluwer, Dordrecht.* 341-366.

71. Wynn-Williams, D.D. and Edwards, H.G.M. (2000). Proximal analysis of regolith habitats and protective biomolecules in situ by laser Raman spectroscopy: Overview of terrestrial Antarctic habitats and Mars analogs. *Icarus.* 144(2): 486-503.

72. Wynn-Williams, D.D., Russell, N.C. and Edward, H.G.M. (1997). Moisture and habitat structure as regulators for microalgal colonists in diverse Antarctic terrestrial habitats. In: Lyon WB, Howard-Williams C, Hawes I (eds) Ecosystem Processes in Antarctic Ice-free Landscapes. *Balkema, Rotterdam.* 77-88.

73. Yurkov, V.V. and Beatty, J.T. (1998). Aerobic anoxygenic phototrophic bacteria. *Microbiol Mol Biol Rev.* 62: 695-724.

7

Algal Biotechnology: Potential Source for the Future Energy Demand

Ravindra Singh[1] and Sadhana Chaurasia[2*]

[1]Department of Biological Sciences,
Mahatma Gandhi Chitrakoot Gramodaya Vishwavidyalaya,
Chitrakoot, Satna-485334, Madhya Pradesh, India
[2]Department of Energy and Environment,
Mahatma Gandhi Chitrakoot Gramodaya Vishwavidyalaya,
Chitrakoot, Satna-485334, Madhya Pradesh, India
E-mail: rsinghmgcgv@gmail.com, sadhanamgcgv@gmail.com

Abstract

Fossil fuel energy resources are depleting rapidly.Hence it is necessary to look for alternative fuels, which can be produced from materials available within the country. Bio fuels form algae are gaining much more attention due to high lipid content and fast growth rate but the bio fuel technologies are needed to improve for a commercial-scale production. This paper reviews on potential of algae and various methods of bio fuels production.

Keywords: Algae, Open pond, Photo bioreactor, Transesterification

Introduction

The rapid depletion of fossil fuels together with the uncertain global climate, increasing the demand for alternate energy sources which can be substitute of fossil fuels. It is reported that the present petroleum consumption is 10^5 times faster than the nature can create [1,2] and at this rate of consumption, the world's fossil fuel reserves will be diminished by 2050 [1,3]. Today intense interest has focused on the bio fuel from photosynthetic plants such as, soybean, rapeseed, sunflower, palm, coconut, jatropha, karanja [4,8]. But bio fuels production from these crops has become a major controversy due to food versus fuel competition [9]. In current scenario microalgae are gaining much more attention as a bio fuel source of energy over traditional crops due to the following reasons:

(1) High lipid content, fast growth rates, makes microalgae an alternative to terrestrial energy crops for biodiesel production [4,10].

(2) Microalgae require non-arable land for their cultivation and can utilise industrial flue gas as carbon source and moreover it can be harvested daily [7].

(3) Another advantage of algal bio fuels is CO_2 sequestration, because algae use CO_2 in its photosynthesis process more efficiently than any other biomass [11-13].

Due to these reasons algal biotechnology is an emerging field of bio-energy.

Potential of algae

Algae have higher growth rates than terrestrial plants, allowing a large quantity of biomass to be produced in a shorter time. Algae growth rates of 10 to 50 g m^{-2} d^{-1} (grams of algal mass per square meter per day) have been reported [14,15]. The main components of a typical algal feedstock are proteins, carbohydrates, lipids, and other valuable components, e.g. pigment, anti-oxidants, fatty acids, vitamins etc. Compared to terrestrial plants such as corn and soy, algae have shorter harvest times because they can double their mass every 24 hours [14,5]. These short harvest times allow for much more efficient and rapid production of algae compared to corn or soy crops. The yields of different oil producing feedstock can be examined, as shown in Table 1.

Table 1: Amount of oil produced by various feedstock's' [14, 16]

Feedstock	Liters/Hectare
Castor	1413
Sunflower	952
Palm	5950
Soya bean	446
Coconut	2689
Algae	100000

Production of micro algal biomass

Microalgae are photosynthetic, photosynthetic growth of micro algal biomass require a light source, carbon dioxide, water, inorganic salts and temperature of 15 and 30°C The growth medium for algae must contain essential nutrients such as nitrogen, phosphorus, iron and sometimes silicon [17,18] for algal growth. Algae are traditionally cultivated either in open ponds, known as high rate ponds (HRP), or in enclosed systems known as photo bioreactors. Each system has its own advantages and disadvantages.

(1) Open Raceways

Open ponds do not allow microalgae to use carbon dioxide as efficiently and limits biomass production [17,5]. In this system, the shallow pond is usually with about 1 foot deep; algae are cultured under conditions identical to the natural environment. The pond is usually designed in a "raceway" or "track" configuration, in which a paddlewheel provides circulation in mixing of the algal cells and nutrients.

Although open ponds cost less to build and operate than enclosed photo bioreactors, this culture system has its intrinsic disadvantages. Since these are open air systems, they often experience a lot of water loss due to evaporation. Biomass productivity is also limited by contamination with unwanted algal species as well as organisms that feed on algae. In addition, optimal culture conditions are difficult to maintain in open ponds and recovering the biomass from such a dilute cell yield is expensive [17,19].

(2) Closed photo-bioreactors

Photo bioreactors are another method to cultivate microalgae. Photo bioreactors can overcome the problems of contamination and evaporation encountered in open ponds [17,20,21]. The biomass productivity of photo bioreactors can be 13 times more than that of a traditional raceway pond on average [10]. There are many different shapes of bioreactors, but they usually fall into two broad categories: 1) use of natural light or 2) use of artificial illumination. Enclosed photo bioreactors are often tubular to allow for a greater amount of light penetration. Thus tubes, whether helical or straight provides more surface area to volume ratio for easy growth of algae. However, enclosed photo bioreactors also have some disadvantages. Variations in light and temperature are common in all photoautotrophic systems, causing suboptimal growth of the microalgae. The scale-up is very difficult in these systems, and warrants a high cost to do so [22].

Bio-fuels

Bio fuels are referred to liquid, gas and solid fuels, produced from biomass. They may be derived from forest, agricultural or fishery products or municipal wastes, also including by-products and wastes originated from agro-industry, food industry and food services. A variety of fuels can be produced from biomass such as ethanol, methanol, biodiesel, Fischer-Trops diesel, hydrogen and methane [23,24].

First generation bio-fuels

Bio fuels derived from edible oils such as rapeseed [25,26] soybeans [25,27-29] palm Oil [30-34] and sunflower [26,35,36] are considered to be first generation bio-fuel feedstock because they were the first crops to be used to produce biodiesel. The use of these first generation biodiesel sources has generated many problems, mainly due to their impact on global food markets and food security [37]. For example, palm and soy are crops whose oils are a vital part of human food. Diverting these food crops to produce oil in the large-scale production of biodiesel could bring imbalance to the global food market [38], and as a consequence, the world could suddenly face a 'food versus fuel crisis.

Second-generation bio-fuels

Second-generation bio fuels derived from lignocellulosic agriculture and forest residues and from non-food crop feedstock a such as jatropha oil, waste cooking oil and animal fats do not affect food security and have significant advantages over first generation oil crops. However sustainability of second generation bio fuels is not favourable due to concern over competing land use or required land use changes [37].

Third-generation bio-fuels

Third-generation bio fuels specifically derived from microbes and microalgae are considered to be most promising alternative resources that is devoid of the major drawbacks associated with first and second-generation bio fuels due to their high photosynthetic efficiency to produce biomass and their higher growth rates and productivity compared to conventional crops [55,39].

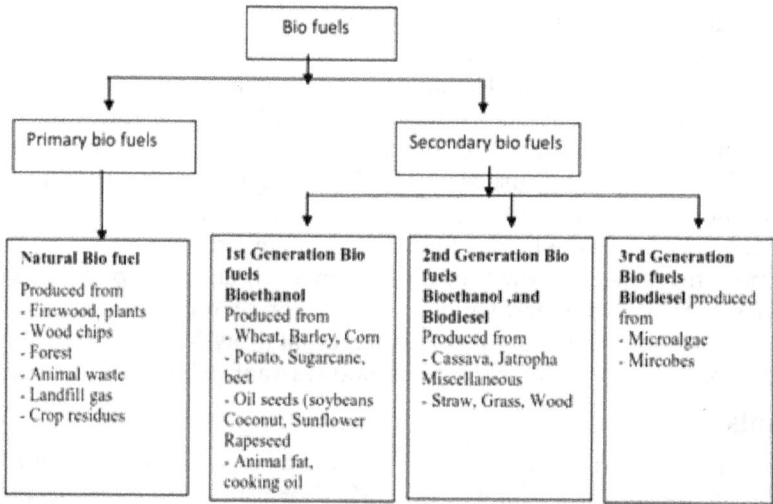

Figure 1: Bio-fuels classification.

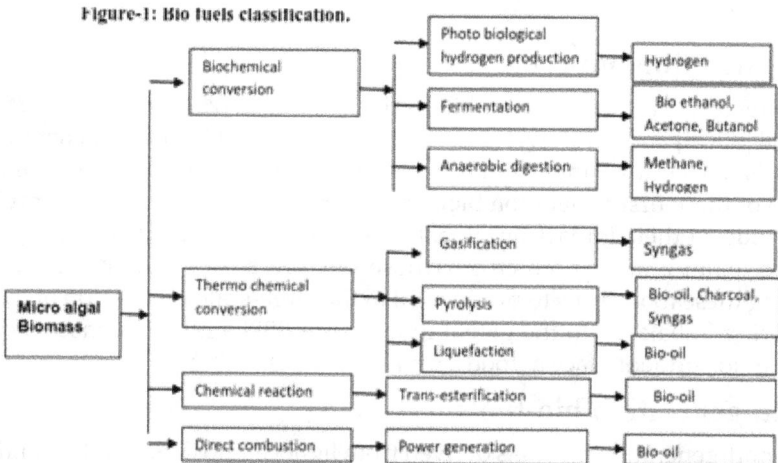

Figure 2: Flow chart of Bio fuel production processes from microalgae biomass

Oil extraction from microalgae

Expeller/oil pressing

Expeller/oil pressing is a mechanical method for the extraction of oil using pressure. Micro algal biomass needs to be dried at high pressure for the optimal performance of the process. Expeller/oil pressing can extract up to 75% of oil. [43,44]

Ultrasonic extraction

Ultrasonic extraction uses intense ultrasonic waves to create tiny cavitations bubbles. The collapse of these bubbles near the cell walls creates shock waves, which rupture the cell walls of microalgae resulting in the release of algal oil into the hexane solvent. After the sonification process the hexane algae mixture is centrifuged to separate the algae from hexane. The hexane thus obtained has dissolved oil which has to be separated. Hexane has a lower boiling point than oil and hence can be evaporated at the boiling point of hexane saving behind algal oil [45]. More than 90% extraction of fatty acids and pigments from *Scenedesmus obliquus*, have been reported using this method [46].

The Soxhlet extraction method

The Soxhlet method is the most commonly used method for the extraction of oils from dry microalgal biomass, in which organic solvents such as hexane, benzene, cyclohexane, acetone and chloroform are used in repeated washings of microalgae under reflux in special glassware called a Soxhlet extractor. This method can be applied to any low oil content materials [43,46].

Supercritical fluid extraction

In supercritical fluid extraction, high pressures and temperatures are used for the rupturing the microalgal cells. Carbon dioxide is the most commonly used supercritical solvent because it does not lead to contamination or thermal degradation of the compounds. In this process, CO_2 is compressed beyond its supercritical state (31°C, 74 bar). CO_2 at this state is brought into contact with microalgae and penetrates into microalgal pores due to its high diffusion rate. When carbon dioxide is depressurized, the substances of interest are efficiently collected with less solvent residues as compare to other extraction methods [46,47,48].

Table 2: Advantage and limitations of various extraction methods for algal oil

Extraction methods	Advantages	Limitations	References
Oil press	Easy to use, no solvent involved	Large amount of sample required, slow process	[51]
Solvent extraction	Solvent used are relatively inexpensive; Reproducible	Most organic solvents are highly flammable and/ or toxic; solvent recovery is expensive and energy intensive; large volume of solvent needed	[52,53]

Extraction methods	Advantages	Limitations	References
Supercritical fluid extraction	Non-toxicity (absence of organic solvent in residue or extracts), 'green solvent' used; non-flammable, and simple in operation	Often fails in quantitative extraction of polar analyses from solid matrices, insufficient interaction between supercritical CO_2 and the samples	[54,55]
Ultrasound	Reduced extraction time; reduced solvent consumption; greater penetration of solvent into cellular materials; improves the release of cell contents into the bulk medium	High power consumption; difficult to scale up	[56,57]

Algal biodiesel

Algal oil can be converted into biodiesel by the transesterification process biodiesel is composed of methyl or ethyl esters produced from vegetable oil or animal oil and has fuel properties similar to diesel fuel which offers many benefits.

Transesterification of algae oil

Biodiesel production from microalgae can be done using several well-known industrial processes, the most common of which is base catalyzed transesterification with alcohol. Parent oil used in making biodiesel consists of triglycerides (Fig.-1) in which three fatty acid molecules are esterified with a molecule of glycerol. In making biodiesel, triglycerides are reacted with methanol in a reaction known as transesterification or alcoholysis. Transestrification produces methyl esters of fatty acids that are biodiesel, and glycerol. The reaction occurs stepwise: triglycerides are first converted to diglycerides, then to monoglycerides and finally to glycerol [17].

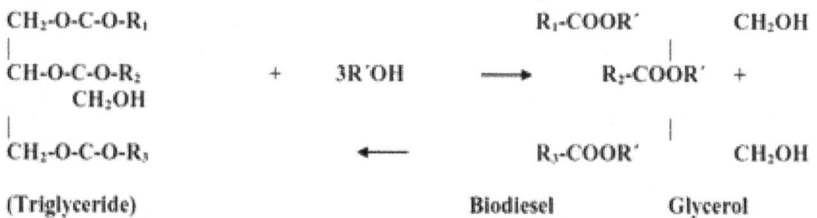

$$
\begin{array}{lll}
CH_2\text{-}O\text{-}C\text{-}O\text{-}R_1 & R_1\text{-}COOR' & CH_2OH \\
\quad | & \quad | & \\
CH\text{-}O\text{-}C\text{-}O\text{-}R_2 \quad + \quad 3R'OH \quad \longrightarrow & R_2\text{-}COOR' \quad + & \\
\quad\quad CH_2OH & & \\
\quad | & \quad | & \\
CH_2\text{-}O\text{-}C\text{-}O\text{-}R_3 \quad\quad \longleftarrow & R_3\text{-}COOR' & CH_2OH \\
& & \\
(Triglyceride) & Biodiesel & Glycerol
\end{array}
$$

Bio-ethanol

In general, two methods are normally adopted for the production of bio ethanol from biomass. The first one is a biochemical process, i.e. fermentation and the other is by thermo-chemical process or gasification. The fermentation process produces ethanol and CO_2 in addition to methane by the anaerobic digestion of the remaining algal biomass slurry left after fermentation, which can further be converted to produce electricity. A number of advantages have been reported in the production of bio ethanol from algae. Fermentation process involves less intake of energy and the process is much simpler in comparison to the biodiesel production system. In addition CO_2 produced as a by-product from the fermentation process can be recycled as carbon sources for microalgae cultivation, thus reducing greenhouse gas emissions [49].

Hydrogen production routes from biomass

The methods available for the hydrogen production from biomass can be divided into two main categories: thermo chemical and biological routes. Hydrogen can be produced from bio renewable feedstock via thermo chemical conversion processes such as pyrolysis, gasification, steam gasification, steam reforming of bio-oils, and supercritical water gasification. Biological production of hydrogen can be classified into the following groups: (i) bio-photolysis of water using green algae and blue-green algae (cyanobacteria), (ii) photo-fermentation, (iii) dark-fermentation, and (iv) hybrid reactor system. The advantage of the thermo chemical process is that its overall efficiency (thermal to hydrogen) is higher (hw52%) and production cost is lower.

Algal Bio-methane Production

The production of biogas from biomass is gaining increasing importance worldwide. Recent studies of the Institut fur Energetik und Umwelt Leipzig predict that a significant part of the methane demand in Europe can be satisfied with biogas. However, the knowledge of the biological processes which take place in a biogas production facility is limited today; therefore research in this field is needed and important to improve the bio methane production process. Bio methane can be made from a very wide range of biomass crops as well as from a range of crop residues. A major limiting factor for the future growth of bio methane production from plant sources is the availability of photo synthetically grown biomass. Currently, a 500 KW bio methane plant requires approx 10,000–12,000 tons of biomass feedstock per year with maize currently being the major crop plant feedstock. Using cereals and sunflowers, typical yields between 2,000 to 4,500 m^3 bio methane per hectare per year have been reported. Yields from maize are higher and vary in dependence of the species and the time of harvesting between 5,700 and 12,400 m^3 of bio methane per year per hectare.

Fermentation

Commercially ethanol is produced in different countries from and starch crops and sugar crops by the process of fermentation. Corn contains maximum 70% starch, and globally one of most valuable feedstock for ethanol based bio industry. Large variety of carbohydrates is used for ethanol production. Fermentation of sucrose is achieved by utilising commercial yeast such as, invertase, *Saccharomyces ceveresiae* and zymase enzymes. The hydrolysis of sucrose can be performed in the presence of invertase enzyme that converts it to glucose and fructose.

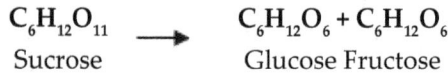

$$C_6H_{12}O_{11} \longrightarrow C_6H_{12}O_6 + C_6H_{12}O_6$$

Sucrose   Glucose   Fructose

Second, enzyme zymase converts the fructose and glucose into ethyl alcohol.

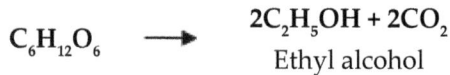

$$C_6H_{12}O_6 \longrightarrow 2C_2H_5OH + 2CO_2$$

Ethyl alcohol

Conclusion

Algae have immense potentials for bio fuels production. However, these potentials largely depend on utilisation of technology [40]. A number of researches are going on genetically modifying algae to speed growth, increasing lipid content and favourably changing the economics of the algae-based bio fuels [58]. More innovations are still needed for the development of technologies which reduce costs while increasing the yields. Crucial economic challenge for algae producers is to discover low cost oil extraction and harvesting methods. With the advent of cheaper photo bioreactors (PBRs), these costs are likely to come down significantly in the next few years. In the present scenario, reducing these costs is critical to algae bio fuel companies for its successful commercial implementation [49].

References

1. Ahmad, A.L., Yasin, N.H., Derek, C.J.C. and Lim, J.K. (2011). Microalgae as a sustainable energy source for biodiesel production: A review. *Renewable and Sustainable Energy Reviews*. 15: 584-593.

2. Alam, F., Abhijit, D., Rasjidin, R., Mobin, S., Moria, H. and Baqui, A. (2012). Biofuels from Algae-is it viable alternative? *Procedia Engineering*. 49: 221-227.

3. Alcantara, R., Amores, J., Canoira, L., Fidalgo, E., Franco, M.J. and Navarro, A. (2000). Catalytic production of biodiesel soy-bean oil, used frying oil and tallow. *Biomass Bioenergy*. 18: 515-27.

4. Applied Microbiology and Microbial Biotechnology, Mendez-Vilas A (ed.). *Formatex*. 1355-1366.

5. Bajjhaiya, A.K., Mandotra, S.K., Suseela, M.R. and Toppo, K.R. (2010). Algal Biodiesel: the next generation biofuels for India. *Asian J. Exp. Biol. Sci.* 1(4): 728-739.

6. Balat, H. and Kırtay, E. (2010). Hydrogen from biomass-Present scenario and future Prospects. *International journal of hydrogen energy*. 35: 7416-742 6.

7. Beer, L.L., Boyd, E.S., Peters, J.W. and Posewitz, M.C. (2009). Engineering algae for biohydrogen and biofuel production. *Curr. Opin. Biotechnol.* 29: 264-71.

8. Brennan, L. and Owende, P. (2010). Biofuels from microalgae – a review of technologies for production, processing, and extractions of biofuels and co-products. *Renew. Sust. Energ. Rev.* 14: 557-577.

9. Canakci, M., Azsezen, A.N., Arcaklioglu, E. and Erdil, A. (2009). Prediction of performance and exhaust emissions of a diesel engine fueled with biodiesel produced from waste frying palm oil. *Expert Syst. Appl.* 36: 9268-80.

10. Celikten, I., Koca, A. and Arslan, M.A. (2010). Comparison of performance and emissions of diesel fuel, rapeseed and soybean oil methyl esters injected at different pressures. *Renew Energy.* 35: 814-20.

11. Chaumont, D. (1993). Biotechnology of algal biomass production: a review of systems for outdoor mass culture. Journal of Applied Phycology. 5: 593-604.

12. Chisti, Y. (2007). Biodiesel from microalgae. *Biotechnol Adv.* 25: 294-306.

13. Demirbas, A. (2008). Comparison of transesterification methods for production of biodiesel from vegetable oils and fats. *Energy Convers Manage.* 49: 125-130.

14. Demirbas, A. (2009). Global renewable energy projections. *Energy Source.* 4(2): 212-224.

15. Dragone, G., Fernandes, B., Vicente, A.A. and Teixeira, J.A. (2010). Third generation biofuels from microalgae in Current Research, Technology and Education Topics in

16. Fabiana, P., Sole, M., Garcia, J. and Ferrer, I. (2013). Biogas production from microalgae grown in wastewater: Effect of microwave pretreatment. *Applied Energy.*108: 168-175.

17. Galloway, J.A., Koester, K.J., Paasch, B.J. and Macosko, C.W. (2004). Effect of sample size on solvent extraction for detecting cocontinuity in polymer blends. *Polymer.* 45: 423-8.

18. Gendy, T. (2013). Commercialization potential aspects of microalgae for biofuel production: An overview. *Egyptian Journal of Petroleum.* 22: 43-51.

19. Gude, V. and Patil P. (2013). Microwave energy potential for biodiesel Production. *Sustainable Chemical Processes.* 1: 51-31.

20. Gui, M.M., Lee, K.T. and Bhatia, S. (2008). Feasibility of edible oil vs. non-edible oil vs. Waste edible oil as biodiesel feedstock. *Energy.* 33: 1646-53.

21. Haas, M.J. (2005). Improving the economics of biodiesel production through the use of low value lipids as feedstocks: vegetable oil soapstock. *Fuel Process Technol.* 86: 1087-96.

22. Haik, Y.S., Mohamed, Y.E. and Abdulrehman, T. (2011). Combustion of algae oil methyl ester in an indirect injection diesel engine. *Energy.* 36: 1827-1835.

23. Hall, D.O., Mvnick, H.E. and Williams, R.H. (1991). Cooling the greenhouse with bioenergy. *Nature.* 353: 11-2.

24. Harun, R., Singh, M., Forde, Gareth, M. And Danquah Michael, K. (2010). Bioprocess engineering of microalgae to produce a variety of consumer products. *Renew Sust. Ener. Rev.* 14: 1037-47.

25. Herrero, M., Ibanez, E., Senorans, J. and Cifuentes, A. (2004). Pressurized liquid extracts from Spirulina platensis microalga: determination of their antioxidant activity And preliminary analysis by micellar electrokinetic chromatography. *J Chromatogr A.* 1047: 195-203.

26. Kalam, M.A. and Masjuki, H.H. (2002). Biodiesel from palm oil – an analysis of its properties and potential. *Biomass Bioenergy.* 23: 471-9.

27. Kansedo, J., Lee, K.T. and Bhatia, S. (2009). Biodiesel production from palm oil via heterogeneous transesterification. *Biomass Bioenergy.* 33: 271-6.

28. Kiran, B., Kumar, R. and Deshmukh, D. (2014). Perspectives of microalgal biofuels as a renewable source of energy. *Energy Conversion and Management.* 1-17.

29. Lang, X, Dalai, A.K., Bakhshi, N.N., Reaney, M.J. and Hertz, P.B. (2001). Preparation and characterization of bio-diesels from various bio-oils. *Bioresour Technol.* 80: 53-62.

30. Luque-Garcia, J.L., Luque, De. and Castro, M.D. (2003). Ultrasound: a powerful tool for leaching. TrAC *Trends Anal Chem.* 22: 41-7.

31. Macias-Sanchez, M.D., Mantell, C., Rodriguez, M., Martinez, De. La. Ossa, E., Lubian, L.M. and Montero, O. (2005). Supercritical fluid extraction of carotenoids and chlorophyll a from *Nannochloropsis gaditana. J Food Eng.* 66: 245-51.

32. Martin, P.D. (1993). Sonochemistry in industry. Progress and prospects. *Chem Ind (London).* 7: 233-236.

33. Mata Teresa, M., Martins Antonio, A. and Caetano Nidia, S. (2010). Microalgae for biodiesel production and other applications: a review. *Renew Sust. Energy Rev.* 14: 217-32.

34. Michael, B.J. (2009). Microalgal Biodiesel Production through a Novel Attached. Culture System and Conversion Parameters.

35. Minowa, T., Yokoyama, A.Y., Kishimoto, M. and Okakurat, T. (1995). Oil production from algal cells of Dunaliella tertiolecta by direct thermochemical liquefaction. *Fuel.* 74(12): 1735-8.

36. Molina, E.M. (1999). Scale-up of tubular photobioreactors. 8th Tri Annual International Conference on Applied Algology (8th ICAA), Montecatini Terme, Italy, Kluwer Academic Publ.

37. Molina, E.M. and Fernandez, F.G. Scale-up of tubular photobioreactors. 8th Tri Annual International Conference on Applied Algology (1999) (8th ICAA), Montecatini Terme, Italy, Kluwer Academic. Publ.

38. Molina, G.E, Acien-Fernandez, F.G., Garcia-Camacho, F., Camacho-Rubio, F. and Chisti, Y. (2000). Scale-up of tubular photobioreactors. *J Appl Phycol.* 12: 355-368.

39. Moscoso, Jose Luis Garcia,Obeid, Wassim, Sandeep Kumar, Hatche Patrick, G. (2013). Flash hydrolysis of microalgae (*Scenedesmus* sp.) for protein extraction and production of biofuels intermediates. *J. of Supercritical Fluids*. 82: 183–190.

40. Mostafa, S. (2010). Microbiological aspects of biofuel production: current status and future Prospect. *Journal of Advanced Research*. 103-111.

41. Nigam, P. and Singh, A. (2011). Production of liquid biofuels from renewable resources. *Progress in Energy and Combustion Science*. 37: 52-68.

42. Oilgae. (2008). www.oilgae.com

43. Ooi, Y.S., Zakaria, R., Mohamed, A.R. and Bhatia, S. (2004). Catalytic conversion of palm oil Based fatty acid mixture to liquid fuel. *Biomass Bioenergy*. 27(5): 477-84.

44. Pathak, V., Singh, R. and Gautam, P. (2014). Algal Biodiesel as an Emerging Source of Energy: A Review. *International Journal of Research (IJR)*. 1(6): 2348-6848.

45. Pathak, V., Singh, R. and Gautam, P. (2014). Algal Biodiesel as an Emerging Source of Energy: A Review. *International Journal of Research*. 1(6): 1-8.

46. Pathak, V., Singh, R. and Gautam, P. (2015). Microalgae as Emerging source of Energy: A Review. *Research Journal of Chemical Sciences*. 5(3): 1-5.

47. Pawliszyn, J. (1993). Kinetic model of supercritical fluid extraction. *J Chromatogr Sci*. 31: 31-7.

48. Rattanaphra, D. and Srinophakun, P. (2010). Biodiesel production from crude sunflower oil and crude jatropha oil using immobilized lipase. *J. Chem. Eng. Jpn*. 43(1): 104-8.

49. Reddy, Harvind K. Muppaneni, Tapaswy, Patil, Prafulla D.Ponnusamy, Sundaravadivelnathan Cooke, Peter, Schaub, Tanner and Deng, Shuguang. (2014). Direct conversion of wet algae to crude biodiesel under supercritical ethanol conditions. *Fuel*. 115: 720-726.

50. Satyanarayana, K.G. and Mariano, A.B. (2011). Vargas JVC: A review on microalgae, a versatile source for sustainable energy and materials. *Int. J. Energy Res*. 35: 291–311.

51. Schenk Peer, M. (2008). Second Generation Biofuels: High-Efficiency Microalgae for Biodiesel Production. *Bioenerg. Res*. 20-43.

52. Shay, E.G. (1993). Diesel fuel from vegetables-oils-status and opportunities. *Biomass Bioenergy*. 4: 227-42.

53. Sheehan, J. (1999). A Look Back at the U.S. Department of Energy's Aquatic Species Program – biodiesel from algae. Report NREL/TP-580-24190, National Renewable Energy Laboratory, Golden, CO.

54. Siler-Marinkovic, S. and Tomasevic, A. (1998). Transesterification of sunflower oil in situ. *Fuel*. 77(12): 1389-1392.

55. Singh, J. and Sai, G. (2010). Commercialization potential of microalgae for biofuels production. *Renewable and Sustainable Energy Reviews*. 14: 2596-2610.

56. Sivakumar, P. (2014). Mass cultivation of microalgae and extraction of total hydrocarbons: A kinetic and thermodynamic study. *Fuel.* 119: 308-312.

57. Smith Val, H. (2009). Reviewed on The ecology of algal biodiesel production. *Trends in Ecology and Evolution.* 25(5): 1-9.

58. Spolaore, P., Joannis-Cassan, C., Duran, E. and Isambert, A. (2006). Commercial applications of microalgae. *J. Biosci. Bioeng.* 101: 87-96.

59. Sudhakar, K. and Premalatha, M. (2012). Micro-algal Technology for Sustainable Energy Production: State of the Art. *Journal of Sustainable Energy & Environment.* 3: 59-62.

60. Sumathi, S., Chai, S.P. and Mohamed, A.R. (2008). Utilization of oil palm as a source of renewable energy in Malaysia. *Renew Sust. Energy Rev.* 12: 2404-2421.

61. Topare, N.S., Raut, S.J., Renge, V.C., Khedakar, S.V., Chavan, V.P. and Bhagat, S.L. (2011). Extraction of oil from algae by solvent extraction and Oil expeller method. *Int J Chem. Sci.* 9(4): 1746-50.

62. Wang, B., Li, Y., Wu, N. and Lan, C. (2008). CO_2 bio-mitigation using microalgae. *Applied Microbiology and Biotechnology.* 79: 707-718.

8

Approaches in Studying the Diversity of Arbuscular Mycorrhizal Fungi

Harbans Kaur Kehri*[1] and Ovaid Akhtar[2]

*Sadasivan Myopathology Laboratory, Department of Botany,
University of Allahabad, Allahabad-211002, Uttar Pradesh, India
E-mail: kehrihk@gmail.com*

Abstract

Arbuscular Mycorrhizal (AM) fungi are integral component of the soil ecosystem. They are found in symbiotic association with roots of higher plants. Being endophytic and obligate nature it is very difficult to culture in auxenic medium. This nature hampers the approaches in studying the diversity and ecology of AM fungi. In recent past the morphological methodologies has been developed so as to study these important communities within the root and in soils. The morphological approaches are solely based on the characterization of azygospores produced in the soils and in the roots. However, these methods have some problems and limitations in characterization of AM fungal spores. To overcome this, scientists have applied molecular approaches in understanding the taxonomy and ecology of AM fungi. It has now been established to use SSU-ITS-LSU region of the nuclear ribosomal gene (~1.5-1.8 kb) as potential barcode region for the identification of AM fungi at species level. This literature encompasses morphological and molecular approaches in the field of AM fungi and relevance of the two approaches has been discussed.

Key words: Arbuscular Mycorrhizal (AM) fungi, Azygospores, Morphological methods, PCR, Primers.

Introduction

Arbuscular Mycorrhizal (AM) fungi are the obligate symbiont of the roots of higher plants since the origin of the land plants. Perhaps when the first land plant has moved on the ground, AM fungi infected them and might have established symbiotic association. However, being endophytic nature this fungal group was hidden to the human eyes and remained in dark side. It was Franciszek Kamienski (1881), who discovered the fungus in association with the roots of *Monotropa hypopitys* L. Four years later to this discovery Frank (1885) coined the term *"Mycorrhiza"* for this symbiotic association.

AM fungi are ubiquitous and found everywhere, where plants can grow. They were reported to colonize the early land plants in the pre-historic time during the devonian period (400 million years back) and were integral components of the palaeoecosystem. Taylor *et al.*, (1995) reported the vesicular arbuscular mycorrhizal fungi in the fossil axis of the Rhynie chert plant. These fungi are reported to have structures like modern AM fungi and named as *Glomites rhyniensis*. Since the origin of the land plant this group of fungi evolved with the evolution in land plant and now it is well established that up to 90% of the plants, including angiosperms, gymnosperms and pteridophytes having roots, as well as the gametophytes of some mosses, lycopods, and *Psilotalus*, which do not have true roots are mycorrhizal (Smith and Read, 2008; Brundrett, 2009).

AM fungi belong to the glomeromycota (Schüßler et al., 2001) spatially consist two types of structures viz., intraradical and extraradical. The intraradical structure consists of arbuscules, vesicles, spores and hyphae. The extraradical structure consists of hyphae, spores and in some genera auxiliary cells. Spores are the asexual structures formed singly or in groups and some time in compact aggregates with a definite peridium called as sporocarp. Spore is main component of AM fungi taken into consideration for identification and classification (Stürmer and Morton, 1999; Walker, 1983). Vesicles are the reserve food storage organs and contains lipid granules. These vesicles may act like propagules and can cause infection. In genera of the family gigasporaceae, exraradical auxiliary cells are formed in place of vesicles. Arbuscules are fan-shaped or brush like structures, formed in deeper cells of the cortex and provide the active sites for exchange of various molecules across. AM fungi are obligate biotrophs and can grow only on feeding of photosynthates of their alive host plant.

Functionally, AM fungi are of great importance as they play diverse roles (Fig. 1) in the establishment and survival of plants in various ecosystems. It increases the root absorptive area and hence the plant nutrition (Bieleski, 1973), actively influence the succession of plant communities (Janos, 1980), equalize the level of nutrition of co-existing plants by formation of hyphal bridges transferring nutrients between them (Newman, 1988). AM fungi improve soil structure through binding sand grains into aggregates by extraradical hyphae (Koske *et al.*, 1975; Sutton and Sheppard, 1976). AM fungi influence the primary productivity in terrestrial ecosystem by influencing the global phosphate and carbon cycling (Fitter, 2005). They play an essential role in enhancing plant growth in semi-arid agro-ecosystem (McGee, 1989), particularly for plants grown in eroded soils (Herrera *et al.*, 1993).

AM fungi have been suggested as useful inoculants for indigenous seedling production and reforestation (Wubet *et al.*, 2003). Additionally, AM fungi increased the tolerance of plants to various toxic metals (Dehn and Schüepp, 1989; Griffioen and Ernst, 1989), water stresses (Stahl and Smith, 1984), as well as pathogenic fungi and nematodes (Schönbeck, 1978). Although even low colonization of plant roots by AM fungi can alleviate such stresses (Pongrac *et al.*, 2009), the effect of influence of AM fungi may differ, because different species or even strains of a given species of AM fungi may variously affect plants (Abbott and Robson, 1981; Kaldorf *et al.*, 1999; Maherali and Klironomos, 2007; Sýkorová *et al.*, 2012).

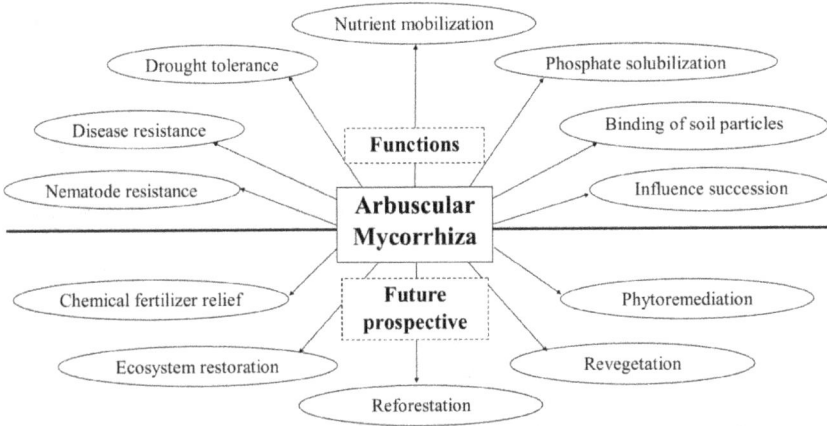

Figure 1: Benefits of AM fungi to host plants and future perspectives in ecosystem restoration.

The azygospore is only structure which transfers the genetic information vertically. Horizontal gene transfer has not been reported yet in AM fungi, hence despite of ancient origin, less number of species have been evolved and the characterized species are even very few in number. The obligated nature of AM fungi has limited the study. Hence, numerous unsuccessful attempts of application of AM fungi probably partly resulted from erroneous species identification and the difficult nature of AM fungal taxonomy (Schüßler *et al.*, 2011; Krüger *et al.*, 2012). This literature contains all the methodologies developed till date for the taxonomic and ecological study of AM fungi including anatomical, histochemical and relatively recent molecular approaches.

Morphological approaches for study of AM fungal diversity

AM fungi are the obligate biotrophs and this property hampered the culture and multiplication in artificial synthetic media. The traditional methods and approaches for identification and classification are solely based on the morphological nature of extramatrical spores (Stürmer and Morton, 1997; Walker, 1983). Spore morphology includes shape, size and color of the spore at maturity, surface ornamentation, composition and thickening of composite wall layers, subtending hyphae, their shape, septation (if any), etc. Beside the extrametrical spores, intraradical hyphae, vesicle/auxiliary cells are also considered as important parameters for the identification of AM fungi up to family level (Merryweather and Fitter, 1998).

The morphological approaches developed and applied till date for the study of AM fungi can be described under the following heads.

Extraction of AM fungal spores from soil

For the study of AM fungal spores the spores are firstly extracted through wet sieving and decanting method (Gerdeman and Nicolson, 1963) either from field soil or from pot culture (Walker, 1999) (Fig. 2). In this method soil suspension containing spores are allowed to pass through sieves of various pore size and spores are recovered on filter paper for observation. Later, spores of AM fungi were collected by using gelatin column (Mosse and Jones, 1968). Furlan and Fortin (1975) used floatation-bubbling method for the isolation of AM fungal spores by using 50% aqueous glycerol solution. 2M sucrose solution was used by Smith and Skipper (1979) for extraction of spores from decanted soil-spore suspension by centrifugation at 2000 rpm for 10 minutes. For the mass collection of spores form soil 30% sucrose solution was used by Mertz *et al.,* (1979). After the method of wet sieving and decanting, a number of methods have been developed for the collection of AM fungal spores from soils but all the methods necessarily are the modification of earlier one and thus, wet sieving and decanting method is still used as the first step for isolation of mycorrhizal spores from soil. By using any above method or by combination of more than one method, one can harvest extrametrical (extraradical) spores for the study of fungi.

Study of spores

After extraction process, spores are manually picked with the help of needle or micropipette and mounted on to a glass slide in polyvinyl lacto glycerol (PVLG) (Omar *et al.,* 1979) or in PVLG + Melzer's reagent.

Application of gentle pressure over the coverslip ruptures the spores and wall layers get separated. The slides prepared in this way are permanent and can be stored for years. However, heating of slide in oven at 50ºC for 24 hours can give better view under microscope, as the heating treatment removes the water content and the mounting medium becomes less thick, which allows the more details in one plane. In this way, photograph taken can have more details of the spores. Following are the important microscopic characters, which are taken into consideration while studying the AM fungal spores.

Shape and Size

AM fungi show great variability in their spore shape, for example, Trappe (1982) described the shape from globose, ellipsoid, obovoid, reniform, irregular and narrow clavate to broadly clavate for taxonomy of the nine genera of endogonaceae. Determination of shape of AM fungal spores has some limitations and need very care while describing it. As the AM fungal spores/azygospores/ vegetative spore borne at the tip of vegetative mycelium through bulging, the shape which a spore will take depends basically upon the genus. However, these may vary depending upon the surroundings of spores, where they form, and up to some extent on the stages of development. For example, some genera of *Glomus* form spore both in free and aggregate form and shape of the two types of same genus vary. Moreover, the genera, which forms spores both in roots as well as in soil also varies in their shape. Due to the lack of space and compactness, the shape of spore inside the roots becomes irregular due to the pressure of adjacent spores. Likewise, AM fungal spores show higher degree of variability. It may range from less than 20⊙m (in Acaulosporaceae) to more than 500⊙m (in sporocarpic genera) in diameter at maturity (Trappe, 1982; Schenck and Perez, 1990). Beside the genus type, the size is smaller at maturity whereas, as the spore gets mature its size increases.

Study of the shape and size of spores needs expertise and familiarity with diverse group of AM fungi and their developing stages. The higher the number of spores of one type taken into consideration, the accurate will be the determination of shape and size and hence identification.

Color

Depending upon the fungal type and stages of development, spores of AM fungi may vary in color. According to the Trappe (1982) they may of white to gray, yellow, brown, orange to red, violet and nearly black in color. The determination of color of spore is very difficult as color of spore changes in the same genus, as it ages. So researchers should be familiar with age of spore whether the spore is in mature stage. For color determination, one should also be familiar with the different types of colors and the properties of light which are allowed to pass through the spores mounted on glass slides. For example, tungsten filament with Kohlar illumination without any filter has been used for the determination of color of spores by Trappe (1982). Now a days for better resolution all the research microscopes are generally fitted with blue color filter. Hence, the color of spores seen in modern microscope may vary with those written in the old literature. A standard color chart (Kornerup and Wanscher, 1983) is often appropriate but is problematic issue due to the diverse combinations. A more convenient and accurate method is to determine the color in descriptive form of CMYK format. Any photo editing computer program can have the function to write the color of spore taken as photograph in the form of CMYK format (#-#-#-#). '#' can be any number between the '0' to '100' and the combination of these four numbers produces all the color. This descriptive form of color determination in supplement with color name has been applied by Morton (2002).

Surface ornamentation

Surface ornamentation of AM fungal spores at young stage remain almost smooth. As the spore ages, a definite pattern is developed on the surface of spores, however, the pattern which is to be developed on the surface is highly constant within the same species and characteristics to the fungal morphotypes. The surface ornamentation at maturity may be smooth, roughened, outer layer sloughing away (Berch and Koske, 1986; Blaszkowski *et al.*, 2001), pits, warts, covered with spines, with polygonal projection, reticulation, cerebriform fold, spiraling ridges, etc.

Wall layers

Number of wall sub-layers (laminae) and its thickness has great importance in identification of AM fungal spores. Wall sub-layer is also highly variable among the species and constant with in the same species and hence used in the identification of morphotypes (Walker, 1983, 1986; Morton, 2002; Morton and Redecker, 2001). For studying the wall sub-layer, it is recommended to mount the spores in PVLG+Melzer's reagent. The iodine in melzer's reagent shows dextrinoid reaction with the laminae and depending on the composition of different laminae, it take purple, light brown, brown to red in color.

Hyphal attachment

As the AM fungi produce spore at the tip of a parental hyphae, hence except in the family acaulosporaceae, the spores are found attached to the subtending hyphae. The type of hyphal attachment and its microscopic details such as, shape, continuity of spore was layer, separation of spore content, constriction, etc., are considered as important morphological parameter for studying morphotypes. In the family acaulosporaceae spore are formed from the saccules, hence no hyphal attachment is there and in few genera of *Glomus* (for example, *G. multicaule*) more than one subtending hyphae are found (Gerdemann and Bakshi, 1976). This is mainly due to the intercalary nature of spore rather than terminal.

Study of AM association

The intramatrical structures are also being taken into consideration for the taxonomic study of AM fungi and these characters give resolution at family level (Merryweather and Fitter, 1998). For studying intraradical mycorrhizal structures it is essential to examine intact infected root fragments. Due to the endosybiontic and obligate nature, this fungi could not be studied in auxenic culture. For this purpose, it is mandatory to study the root fragments colonized by AM fungi.

Staining of AM fungi in roots

Phillips and Hayman (1970) first of all developed a staining procedure for studying intramatrical structure of AM fungi, in which they cleared the infected root fragments in 10% KOH solution overnight and then stained in 0.05% trypan blue or aniline blue or cotton blue prepared in lasctophenol, acetic acid and glycerol. In this method root fragments are allowed to react with clearing and staining reagent in at least for 24 hours. Later on time of treatment was reduced by heat treatment at 90º C for 1 hour. Koske and Gemma (1989) described a new method for staining of AM fungal structure in root, in which they used 0.05% acid fuchsin. Ink-vinegar method of staining (Vierheilig *et al.*, 1998) has also been applied in AM fungal research.

Study of AM association in roots

For studying the intensity of AM fungal colonization in roots, four statistical methods has been compared by Giovannetti and Mosse (1980), of which slide method has gain more popularity in mycorrhizal research. Later McGonigle *et al.*, (1990) have published a new statistical method for estimation of intensity of AM association in roots. These methods provided the useful tool for ecological studies of AM fungi.

SEM approach for studying AM fungi

Although little work has been done with reference to the scanning electron microscopy (SEM) of AM fungi (Orłowska *et al.*, 2002), however, the application of

SEM in AM fungal research could be ad-on in understanding the ultrastructural details of spore surface. For the preparation of material for the SEM study, spores of excised roots are dehydrated first either by air drying or with the treatment of 2% osmium tetroxide. Then, spore are passed through graded ethanol series and finally mounted on stub. Coating with carbon or gold make the material electron dense.

Limitations of morphological approaches for AM fungal study

Since the upsurge in AM fungal research, microbial taxonomists have done a lot of work related to the morphological study of AM fungi and established the references for identification and classification but, morphological characterization is solely based on specific strains of fungi and hence lack of homogeneity in parameters is well known problem still persisting. Many workers have identified and classified the AM fungi on microscopic observation of spores collected form soil (Bethenfalvay and Yoder, 1981; Schenck and Perez, 1990). Approximately 150 species of AM fungi have been described in order Glomales on the basis of morphological characteristics of their azygospores (Walker and Trappe, 1993). At present the phylum Glomeromycota comprises three classes, five orders, 15 families, 34 genera and 250 species (Oehl *et al.*, 2011a,b; Schüßler and Walker, 2010).

The morphological characterization of AM fungi in ecological and phylogenetic studies on the basis of spores has several cons. The most physiologically active and important part of symbiosis are the arbuscules, which are formed within the inner cortex. Hence, functionally important part of the root-AM fungal symbiosis is the infected length of root. It is well established fact, that more than one type of fungi can infect the root simultaneously, but it is not always essential to produce extraradical spores by all symbiotically active fungi. The time of spore formation varies from species to species. Spores are the resting stages and not always reflect those species that are physiologically active at the time (Sanders, 2004). Moreover, some fungi do not produce azygospores in certain environmental conditions. Environmental factors play a significant role in the dynamics of AM fungi, especially on the formation of extraradical spores. Behavior of AM fungi is affected by soil pH (Wang *et al.*, 1985), nutrient level (Mosse *et al.*, 1981) and interactions with other micro-organisms (Bagyaraj, 1984). Hence, extraradical spores do not necessarily present whole actively participating AM fungi in symbioses. Under certain conditions or during certain seasons of the year, some AM fungi may produce many spores and therefore appear to be dominant root colonizers, whereas under different conditions, they may not sporulate at all. Furthermore, the dynamics of spore production versus root colonization may differ among species (Bever *et al.*, 1996). Non-sporulating species may not be detected at all whereas prolific spore-producers dominate our views of AM fungal ecology. The production of spores is not always correlated with root colonization (Clapp *et al.*, 1995; Merry weather and Fitter, 1998). When no spores are formed, the intraradical structures of AM fungi at best allow identification only up to the family level (Merry weather and Fitter, 1998).

For ecological studies of AM fungi in the rhizosphere, it is essential to study the intact root fragment housing active AM fungi. Moreover, several lineages of AM fungi have been discovered that do not stain at all within the roots, or only very weakly using the standard dyes (Morton and Redecker, 2001). Thus, the species composition of active AM populations within roots can only be analyzed by molecular methods. Molecular techniques, especially nucleic acid based approaches, have the potential to revolutionize our understanding of AM fungi. For these reasons, several PCR-based detection methods have been developed in recent years and some have already been applied under field conditions (Clapp *et al.*, 1995; Helgason *et al.*, 1998; Hijri *et al.*, 2006).

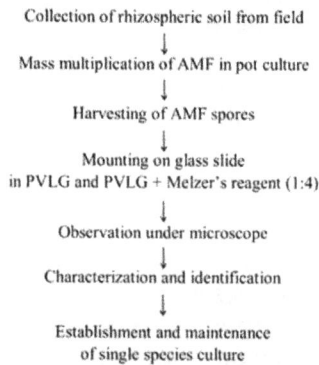

Collection of rhizospheric soil from field
↓
Mass multiplication of AMF in pot culture
↓
Harvesting of AMF spores
↓
Mounting on glass slide
in PVLG and PVLG + Melzer's reagent (1:4)
↓
Observation under microscope
↓
Characterization and identification
↓
Establishment and maintenance
of single species culture

Figure 2: Schematic representation of morphological approach for study of AM fungal spore.

Molecular approaches on AM fungal study

With the development in the field of molecular biology, in last three decades AM fungal taxonomists have focused towards the study of AM fungi on basis of variations in the genetic region. Methods have been developed and applied (Clapp *et al.*, 1995; Helgason *et al.*, 1998) to study the AM fungal diversity by using colonized root fragment, isolated spores as well as directly from soils (Fig. 3). The nuclear ribosomal DNA (nrDNA) has been well established as a molecular marker for characterization of AM fungi (Clapp *et al.*, 2002). nrDNA (Fig. 4) are present in multiple repeat and consists of conserved gene for small subunit – SSU and large subunit – LSU. These exon sequences are separated by non-functional spacer called – Internal Transcribed Spacer (ITS). Spacer sequence ITS1 separated between 18s and 5.8s segment of rDNA while ITS2 separates the 5.8s and 28s segment. These regions – SSU, ITS and LSU alone as well as in combination, are important tool for revealing phylogenetic relationships and developing molecular probes to identify glomeromycotan fungi (Redecker, 2000; Schüßler, 1999; van Tuinen *et al.*, 1998).

The first step in studying molecular diversity of AM fungi is selection of suitable material for the DNA extraction. DNA are extracted from host root or isolated spores from soil. In both the processes, it is essential to amplify marker region through PCR. Root fragments are good materials while studying the AM fungal diversity of an ecosystem, however, for identification of AM fungal genus from pot culture, healthy spores are best.

Extraction of AM fungal DNA

For working with one species culture or trap culture with good amount of spore, few spores of one type or selected and crushed in molecular grade water or in PCR buffer (Lanfranco *et al.*, 2001). By crushing DNA released into the solution and this is utilized as template for PCR (Redecker *et al.*, 1997). DNA from root samples are extracted by CTAB method of Li (2000).

Amplification of AM fungal DNA by PCR

For sequencing, multiple copies of a barcode region of DNA is the prerequisite. Hence, polymerase chain reaction (PCR) is mandatory step in molecular research of AM fungi. Following are the variant of PCR frequently applied to the AM fungal nrDNA multiplication.

Nested-PCR

In recent few years, to multiply the very minute quantity of DNA content extracted either from spore or from colonized root, nested PCR, a two-step highly sensitive amplification process has been applied in studying AM fungi. In nested PCR, first amplification is based on crude DNA from the samples and universal sets fungal primers are used where as in second step, DNA template is the product of first step and more specific primers of glomeromycotan fungi are utilized (Renker *et al.*, 2003; Kjøller and Rosendahl, 2000; Rosendahl and Matzen, 2008).

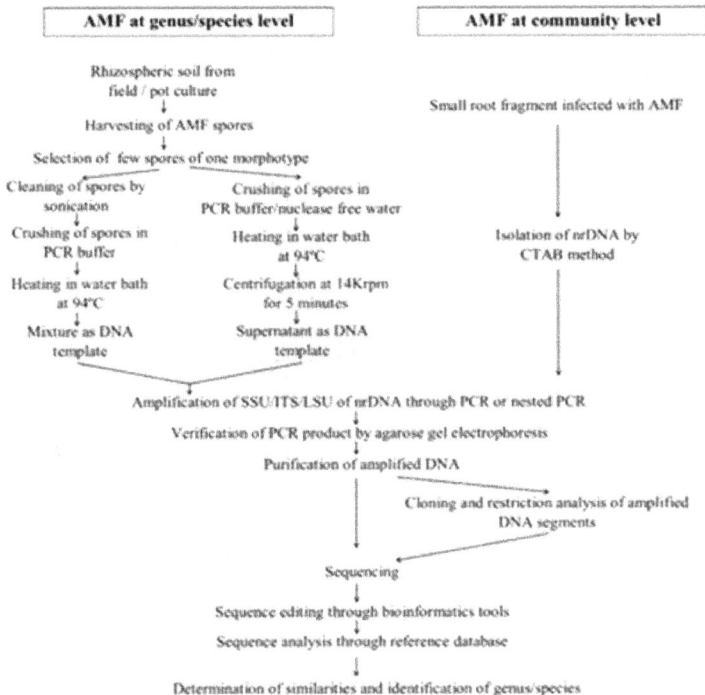

Figure 3: Schematic representation of molecular approaches for the study of AM fungi.

PCR-denaturing gradient gel electrophoresis

Another version of PCR is PCR-denaturing gradient gel electrophoresis (DGGE) which was initially developed to study mutation has now-a-days become one of the most applied culture-independent techniques to study the community structure of microorganisms (Muyzer and Smala, 1998). In recent years, PCR-DGGE has been successfully applied by various workers (J. C. Santos *et al.*, 2006; Wei *et al.*, 2014, 2015) in analyzing the AM fungi of various habitats.

PCR-RFLP

PCR coupled with restriction fragment length polymorphism (RFLP) has been applied in mycorrhizal research to identify strains of mycorrhizal fungi and also to differentiate and identify mycorrhizal symbionts unambiguously. Redecker *et al.*, (1997) demonstrated that PCR/restriction analysis of internal transcribed spacers (ITS) of ribosomal DNA allowed species of the Glomales to be distinguished using a minute amount of fungal biomass, and that the generated fragment patterns were highly reproducible. The PCR/RFLP polymorphism of the ITS region is generally regarded as appropriate to differentiate AM fungi at the species level (Renker *et al.*, 2003, 2006; Hempel *et al.*, 2007). Application of 'next-generation' sequencing (NGS) and single molecule real time sequencing (SMRT) technologies to AM fungal genetics possesses the potentiality to revolutionize the genome sequencing.

Primers for amplification of barcode region

The analysis of nuclear ribosomal genes (nrDNA) allows the concurrent elucidation of phylogeny and primer design for specific phylogenetic groups of AM fungi. Various sets of primer have been designed from time to time as the knowledge on sequences of AM fungi had become available. These primers basically vary in the specificity of nrDNA regions (Fig. 4).

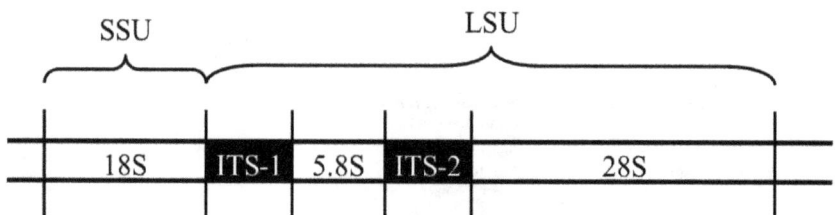

Figure 4: Various regions of ribosomal DNA used as barcode in AM fungal research.

Universal eukaryotic primers

Universal eukaryotic primers NS5 and ITS4 were the first primer set successfully tested for AM fungi by White *et al.*, (1990). NS31 primer has been designed by Simon *et al.*, (1992). Another eukaryotic primer, VANS1 has been designed by Simon *et al.*, (1992) on basis of analysis of three Glomalean 18S subunit sequences available at that time. Recent studies have shown that the annealing site of VANS1 is not well conserved within the Glomales (Clapp *et al.*, 1999) and that several newly characterized ancestral lineages of the Glomales do not have this site at all (Redecker *et al.*, 2000). J. C. Santos *et al.* (2006) successfully applied the universal eukaryotic primer set AM1 (Helgason *et al.*, 1998) and the NS31-GC (Kowalchuk *et al.*, 1997). NS31-GC corresponds to the universal eukaryotic NS31 (Simon *et al.*, 1992) plus a GC-clamp sequence.

Fungi specific Primers

Other primers reported were specific only for single isolates or species and, therefore, are of restricted applicability (van Tuinen *et al.*, 1998) or do not exclude other fungi (Helgason *et al.*, 1998). These primers also had some mismatch.

Glomeromycota specific primers

Later some other sets of primer has been designed for certain groups of glomeromycotan fungi to overcome this problem (Kjøller and Rosendahl, 2000; Redecker, 2000; Wubet *et al.*, 2003; Gamper and Leuchtmann, 2007). Redecker (2000) has successfully tested the new combinations of the previously designed primers. In nested PCR he used NS5 and ITS4 (White *et al.*, 1990) universal primer set in first step. In second step (nested PCR) he used GLOM5.8R, GIGA5.8R (White *et al.*, 1990) in combination with ITS1F (Gardes and Bruns, 1993) in first variant and in the second variant forward specific primers (ARCH1311, ACAU1660, LETC1670) combined with ITS4 were used.

Primers for partial nrDNA of glomeromycota

The ITS region was used for AM fungi to separate species e.g. in the *Ambisporaceae* (Walker *et al.*, 2007) and in combination with the LSU rRNA gene for *Diversisporaceae* (Gamper *et al.*, 2009), but species recognition of the ITS alone for very closely related species e.g. for *Rhizophagus intraradices* FL208 and *Rhizophagus irregularis* DAOM197198 is not always robust, due to high intraspecific variability.

The most comprehensive taxon sampling for Glomeromycota covers the SSU rDNA region (Schüßler *et al.*, 2001), for which a new AM fungi specific primer pair AML1 and AML2 (Lee *et al.*, 2008) was published. AML1 and AML2 was the first primer which covers the all reported AM fungi at that time and widely used for characterization of AM fungi in the field (Lee *et al.*, 2008; Beck *et al.*, 2007; Öpik *et al.*, 2008, 2010; Turrini *et al.*, 2008; Long *et al.*, 2010; Ryszka *et al.*, 2010).

Primers sets for ~1.5-1.8 kb SSU-ITS-LSU

However, this set of primer only resolves the AM fungi at generic level. For species level resolution and to develop robust phylogenetic tree some workers, (da Silva *et al.*, 2006; Kruger *et al.*, 2009) considered the SSU rDNA region as unsuited for community analysis at species-level and DNA barcoding. Recently Kruger *et al.* (2009) designed PCR primer set (SSUmAf-LSUmAr, SSUmCf-LSUmBr), which amplify a fragment of ~1.5-1.8 kb covering the 3' SSU, the whole ITS and a part of the LSU rDNA region and resolve the detection of all known AM fungal lineages. SSU-ITS-LSU fragment of 1.5 kb could be now utilized as baseline for AM fungal DNA barcoding, because shorter fragments failed to separate closely related species robustly. However, species identification is only as good as the reference sequence database is (Begerow *et al.*, 2010).

Conclusions

Use of molecular techniques has revolutionized the understanding of AM fungal taxonomy, phylogeny and ecology. Advancement in the molecular approaches have ability to bridge the gap between spore-based morphological taxonomy and what is present in function symbioses. By combined application of classical and modern molecular approaches, detailed behavior of AM fungi could be studied at species and community level, which is otherwise not possible. For studying the AM fungal diversity involved in the symbioses molecular techniques can provide direct and accurate identification tool.

However, less than 5% of the molecular taxonomic units have been characterized and established in culture as species. Hence, most of the similar morphotypes are erroneously identified as single species. As the molecular tools have been developed for the study of AM fungal taxonomy, diversity and ecology the morphological methods should have also to be developed in parallel. But, being classical approach, the latter is hampered due to lack of funding, labor and patience.

For accurate identification research should be carried out in both the molecular and morphological direction. Methods for the multiplication and single species culture should be improved to reduce the time and labor. The molecular method should also be improved for better resolution of species with high efficiency and accuracy. Thus, simultaneous approach will definitely unrevealed the hidden mystery of AM fungal world. By this way, novel species could be identified and the advantages of this fungal group would be utilized in the field of agriculture, reforestation, heavy metal remediation, restoration of ecosystem, etc.

References

1. Abbott, L.K. and Robson, A.D. (1981). Infectivity and effectiveness of five endomycorrhizal fungi: competition with indigenous fungi in field soils. *Aust. J. Agric. Res*. 32: 621-630.
2. Bagyaraj, D.J. (1984). Biological interactions with VA mycorrhizal fungi. In C.L, Powell, and D.I. Bagyaraj, (Eds.), *VA Mycorrhiza*. Boca Raton, FL: CRC Press. 131-153.
3. Beck, A., Haug, I., Oberwinkler, F. and Kottke, I. (2007). Structural characterization and molecular identification of arbuscular mycorrhiza morphotypes of *Alzatea verticillata (Alzateaceae)*, a prominent tree in the tropical mountain rain forest of South Ecuador. *Mycorrhiza*. 17: 607-625.
4. Begerow, D., Nilsson, H., Unterseher, M. and Maier, W. (2010). Current state and perspectives of fungal DNA barcoding and rapid identification procedures. *Applied Microbiology and Biotechnology*. 87(1): 99-108.
5. Berch, S.M. and Koske, R.E. (1986). *Glomus pansihalos*: a new species in the Endogonaceae, Zygomycetes. *Mycologia*. 78: 838-842.
6. Bethenfalvay, G.J. and Yoder, J.F. (1981). The *Glycine max-Glomus fasciculatus-Rhizobium japonicum* Symbiosis; Phosphorus effect on nitrogen fixation and mycorrhizal infection. *Physiologia Plantarum*. 52: 141-145.

7. Bever, J.D., Morton, J., Antonovics, J. and Schultz, P.A. (1996). Host-dependent sporulation and species diversity of mycorrhizal fungi in a mown grassland. *Journal of Ecology.* 75: 1965-1977.

8. Bieleski, R.L. (1973). Phosphate pools, phosphate transport and phosphate availability. *Ann. Rev. Plant Physiol.* 24: 225-252.

9. Blaszkowski, J., Tadych, M. and Madej, T. (2001). *Glomus arenarium,* a new species in Glomales (Zygomycetes). *Acta Soc. Bot. Pol.* 70: 97-101.

10. Brundrett, M. (2009). Mycorrhizal associations and other means of nutrition of vascular plants: understanding the global diversity of host plants by resolving conflicting information and developing reliable means of diagnosis. *Plant and Soil.* 320: 37-77.

11. Clapp, J.P., Rodriguez, A. and Dodd, J.C. (2002). Glomales rRNA gene diversity-all that glistens is not necessarily glomalean? *Mycorrhiza.* 12: 269-270.

12. Clapp, J.P., Young, J.P.W., Merryweather, J.W. and Fitter, A.H. (1995). Diversity of fungal symbionts in arbuscular mycorrhizas from a natural community. *New Phytol.* 130: 259-265.

13. Clapp, J.P., Fitter, A.H. and Young, J.P.W. (1999). Ribosomal small subunit sequence variation within spores of an arbuscular mycorrhizal fungus, *Scutellospora* sp. *Mol Ecol.* 8: 915–921.

14. Da-Silva, G.A., Lumini, E., Maia, L.C. and Bonfante, P.B. (2006). Phylogenetic analysis of *Glomeromycota* by partial LSU rDNA sequences. *Mycorrhiza.* 16: 183-189.

15. Dehn, B. and Schuepp, H. (1989). Influence of VA mycorrhizae on the uptake and distribution of heavy metals in plants. *Agric. Ecosys. Environ.* 29: 79-83.

16. Fitter, A.H. (2005). Darkness Visible: Reflections on Under-ground Ecology. *Journal of Ecology.* 93: 231-243.

17. Frank, A.B. (1885). Über die auf Wurzelsymbiose beruchende Ernarung gewisser Baume durch unterirdische Pilze. *Ber. Deutch Bot. Gessell.* 3: 128-145.

18. Furlan, V. and Fortin, J.A. (1975). "A Flotation-.Bubbling System for Collecting Endogonaceae. Spores From Sieved Soil," *Naturalistic Canadien.* 102:663-667.

19. Gamper, H. and Leuchtmann, A. (2007). Taxon-specific PCR primers to detect two inconspicuous arbuscular mycorrhizal fungi from temperate agricultural grassland. *Mycorrhiza.* 17: 145-152.

20. Gamper, H.A., Walker, C. and Schu¨ßler, A. (2009). Diversispora celata sp. nov: molecular ecology and phylotaxonomy of an inconspicuous Arbuscular mycorrhizal fungus. *New Phytologist.* 182: 495-506.

21. Gardes, M. and Bruns, T.D. (1993). ITS primers with enhanced specificity for Basidiomycetes application to the identification of mycorrhizae and rusts. *Molecular Ecology Notes.* 2: 113-118.

22. Gerdemann, J.W. and Bakshi, B.K. (1976). Endogonaceae of India: Two new species. *Trans. Br. Mycol. Soc.* 66: 340-343.

23. Giovannetti, M. and Mosse, B. (1980). An evaluation of techniques for measuring vesicular arbuscular mycorrhizal infection in roots. *The New Phytologist.* 84: 489-500.

24. Griffioen, W.A.J. and Ernst, W.H.O. (1989). The role of VA mycorrhiza in the heavy metal tolerance of *Agrostis capillaris* L. Agric. *Ecosys. Environm.* 29: 173-177.

25. Helgason, T., Daniell, T.J., Husband, R., Fitter, A.H. and Young, J.P.W. (1998). Ploughing up the wood-wide web? *Nature.* 394: 431.

26. Hempel, S., Renker, C. and Buscot, F. (2007). Differences in the species composition of arbuscular mycorrhizal fungi in spore, root and soil communities in a grassland ecosystem. *Environ. Microbiol.* 9: 1930-1938.

27. Herrera, M.A., Salamanca, C.P. and Barea, J.M. (1993). Inoculation of woody legumes with selected arbuscular mycorrhizal fungi and rhizobia to recover desertified Mediterranean ecosystems. *Applied Environmental Microbiology.* 59: 129-133.

28. Hijri, I., Sýkorová, Z., Oehl, F., Ineichen, K., Mader, P., Wiemken, A. and Redecker, D. (2006). Communities of arbuscular mycorrhizal fungi in Arable soils are not necessarily low in diversity. *Mol. Ecol.* 15: 2277-2289.

29. Janos, D.P. (1980). Mycorrhizae influence tropical succession. *Biotropica.* 12: 56-64.

30. Kaldorf, M.O., Kuhn, A.J., Schröder, W.H., Hildebrandt, U. and Bothe, H. (1999). Selective element deposits in maize colonized by a heavy metal tolerance conferring arbuscular mycorrhizal fungus. *J. Plant Physiol.* 154: 718-728.

31. Kamienski, F. (1881). Die Vegetationsorgane der *Monotropa hypopitys* L. *Bot. Zeitschr.* 39: 225-234.

32. Kjøller, R. and Rosendahl, S. (2000). Detection of arbuscular mycorrhizal fungi (*Glomales*) in roots by nested PCR and SSCP (single stranded conformation polymorphism). *Plant Soil.* 226: 189-196.

33. Kornerup, A. and Wanscher, J.H. (1983). *Methuen handbook of colour.* 3rd Ed. London: Methuen and Co., Ltd.

34. Koske, R.E. and Gemma, J.N. (1989). A modified procedure for staining roots to detect VA mycorrhizas. *Mycological Research.* 92: 486-505.

35. Koske, R.E., Sutton, J.C. and Sheppard, B.R. (1975). Ecology of *Endogone* in Lake Huron sand dunes. *Can. J. Bot.* 53: 87-93.

36. Kowalchuk, G.A., Gerards, S. and Woldendorp, J.W. (1997). Detection and characterization of fungal infections of *Ammophila arenaria* (marram grass) roots by denaturing gradient gel electrophoresis of specifically amplified 18S rDNA. *Applied and Environmental Microbiology.* 63: 3858-3865.

37. Krüger, M., Krüger, C., Walker, C., Stockinger, H. and Schußler, A. (2012). Phylogenetic reference data for systematics and phylotaxonomy

of arbuscular mycorrhizal fungi from phylum to species level. *New Phytol.* 193: 970-984.

38. Krüger, M., Stockinger, H., Krüger, C. and Schußler, A. (2009). DNA-based species-level detection of Arbuscular mycorrhizal fungi: one PCR primer set for all AMF. *New Phytologist.* 183: 212-223.

39. Lanfranco, L., Bolchi, A., Ros, E.C., Ottonello, S. and Bonfante, P. (2002). Differential expression of a metallothionein gene during the presymbiotic *versus* the symbiotic phase of an arbuscular mycorrhizal fungus. *Plant Physiology.* 130: 58-67.

40. Lee, J., Lee, S. and Young, J.P.W. (2008). Improved PCR primers for the detection and identification of Arbuscular mycorrhizal fungi. *FEMS Microbiology Ecology.* 65: 339-349.

41. Li, M.G. (2000). Operating Principles and Techniques of Plant Gene. *Tianjin Science and echnology* Press, Tianjin, China.

42. Long, L.K., Yao, Q., Guo, J., Yang, R.H., Huang, Y.H. and Zhu, H.H. (2010). Molecular community analysis of Arbuscular mycorrhizal fungi associated with five selected plant species from heavy metal polluted soils. *European Journal of Soil Biology.* 46: 288-294.

43. Maherali, H. and Klironomos, J.M. (2007). Influence of phylogeny on fungal community assembly and ecosystem functioning. *Science.* 316: 1746-1748.

44. McGee, P. (1989). Variation in propagule numbers of vesicular–arbuscular mycorrhizal fungi in a semi-arid soil. *New Phytopathology.* 92: 28-33.

45. McGonigle, T.P., Miller, M.H., Evans, D.G., Fairchild, G.L. and Swan, J.A. (1990). A new method which gives an objective measure of colonization of roots by vesicular-arbuscular mycorrhizal fungi. *New Phytol.* 115: 495-501.

46. Merryweather, J. and Fitter, A. (1998). The arbuscular mycorrhizal fungi of *Hyacinthoides non-scripta*: I. Diversity of fungal taxa. *New Phytol.* 138: 117-129.

47. Mertz, Jr.S.M., Heithans, J.J.III. and Bush, R.L. (1979). Mass production of axenic spores of the endomycorrhizal fungus Gigaspora margarita. *Transaction of British Mycol. Soc.* 72: 167-169.

48. Morton, J.B. (2002). International Culture Collection of Arbuscular and Vesicular-Arbuscular Mycorrhizal Fungi. West Virginia University.

49. Morton, J.B. and Redecker, D. (2001). Two families of Glomales, Archaeosporaceae and Paraglomaceae, with two new genera *Archaeospora* and *Paraglomus,* based on concordant molecular and morphological characters. *Mycologia.* 93: 181-195.

50. Mosse, B. and Jones, G.W. (1968). Separation of endogone spores from organic soil debris by differential sedimentation on gelatin columns. *Transaction of British Mycological Society.* 51: 604-608.

51. Mosse, B., Stribley, D.O. and LeTacon, F. (1981). Ecology of mycorrhizae and mycorrhizal fungi. *Advances in Microbial Ecology.* 5: 137-210.

52. Muyzer, G. and Smalla, K. (1998). Application of denaturing gradient gel electrophoresis (DGGE) and temperature gradient electrophoresis (TGGE) in microbial ecology. *Antonie Van Leeuwenhoek.* 73: 127-141.

53. Newman, E.I. (1988). Mycorrhizal links between plants: their functioning and ecological significance. *Adv. Ecol. Res.* 18: 243-270.

54. Oehl, F., Silva, D.K.A., Maia, L.C., Ferreira, N.M. and da Silva, G.A. (2011a). *Orbispora* gen. nov., ancestral in the *Scutellosporaceae* of the *Glomeromycetes*. *Mycotaxon.* 116: 161-169.

55. Oehl, F., Silva, G.A., Goto, B.T., Maia, L.C. and Sieverding, E. (2011b). *Glomeromycota*: two new classes and a new order. *Mycotaxon.* 116: 365-379.

56. Omar, M.B., Bollan, L. and Heather, W.A. (1979). A permanent mounting medium for fungi. *Bull. Br. Mycol. Soc.* 13: 31-32.

57. Öpik, M., Moora, M., Zobel, M., Saks, Ü., Wheatley, R., Wright, F. and Daniell, T. (2008). High diversity of arbuscular mycorrhizal fungi in a boreal herb-rich coniferous forest. *New Phytologist.* 179: 867-876.

58. Öpik, M., Vanatoa, A., Vanatoa, E., Moora, M., Davison, J., Kalwij, J.M., Reier, Ü. and Zobel, M. (2010). The online database MaarjAM reveals global and ecosystemic distribution patterns in arbuscular mycorrhizal fungi (*Glomeromycota*). *New Phytologist.* 188: 223-241.

59. Orlowska, E., Zubek, S., Jurkiewicz, A., Szarek, L.G. and Turnau, K. (2002). Influence of restoration on arbuscular mycorrhiza of *Biscutella laevigata* L. (Brassicaceae) and *Plantago lanceolata* L. (Plantaginaceae) from calamine spoil mounds. *Mycorrhiza.* 12:153-160.

60. Phillips, J.M. and Hayman, D.S. (1970). Improved procedures for clearing roots and staining parasitic and vesicular-arbuscular mycorrhizal fungi for rapid assessment of infection. *Transactions of the British Mycological Society.* 55: 158-160.

61. Pongrac, P., Sonjak, S., Vogel-Mikuš, K., Kump, P., Nečemer, M. and Regvar, M. (2009). Roots of metal hyperaccumulating population of *Thlaspi praecox* (Brassicaceae) harbour arbuscular mycorrhizal and other fungi under experimental conditions. *Int. J. Phytorem.* 11: 347-359.

62. Redecker, D. (2000). Specific PCR primers to identify arbuscular mycorrhizal fungi within colonized roots. *Mycorrhiza.* 10: 73-80.

63. Redecker, D., Morton, J.B. and Bruns, T.D. (2000). Ancestral lineages of Arbuscular mycorrhizal fungi (Glomales). *Mol Phylogenet Evol.* 14: 276-284.

64. Redecker, D., Thierfelder, H., Walker, C. and Werner, D. (1997). Restriction analysis of PCR-amplified internal transcribed spacers of ribosomal DNA as a tool for species identification in different genera of the order Glomales. *Appl Environ Microbiol.* 63: 1756-1761.

65. Renker, C., Heinrichs, J., Kaldorf, M. and Buscot, F. (2003). Combining nested PCR and restriction digest of the internal transcribed spacer region to characterize Arbuscular mycorrhizal fungi on roots from the field. *Mycorrhiza.* 13: 191-198.

66. Renker, C., Weißhuhn, K., Kellner, H. and Buscot, F. (2006). Rationalizing molecular analysis of field-collected roots for assessing diversity of arbuscular mycorrhizal fungi: to pool, or not to pool, that is the question. *Mycorrhiza*. 16: 525-531.

67. Rosendahl, S. and Matzen, H.B. (2008). Genetic structure of arbuscular mycorrhizal populations in fallow and cultivated soils. *New Phytol.* 179: 1154-1161.

68. Ryszka, P., Błaszkowski, J., Jurkiewicz, A. and Turnau, K. (2010). Arbuscular mycorrhiza of *Arnica montana* under field conditions-conventional and molecular studies, *Mycorrhiza*. 20: 551-557.

69. Sanders, I.R. (2004). Plant and arbuscular mycorrhizal fungal diversity: are we looking at the relevant levels of diversity and are we using the right techniques? *New Phytologist*. 164: 415-418.

70. Santos, J.C., Finlay, R.D. and Tehler, A. (2006). Molecular analysis of arbuscular mycorrhizal fungi colonising a semi-natural grassland along a fertilisation gradient. *New Phytologist*. 172: 159-168.

71. Schenck, N.C. and Perez (1990). Manual for the identification of VA Mycorrhizal fungi. INVAM, University of Florida, Gainesville, Fk, Pp: 245.

72. Schönbeck, F. (1978). Einfluss der endotrophen Mykorrhiza auf die Krankheitsresistenz höherer Pflanzen. *Z. PflKrank. PflSchutz*. 85: 191-196.

73. Schüßler, A. (1999). Glomales SSU rRNA gene diversity. *New Phytologist*. 144: 205-207.

74. Schußler, A., and Walker, C. (2010). The *Glomeromycota*: a species list with new families and genera. Arthur Schüßler & Christopher Walker, Gloucester. Published in December 2010 in libraries at The Royal Botanic Garden Edinburgh, The Royal Botanic Garden Kew, Botanische Staatssammlung Munich, and Oregon State University. www.amf-phylogeny.com.

75. Schußler, A., Krüger, M. and Walker, C. (2011). Revealing natural relationships among Arbuscular mycorrhizal fungi: culture line BEG47 represents *Diversispora epigaea*, not *Glomus versiforme*. *PLoS ONE*. 6(8): e23333.

76. Schüßler, A., Schwarzott, D. and Walker, C. (2001). A new fungal phylum, the *Glomeromycota*: phylogeny and evolution. *Mycological Research*. 105: 1413-1421.

77. Simon, L., Lalonde, M. and Bruns, T.D. (1992). Specific amplification of 18S fungal ribosomal genes from vesicular–arbuscular endomycorrhizal fungi colonizing roots. *Applied Environmental Microbiology*. 58: 291-295.

78. Smith, G.W. and Skipper, H.D. (1979). Comparison of methods to extract spores of vesicular-arbuscular mycorrhizal fungi. *Soil. Sci. Soc. Amer. J*. 43: 722-725.

79. Smith, S.E. and Read, D.J. (2008). "Mycorrhizal Symbiosis," 3rd Edition, Academic Press, London.

80. Stahl, P.O. and Smith, W.K. (1984). Effects of different geographic isolates of *Glomus* on the water relations of *Agropyron smithii*. *Mycologia*. 76: 261-267.

81. Stürmer, S.L. and Morton, J.B. (1997). Developmental patterns defining morphological characters in spores of four species in *Glomus*. *Mycologia*. 89: 72-81.

82. Stürmer, S.L. and Morton, J.B. (1999). Taxonomic reinterpretation of morphological characters in *Acaulosporaceae* based on developmental patterns. *Mycologia*. 91: 849-857.

83. Sutton, J.C. and Sheppard, B.R. (1976). Aggregation of sand-dune soil by endomycorrhizal fungi. *Can. J. Bot*. 54: 326-333.

84. Sýkorová, Z., Börster, B., Zvolenská, S., Fehrer, J., Gryndler, M., Vosátka, M. and Redecker, D. (2012). Long-term tracing of *Rhizophagus irregularis* isolate BEG140 inoculated on *Phalaris arundinaceae* in a coal mine spoil bank, using mitochondrial large subunit rDNA markers. *Mycorrhiza*. 1: 69-80.

85. Taylor, T.N., Remy, W., Hass, H. and Kerp, H. (1995). Fossil Arbuscular mycorrhizae from the Early Devonian. *Mycologia*. 87: 560-573.

86. Trappe, J.M. (1982). Synoptic keys to the genera and species of zygomycetous mycorrhizal fungi. *Phytopathology*. 72: 1102-1108.

87. Turrini, A., Avio, L., Bavila, C. and Giovannetti, M. (2008). Characterisation of arbuscular mycorrhizal fungi in roots by means of epifluorescence microscopy and molecular methods. *Annals of Microbiology*. 58(1): 157-162.

88. van Tuinen, D., Jacquot, E., Zhao, B., Gollotte, A. and Gianinazzi-Pearson, V. (1998). Characterization of root colonization profiles by a microcosm community of arbuscular mycorrhizal fungi using 25S rDNA-targeted nested PCR. *Molecular Ecology*. 7: 103-111.

89. Vierheilig, H., Coughlan, A.P., Wyss, U. and Piche, Y. (1998). Ink and vinegar, a simple staining technique for arbuscular mycorrhizal fungi. *Appl Environ Microbiol*. 64: 5004-5007.

90. Walker, C. (1983). Taxonomic concepts in the Endogonaceae: spore wall characteristics in species descriptions. *Mycotaxon*. 18: 443-455.

91. Walker, C. (1986). Taxonomic concepts in the Endogonaceae. II. A fifth morphological wall type in endogonaceous spores. *Mycotaxon*. 25: 95-99.

92. Walker, C. (1999). Methods for culturing and isolating arbuscular mycorrhizal fungi. *Mycorrhiza News*. 11: 2-3.

93. Walker, C., and Trappe, J.M. (1993). Names and epithets in the Glomales and Endogonales. *Mycol. Res*. 97: 339-344.

94. Walker, C., Vestberg, M., Demircik, F., Stockinger, H., Saito, M., Sawaki, H., Nishmura, I. and Schu¨ßler, A. (2007). Molecular phylogeny and new taxa in the Archaeosporales (Glomeromycota): Ambispora fennica gen. sp. nov., Ambisporaceae fam. nov., and emendation of Archaeospora and Archaeosporaceae. *Mycological Research*. 111: 137-153.

95. Wang, G.M., Stribley, D.P., Tinker, P.G. and Walker, C. (1985). Soil pH and vesicular-arbuscular mycorrhizae, In A.H. Fitter, (Ed.), *Ecological Interactions in Soil*, (pp. 219–224). Oxford, U.K.: Blackwell Publication.

96. Wei, Y., Hou, H.Li.J., Shangguan, Y., Xu, Y., Zhang, J., Zhao, L. and Wang, W. (2015). Molecular diversity of arbuscular mycorrhizal fungi at a large-scale antimony mining area in southern China. *Journal of Environmental Sciences* 29: 18-26.

97. Wei, Y., Hou, H.Li.J., Shangguan, Y., Xu, Y., Zhang, J., Zhao, L. and Wang, W. (2014). Molecular diversity of arbuscular mycorrhizal fungi associated with a Mn hyperaccumulator-*Phytolacca americana*, in Mn mining area. *Applied Soil Ecology*. 82: 11-17.

98. White, T., Bruns, T., Lee, S. and Taylor, J. (1990). Amplification and direct sequencing of fungal ribosomal RNA genes for phylogenies. In: M.A, Innis, D.H. Gelfand, J.J. Sninsky, T.J. White, (Eds.), PCR protocols: a guide to methods and applications (pp.315-322). San Diego, CA, USA: Academic Press.

99. Wubet, T., Kottke, I., Teketay, D. and Oberwinkler, F. (2003). Mycorrhizal status of indigenous trees in dry Afromontane forests of Ethiopia. *Forest Ecol Manag*. 179: 387-399.

Diversity and Function of Am Fungi in Heavy Metal Contaminated Soils

Harbans Kaur Kehri*, Ifra Zoomi and Pragya Srivastava

Department of Botany, University of Allahabad,
Allahabad-211002, Uttar Pradesh, India
E-mail: kehrihk@gmail.com

Abstract

Arbuscular mycorrhizal fungi (AMF) form mutualistic symbiotic association with more than 90% of terrestrial plant species. They are widely recognized as enhancing plant growth on severely disturbed soils, including those contaminated with heavy metals (HMs). Diversity of AMF population is modified in metal-polluted soils, even in those with metal concentrations that are below the upper limits. Spore density, species richness and increase in species dominance are all inversely related to heavy metal concentration. However, there are certain AMF species which are reported to be present in the roots of plants growing on metal contaminated soils and played an important role in metal tolerance and accumulation.

Key words: Arbuscular mycorrhizal fungi (AMF), Heavy metals (HMs), Heavy metal tolerance

Introduction

Heavy metals (HMs) are naturally present in the soil and are continuously being added to the soil by atmospheric deposition as well as through anthropogenic activities. Some of these activities are; use of chemical fertilizers, long-term application of sewage and industrial effluents, waste disposal and incineration, production of batteries and mining and smelting of metals (Shen *et al.,* 2002). These activities increase the concentration of HM in the soil. In uncontaminated soil the average concentrations of metals are, e.g., Zn 80 ppm, Cd 0.1-0.5 ppm, and Pb 15 ppm. Whereas in contaminated soil the higher concentrations of metals were found, e.g., Zn: >20,000 ppm, Cd: >14,000 ppm, and Pb: >7,000ppm (http://www.

speclab.com/elements/). However, considerable increase of heavy metals, changes the physico-chemical properties of soil which causes acidification i.e., decrease the pH (Koomen *et al.*, 1990) and increases the availability of metals in the soil (Birch and Bachofen, 1990). Presence of HMs in the soil for long period is detrimental to soil health due to their mobility, solubility and non-degradable nature. Mobility and solubility of metals adversely affect growth and development of plants by inhibiting the metabolic enzymatic activities (Foy *et al.*, 1978) and compromise sustainable food production (Pandolfini *et al.*, 1997; Keller *et al.*, 2002; Voegelin *et al.*, 2003; Kabata-Pendias and Mukherjee, 2007).

Soil microorganisms play an important role in the mobilization and mineralization of heavy metals, thereby changing their accessibility to plants (Birch and Bachofen, 1990). However, significant increase of heavy metals in soil, changes the soil physico-chemical properties which could be toxic for soil microorganisms (Chaudri *et al.*, 1993).

Arbuscular mycorrhizal fungi (AMF) are one of the important components of soil biodiversity that establishes mutualistic symbiotic association with more than 90% of higher plants (Barea and Jeffries, 1995). Plants having mycorrhizal association are potentially benefited from mineral nutrition (Smith and read, 1997) because it helps in recycling of nutrients (Jeffries and Barea, 1994). AM fungi can contribute to plant growth, particularly by increasing accessibility of water and relatively immobile minerals such as phosphorus (Vivas *et al.*, 2003; Yao *et al.*, 2003). Accessibility of immobile minerals is due to the presence of extramatrical hyphae which serve as the extension of roots, thereby provides a wider zone for exploration beyond the root hair zone (Khan *et al.*, 2000; Malcova *et al.*, 2003). These extramatrical hyphae also secreates glomalin in rhozosperic zone of plants which is a glycoprotein (Rillig *et al.*, 2001), it binds the soil particles into stable aggregates, improving soil texture that resists the plant from soil erosion (Steinberg and Rillig, 2003). Arbuscular mycorrhizal fungi occur in most of the ecosystems, including HM contaminated soil (Vallino *et al.*, 2006; Zarei *et al.*, 2008; Wu *et al.*, 2010) and played an important role in metal accumulation and tolerance (Jamal *et al.*, 2002). Not only their presence but their diversity is also a key determining factor for productivity (Van der Heijden, *et al.*, 1998) and ecosystem functioning (Van der Heijden and Scheublin, 2007). However, function of AMF species is not same in all the ecosystems but it varies according to the host plants and soil properties (Hildebrandt *et al.*, 1999; Kelly *et al.*, 2005; Redon *et al.*, 2009). Del Val *et al.* (1999); Regvar *et al.* (2003); Vallino *et al.* (2006) reported that soil contaminated with HM influenced the AMF abundance and its diversity. Therefore, knowledge about the diversity and function of AMF in heavy metal contaminated soils is essential; this gives us a clear understanding of impact of HM contamination on AMF community composition and it could be used for better management and restoration of disturbed (heavy metals contaminated) ecosystem.

Heavy Metals and AMF

Metals having specific molecular weight higher than 5cm^3 are considered as heavy metal (HM) (Holleman and Wiberg, 1985; Weast, 1984). Metals like Cu, Zn. Fe, Mn and Ni are essential which are required by the plants for their growth and development (Zenk, 1996). Contrary to this Cd, As, Hg, and Pb are not essential metals (Mertz, 1981). Non-essential metals when present in soil above the permissible limit they considered as contaminants. Sources of heavy metals include; naturally by atmospheric deposition and by anthropogenic activities, such as mining and smelting of metals, use of chemical fertilizers and pesticides, waste disposal and incineration, vehicles exhaust and by municipal and industrial effluents (Figure 1).

When HMs retained in the soil by uncontrolled additions, it hampers the key biochemical processes which alter ecological balance. It can exhibit a range of toxicities toward soil microorganisms (McGrath *et al.,* 1995; Giller *et al.,* 1998; Dai *et al.,* 2004). However, HMs exposure may lead to the development of metal tolerant microbial population. AMF isolates, particularly the ecotypes found in HM contaminated soils can tolerate and accumulate HMs (Gildon and Tinker, 1981; Weissenhorn *et al.,* 1993, 1994; Joner and Leyval, 1997; Leyval *et al.,* 1997; Smith and Read, 1997 Jamal *et al.,* 2002; Zhu *et al.,* 2001). Gildon and Tinker (1981) isolated AMF strain from contaminated soil which tolerate upto 100mg/kg of Zn. Likewise, Weisenhorn *et al.* (1993) isolated two Cd tolerant species of AMF from highly contaminated soil. Sambandan *et al.* (1992) reported 15 AMF species from heavy metal-contaminated soils from India. Of the 15 AMF species isolated, *Glomus geosporum* was abundant and percentage colonization ranged from 22 to 71% and spore count was as high as 622 per 100 g soil.

Mycorrhizal fungi have also been shown to be associated with metallophyte plants on highly polluted soils, where only adapted plants such as *Viola calaminaria* (yellow-violet) can grow (Tonin *et al.,* 2001). This yellow-violet plant is described as an absolute metallophyte plant, which usually colonizes Zn and Pb-rich soils, and accumulates remarkably low levels of metal or none at all, despite the elevated levels of metal in these soils (Baker, 1981). A *Glomus* sp. isolated from the roots of the violet plant improved maize growth in a polluted soil (Hildebrandt *et al.* 1999) and reduced heavy metal concentrations root and shoot in comparison to common Glomus isolate or non-colonized controls. Griffioen *et al.* (1994) reported *Scutellospora dipurpurascens* from the rhizosphere of *Agrostis capillaris* growing in contaminated surroundings of a zinc refinery in the Netherlands. Likewise, Klauberg-Filho *et al.* (2002) conducted a study to evaluate AMF occurrence and diversity in soils from four locations cultivated with grass species and contaminated with heavy metals as a result of Zn extraction and industrialization. A total of 21 species were identified belonging to the following genera; *Acaulospora* (7 sps.), *Scutellospora* (6 sps.), *Glomus* (5 sps.), *Gigaspora* (2 sps.). The most frequent species identified were *Glomus occultum*, *Acaulospora morrowtae*, *A. mellea*, *G.* intraradices, *G. clarum* and *Scutellospora pellucida*. Spore density, species richness and increase in species dominance were all inversely related to heavy metal concentration.

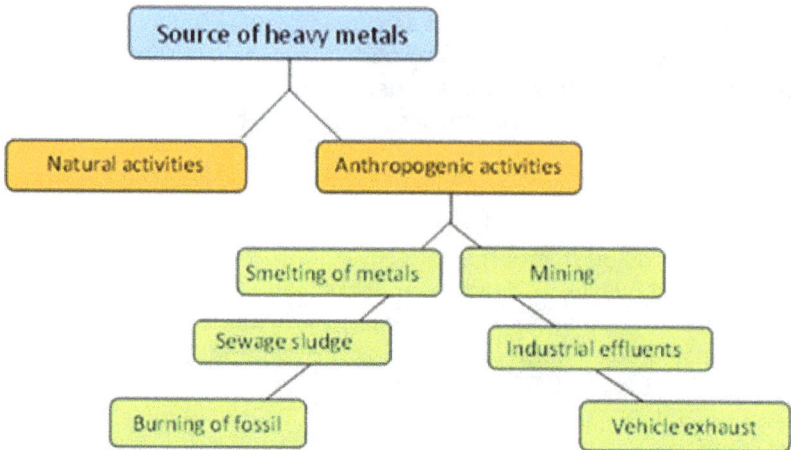

Figure 1: Sources of heavy metal contamination in soil

Diversity of AMF in effluent irrigated soil

There is a gradual decline in availability of fresh water to be used for irrigation all over the World. As a consequence, the use of sewage and other industrial effluents for irrigating agricultural lands is on the rise particularly in peri-urban areas of developing countries. Waste water is not only offers an alternative for irrigation source, but also contain heavy metals such as Cu, Cd, Zn and Pb etc. Although the concentration of heavy metals in effluent is low, but toxic levels increase as a result of long-term application (Sauve *et al.*, 1996; Rattan *et al.*, 2002). Long term irrigation with sewage water can induce changes in the quality of soil and trace element inputs are sustained over long periods (Zhang *et al.*, 2008). These heavy metals often accumulate in the top layer of soil, and are therefore, accessible for uptake by plant roots which are the principal entry points of metals into the plants. It has been reported that AMF community get reduced when natural ecosystem is converted into agro-ecosystem and this reduces in community because of intensification of agricultural land (Oehl *et al.*, 2003; Jefwa *et al.*, 2012)

Arbuscular mycorrhizal fungi are main component of agro-ecosystems. They played an important role in the maintenance of soil structure that prevents soil. Not only their presence but also their diversity is of importance for maintenance of agro-ecosystem. According to Liang *et al.* (2009) AMF could promote maize growth and decrease the uptake of heavy metals (Pb, Cd and Zn) at higher concentrations thus, protecting the plants from the toxicity. However, AM spore abundance and root colonization influenced by Cu, Zn and Fe concentrations present in soil (Datta and Kulkarni, 2012). Long term application of sewage effluents in the Mezquital Valley lead to decrease in AMF diversity (Ortega-Larrocea *et al.*, 2007). Similarly, in a semiarid orange-tree orchard, treated municipal effluents decreases the diversity of AMF as compare to fresh water irrigation but the soil microbial activities was same. However, no negative effects were observed on productivity and vitality of crop (Alguacil *et al.*, 2012). *Glomus mosseae* was reported to be the most abundant ecotypes in effluent irrigated agriculture soil (Ortega-Larrocea *et al.*, 2007).

Elevated level of Cr in soil has already been reported froms treated tannery effluents. AM fungi belonging to genus *Glomus* is dominant in contaminated soil (Vallino *et al.*, 2006). It has enhanced plant accumulation and tolerance to Cr (Davies *et al.*, 2001). Khade and Adholeya, (2009) recorded six species of AMF belonging to two genera *viz.*, *Glomus* and *Scutellospora*. *G. claroideum* was common in all naturally occurring plants in non-contaminated soil while *Scutellospora corolloidea* was restricted to only one plant in Cr contaminated soil of adjoining Kanpur Tanneries, Uttar Pradesh, India. Likewise, single species also recorded in Cu, Zn, Pb, Mn, Cd and Ni enriched soils (Griffioen, 1994) and *G. intraradices* by (Nogales *et al.*, 2012) this suggest low diversity of AMF in Cr contaminated soils. According to Diaz *et al.* (1996) AMF species from heavy metal-contaminated site seem to be more tolerant to than from non-contaminated sites.

In sewage sludge contaminated soil

Application of sewage sludge to the soil reduces the problem of disposal but, it significantly increases the heavy metal content in the soil and can pose long term environmental problem (Sauve *et al.*, 1996). Metals derived from sewage sludge strongly influenced the bioavalibility and speciation of metals (Lester, 1987). The toxicity of metals in the soil also depends on boiavaliblity i.e., transfer of metal to the soil organisms (Berthelin *et al.*, 1995). AMF can achieve resistance to heavy metals by avoidance and tolerance. There are certain strains of *Glomus* which are more tolerant to heavy metals. According to Weissenhorn *et al.* (1993) soil contaminated by application of sewage sludge contained more Cd tolerant AMF than from non-contaminated soil. Spores of *Glomus* spp. isolated from sewage sludge contaminated with Zn exhibit more tolerance to HMs than the AMF isolate from other pollutant such as $CdNO_3$. However, application of sludge with increasing concentration of heavy metals significantly decreases the size and diversity of AMF populations in soil (Del Val *et al.*, 1999). Total number of AMF species decreases with the increase in metals concentration. Six AMF ecotypes (*Glomus mossea*, *G. claoideum*, *G.* sp. III, *G.* sp. IV, *G.* sp. V and *G.* sp. VI) were found in soil amended with sewage sludge, suggesting a certain adaptation in AMF species to resist the metal stress. *Glomus Claoideum*, another ecotype was isolated from plots receiving 300m³ ha⁻¹ yr⁻¹ of contaminated sludge (Del Val *et al.*, 1999). Elevated levels of heavy metals delays reduce and even completely eliminate AM colonization and germination of spore in the field (Gildon and Tinker, 1981). Similarly, a negative correlation between Zn concentrations and AM colonization has also been reported in soil treated with urban industrial sludge (Boyle and Paul, 1988). However, no correlation was observed between AMF diversity and increasing concentration of Cu (Griffioen *et al.*, 1994), Zn and Cd (Weissenhorn and Leyval, 1994) in sludge-amended agricultural soils. *Glomus* species preferentially found in soils with the highest level of contamination indicates that they are well adapted to survive in stress condition (Sonjak *et al.,* 2009; Zarei *et al.,* 2008). These adapted and tolerant species can be used as an ideal candidate for improving the soil quality in metal contaminated agro-ecosystem (Weissenhorn *et al.,* 1993).

Diversity of AMF in Mining and Smelting sites

Mining is an excavation of valuable minerals from the Earth's crust and responsible for largest release of heavy metals into the environment of any industry. It changes soil physical properties, altering microbial communities and leads to destruction of vegetation (Bell *et al.,* 2001). However, AM fungi are reported to improve the ability of plants to cope with environmental stresses (Sylvia and Williams, 1992). Their presence has also been recorded in mining area (Draft and Hacskaylo, 1976; Jasper *et al.,* 1989; Lambert and Code, 1980; Diaz and Honrubia, 1993; Helm and Carling, 1993).

Arbuscular mycorrhizal fungi contribute metal excluder barrier and improving nutritional status (Turnau *et al.,* 1993; Weissenhorn *et al.,* 1995). Gladstone *et al.* (2005) investigated the occurrence AMF in a copper mining soil and found that 21 AMF species (fifteen spp. of *Glomus,* one of each of *Acaulospora, Archaeospora, Entrophospora, Gigaspora, Paraglomus and Scutellospora*) which indicates a tolerance to heavy metals (Pawlowska *et al.,* 2000; Turnau *et al.,* 2001). Likewise, Gonzalez-Chavez *et al.* (2002) reported the various arbuscular mycorrhizal fungi from arsenic mine soil with dominant genera *Glomus* (*Glomus caledonium, Glomus claroideum, Glomus constrictum, Glomus fasciculatum, Glomus intraradices,* and two unidentified *Glomus* species), followed by *Acaulospora* (*Acaulospora delicate assnd Acaulospora undulata*), and *Enterophospora* (*Enterophospora infrequens*). Yuan *et al.* (2010) reported *Acaulospora scrobiculata, Glomus aggregatum, Sclerocystis clavispora, Glomus glomerulatum, Funneliformis mosseae, Paraglomus occultum, and Scutellospora aurigloba* were the most common species and possessed higher spore density in the Sb mine slagheaps site than in the tailing dam with smelting waste and wastewater site. This indicates that moderate heavy metal stress can stimulate spore formation and more diverse community whereas high metal concentration will show inhibition. Number of AMF spores shows negative correlation with levels of Cu, Fe, and P which indicates that high level of HMs adversely affect the AMF propagules (Silva *et al.,* 2001).

Mining degraded areas showed that the AMF were reduced or totally eliminated from these areas (Allen, 1991). The reduction of plant diversity, the high pH and the amount of metals in the soil, possibly were the most important factors that strongly influence the composition of AMF communities in a Pb-Zn mine area at the Anguran open pit mine in Iran (Zarei *et al.,* 2010). They studied the abundance and diversity of AMF associated with dominant plant species using an established primer set for AMF in the internal transcribed spacer (ITS) region of rDNA, nine different AMF sequence types were distinguished after phylogenetic analyses, showing remarkable differences in their distribution patterns. Only one sequence type was found in the highly contaminated area and suggested that with decreasing Pb and Zn concentration, the number of AMF sequence types increased. Similarly, *Glomus* was the dominant genus (belong to phylum Glomeromycota) in the AMF community associated with ramie (plant) in Sb (Sonjak *et al.*, 2009) and *Elsholtzia* in Cu mining area (Yang *et al.*, 2010). *G. intraradices* PH5 isolated from Pd contaminated smelter region by (Malcova *et al.*, 2003) and *G. intraradices, G. mosseae, G. claroideum, G. geosporum* and *G. etunicatum* reported in mine and metallurgical smelter sites of central Europe (Vosatka and Dodd, 2002). It is therefore possible that stress caused by high pH and salinity is not as inhibitory to AMF as stress caused by heavy metals.

Mechanism of AMF in reducing heavy metals uptake

Alleviation of HMs toxicity by AMF has been reported by a number of workers (Heggo *et al.*, 1990; Weissenhorn *et al.*, 1995; Tao, 1997). It has been suggested that metal-tolerant AM fungi could protect plants against harmful effects of heavy metals (Sylvia and Williams, 1992; Leyval *et al.*, 1997). Several biological and physical mechanisms have been proposed to explain metal tolerance of AMF. Immobilization of metals in the fungal biomass is one of the mechanism involved (Zhu *et al.*, 2001; Li and Christie, 2000). This may occur due to intracellular precipitation of metallic cations with PO_4^-. The accumulation of heavy metals in the fungal structures as suggested by their high heavy metal-binding capacity (Joner *et al.*, 2000) could represent a biological barrier. AMF can bind heavy metals beyond the plant rhizosphere by releasing an insoluble glycoprotein commonly known as glomalin (Gonzalez-Chavez *et al.*, 2004; Gohre and Paszkowski, 2006). Gonzalez-Chavez *et al.* (2004) reported that 1 g of glomalin could extract up to 4.3 mg Cu, 0.08 mg Cd and 1.12 mg Pb from polluted soils. AMF vesicles also provide additional detoxification mechanisms by storing toxic compounds. In maize, heavy metals are selectively retained in the inner parenchyma cells coinciding with fungal structures (Kaldorf *et al.*, 1999). AMF provide HMs tolerance in plants by changing its polyamine metabolism thus, sequestered metals in the root system of colonized plants (Paradi *et al.*, 2003). However, mycorrhizal status of the plants did not influence shoot concentration of HMs, but concentration in roots was increased in mycorrhizal plants (Joner and Leyval, 1997). A range of factors like fungal properties, inherent heavy metal-uptake capacity of plants and soil absorption or desorption characteristics can influence heavy-metal uptake by mycorrhizal plants. AMF reduces excessive passive uptake of potentially harmful elements through the roots while maintaining an adequate supply of the other elements like N and P through active hyphal uptake (Joner and Leyval, 1997).

Conclusion

Heavy metal contamination found to influence the AMF community structure and its abundance (Del Val *et al.*, 1999). Spore density, species richness and increase in species dominance were all inversely related to heavy metal concentration. However, only one or few AM fungi have been considered as heavy metals tolerant. AMF isolates, particularly the ecotypes found in metal contaminated (agro-ecosystem, mine spoils and metalliferrous sites) soils can tolerate and accumulate HMs, depending on intrinsic and extrinsic factors (Gildon and Tinker, 1981; Weissenhorn *et al.*, 1993, 1994; Joner and Leyval, 1997; Leyval *et al.*, 1997; Smith and Read, 1997; Gonzalez-Chavez *et al.*, 2002; Vivas *et al.*, 2005). *Glomus* was a dominant genus (belonging to Phylum Glomeromycota) in HMs contaminated soils (Sonjak *et al.*, 2009; Yang *et al.*, 2010). The frequency of occurrence and abundance of spores is much higher than other genera. It possesses ability to adapt and resist in stressful conditions (Zarei *et al.*, 2009). AM fungi from contaminated soils have been reported to cope better with HMs-toxicity than those not exposed to such long-term selection pressure (Gildon and Tinker, 1981; Weissenhorn *et al.*, 1993; Malcova *et al.*, 2003).

Table 1: Diversity of Arbuscular mycorrhizal fungi in Heavy metal contaminated soils

Diversity of AMF	Metal-Contaminated soils	References
Glomus mosseae, G. intraradices and *G. claroideum, Glomus* sp. HM-CL4, *Glomus* sp. HM-CL5	Zn contaminated soil	Turnau *et al.* (2001)
Glomus mosseae, G. intraradices and *Glomus* sp.	Zn and Pb mining site	Zarei *et al.* (2008)
6 spp. of *Glomus*	Calamine spoil soil (Cd, Pb and Zn)	Pawlowska *et al.* (1996)
Unidentified 4 spp. of *Glomus* spp. *Scutellospora coralloidea*	Effluent treated soil	Khade and Adholeya (2005)
Glomus viscosum, G. constrictum, G. irregularis, Septoglomus viscosum, Rhizophagus intraradices and unidentified *Glomus* sp.	Sb mining soil in southern China	Yuan *et al.* (2015)
Scutellospora dipurpurascens	Zn refinery soils	Griffioen (1994)
Glomus rosea, G. aggregatum, G. fasciculatum, G. intraradices, Gigaspora gigantea, Sclerocystis microcarpum, S. dussii, S. pachycaylis, Scetellospora aurigloba, S. erythropa, S. nigra and *S. persica*	Mining soil	Raman *et al.* (1993)
Glomus caledonium, G. claroideum, G. constrictum, G. fasciculatum, G. intraradices, Glomus sp.1 and *Glomus* sp.2 *Acaulospora delicate, A. undulate Enterophospora infrequens*	As mine spoil	Gonzalez-Chavez *et al.* (2002)
Gigaspora sp. *Glomus tenue*	Ni contaminated soil	Tarnau *et al.* (2003)
Glomus intraradices PH5	Smelting site contaminated with Pb	Malcova *et al.* (2003)

Diversity of AMF	Metal-Contaminated soils	References
Glomus macrocarpum, G. mosseae, G. fasciculatum, G. fugianum and *G. constrictum*	Sewage sludge amended soil	Gunwal *et al.* (2014)
Acaulospora scrobiculata *G. taiwannense, G. versiforme,* *G. aggregatum, G. ambisporum* *G. diaphanum, G. etunicatum, G. formosanum, G. geosporum, G.intraradices, G. manihotis, G. mosseae* and *Glomus* sp.1	Smelting sites contaminated with As, Pb, Zn, Cd and Cu	Wu *et al.* (2010)
Glomus albidum, G. clarum, *G. diaphanum, G. etunicatum, G. invermaium G. macrocarpum, G. microaggregatum, G. microcarpum G. mosseae, G. sinuosum, G. tortuosum* *Glomus* sp. 1 *Glomus* sp. 2 *Glomus* sp. 3 *Glomus* sp. 4 *Paraglomus occultum,* *Acaulospora scrobiculata* *Archaeospora leptoticha* *Entrophospora infrequens,* *Gigaspora margarita,* *Scutellospora gilmorei*	Mining site contaminated with Cu	Gladstone *et al.* (2005)
Glomus intraradices, G. mosseae, G. claroideum, G. geosporum and *G. etunicatum*	Mine and metallurgical smelter sites	Vosátka and Dodd, (2002)
Glomus mossea, G. Claroideum, Glomus sp. III, *Glomus* sp. IV, *Glomus* sp. V and *Glomus* sp. VI)	Sewage sludge amended soil Cd, Ni, Zn. Cu and Pb	Del Val *et al.* (1999)
Glomus mossea, G. fasciculatum, G. geosporum and *Sclerocystis sinuosa*	Sewage effluent irrigated soil	Selvaraj and Kim (2004)
Glomus mossea, G. constrictum, G. microcarpum, G. sp. and *Sclerocystis sinusa*	Cd, Zn and Ni contaminated soil	Takacs *et al.* (2001)
Acaulospora scrobiculata Glomus geosporum, G. aggregatum,G. microaggregatum, G. tortosum, G. rubiformis,G. intraradices, Scutellospora calospora and *Entrophospora colombiana*	Mining site	Mehrotra and Prakash (2006)

Table 2: Dominant Arbuscular mycorrhizal fungi (AMF) species of heavy metal contaminated soils

Dominant AMF	Sites	References
Glomus mossea	Effluent treated soil, Sewage sludge amended soil and Mining and smelting sites	Del Val *et al.* (1999), Vosátka and Dodd, (2002), Gladstone *et al.* (2005), Wu *et al.* (2010), Gunwal *et al.* (2014), Turnau *et al.* (2001), Zarei *et al.* (2008) Selvaraj and Kim (2004)
Glomus intraradices	Effluent treated soil, Mining and smelting sites	Turnau *et al.* (2001), Zarei *et al.* (2008), Vosátka and Dodd, (2002), Gonzalez-Chavez *et al.* (2002), Malcova *et al.* (2003), Wu *et al.* (2010), Raman *et al.* (1993), Yuan *et al.* (2015)
Glomus fasciculatum	Sewage effluent irrigated soil, Sewage sludge amended soil and Mining sites	Gunwal *et al.* (2014), Gonzalez-Chavez *et al.* (2002), Raman *et al.* (1993), Selvaraj and Kim (2004)
Glomus claroideum	Mining and smelting	Turnau *et al.* (2001), Del Val *et al.* (1999), Vosátka and Dodd, (2002), Gonzalez-Chavez *et al.* (2002)

References

1. Alguacil, M.M., Torrecillas, E., Hernandez, G., Torres, P. and Roldan, A. (2012). *Jatropha curcas* and *Ricinus communis* differentially affect arbuscular mycorrhizal fungi diversity in soil when cultivated for biofuel production in a Guantanamo (Cuba) tropical system. *Proceedings of the EGU Conference on General Assembly*, Volume 14, April 22-27, 2012, Vienna, Austria. 2703-2703.

2. Allen, M.F. (1991). *The ecology of mycorrhizae.* Cambridge Univ., Cambridge.

3. Baker, A.J.M. (1981). Accumulators and excluders-strategies in the response of plants to heavy metals. *Journal of Plant Nutrition.* 3: 643-654.

4. Barea, J.M. and Jeffries, P. (1995). Arbuscular mycorrhizas in sustainable soil plant systems. In B. Hock and A. Varma (Eds.), *Mycorrhiza structure, function, molecular biology and biotechnology.* Heidelberg, Germany: Springer-Verlag. 521-559.

5. Bartolome-Esteban, H. and Schenck, N.C. (1994). Spore germination and hyphal growth of arbuscular mycorrhizal fungi in relation to soil aluminum saturation. *Mycologia.* 86: 217–226.

6. Bell, F.G., Bullock, S.E.T., Halbich, T.F.J. and Lindsey, P. (2001). Environmental impacts associated with an abandoned mine in the Witbank Coalfield, South Africa. *International. Journal of Coal Geology.* 45: 195-216.

7. Berthelin, J., Leyval, C., Laheurte, R. and Degiudici, J. (1991). Involvements of roots and rhizospere microflora in the chemical weathering of soil minerals. In D. Atkinson, (Ed.), *Plant root growth: an ecologieal perspective.* Oxford UK: British Ecologieal Society. 187-200.

8. Birch, L. and Bachofen, R. (1990). Complexing agents from microorganisms. *Experientia.* 46: 827-834.

9. Boyle, M. and Paul, E.A. (1988). Vesicular-arbuscular mycorrhizal associations with barley on sewage-amended plots. *Soil Biology Biochemistry.* 20: 945-948

10. Chaudri, A.M., McGrath, S.P., Giller, K.E., Rietz, E. and Sauerbeck, D. (1993). Enumeration of indigenous *Rhizobium leguminossarum* biovar trifolii in soils previously treated with metal sewaged sludge. *Soil Biology Biochemistry.* 25: 301-309.

11. Chen, D.M., Ellul, S., Herdman, K. and Cairney, J.W.G. (2001). Influence of salinity on biomass production by Australian *Pisolithus* spp. Isolates. *Mycorrhiza.* 11: 231-236.

12. Daft, J. and Hacskaylo, E. (1976). Arbuscular mycorrhiza in the anthracite and bituminous coal waste of Pennsylvania. *Journal of Applied Ecology.* 13: 532-531.

13. Dai, J., Becquer, T., Rouiller, J.H., Reversata, G., Bernhard-Reversata, F., Nahmania, J. and Laville, P. (2004). Heavy metal accumulation by two earthworm species and its relationship to total and DTPA-extractable metals in soils. Soil Biology Biochemistry. 36: 91-98.

14. Datta, P. and Kulkarni, M. (2012). Arbuscular Mycorrhizal Fungal Diversity in Sugarcane Rhizosphere in Relation with Soil Properties. *Nature of Science Biology.* 4(1): 66-74

15. Davies, F.T., Puryear, J.D., Newton, R.J., Egilla, J.N. and Grossi, J.A.S. (2001). Mycorrhizal fungi enhance accumulation and tolerance of chromium in sunflower (*Helianthus annuus*). *Journal of Plant Physiology.* 158: 777-786

16. Del Val, C., Barea, J.M., and Azcon-Aguilar, C. (1999). Diversity of arbuscular mycorrhizal fungus populations in heavy-metal-contaminated soils. *Applied Environmental Microbiology.* 65: 718-723

17. Diaz, G. and Honrubia, M. (1993). Infectivity of mines soils from south-east Spain. II. Mycorrhizal population levels in spoil sites. *Mycorrhiza.* 4: 85-88.

18. Diaz, G., Azcon-Aguilar, C. and Honrubia, M. (1996). Influence of arbuscular mycorrhizae on heavy metal (Zn and Pb) uptake and growth of *Lygeum spartum* and *Anthillis cytisoides. Plant Soil.* 180: 241-249.

19. Foy, C.D., Chaney, R.L. and White, M.C. (1978). The Physiology of Metal Toxicity in Plants. *Annual Review of Plant Physiology.* 29(1): 511.

20. Gildon, A. and Tinker, P.B. (1981). A heavy metal tolerant strain of a mycorrhizal fungus. *Transactions of the British Mycological Society.* 77: 648-649.

21. Giller, K.E., Witter, E. and McGrath, S.P. (1998). Toxicity of heavy metals to microorganism and microbial processes in agricultural soils: A review. Soil Biology Bichemistry. 30(10-11): 1389-1414.

22. Gohre, V. and Paszkowski, U. (2006). Contribution of the arbuscular mycorrhizal symbiosis to heavy metal phytoremediation. *Planta*. 223: 1115-1122.

23. Gonzalez-Chavez, M.C., Carrillo-Gonzalez, R., Wright, S.F. and Nichols, K.A. (2004). The role of glomalin, a protein produced by arbuscular mycorrhizal fungi, in sequestering potentially toxic elements. *Environmental Pollution*. 130: 317-323

24. Gonzalez-Chavez, M.C., Dhaen, J., Vangronsveld and Dodd, J.C. (2002). Copper sorption and accumulation by the extraradical mycelium of different *Glomus* spp. (arbuscular mycorrhizal fungi) isolated from the same polluted soil. *Plant Soil*. 240: 287-297.

25. Griffioen, W.A.J. (1994). Characterization of a heavy metal-tolerant endomycorrhizal fungus from the surroundings of a zinc refinery. *Mycorrhiza*. 4: 197-200.

26. Griffioen, W.A.J., Iestwaart, J.H. and Ernst, W.H.O. (1994). Mycorrhizal infection of *Agrostis capillaris* population on a copper-contaminated soil. *Plant Soil*. 158: 83-89.

27. Heggo, A. and Angle, S. (1990). Effects of vesicular-arbuscular mycorrhizal fungi on heavy metal uptake by soybeans. *Soil Biology Biochemistry*. 22: 865-869.

28. Helm, D. J. and Carling, D. E. (1993). Use of soils transfer for reforestation on abandoned mined lands in Alaska. II. Effect of soil transfers from different successional stages on growth and mycorrhiza formation by *Populus balsamifera* and *Alnus crispa*. *Mycorrhiza*. 3(3): 107-114.

29. Hildebrandt, U., Kaldorf, M. and Bothe, H. (1999). The zinc violet and its colonization by arbuscular mycorrhizal fungi. *Journal of Plant Physiology*. 154: 709-717.

30. Holleman, A.F. and Wiberg, E. (1985). Lehebuch du Anoranischen chemie. *Water de Gruyter, Berlin*. 868.pr.

31. Jamal, A., Ayub, N., Usman, M. and Khan, A.G. (2002). Arbuscular mycorrhizal fungi enhance zinc and nickel uptake from contaminated soil by soyabean and lentil. *International Journal of Phytoremediation*. 4: 205-221.

32. Jasper, D.A. (1994). Bioremediation of agricultural and forestry soils with symbiotic micro-organisms. *Soil Biology Biochemistry*. 32: 1301-1319.

33. Jeffries, P. and Barea, J.M. (1994). Biogeochemical cycling and arbuscular mycorrhizas in the sustainability of plant soil systems. In S. Gianinazzi, and H. Schuepp, (Eds.), *Impact of Arbuscular Mycorrhizas on Sustainable Agriculture and Natural Ecosystems*. Birkhauser, Basel. 101-115.

34. Jefwa, J.M., Okoth, S., Wachira, P., Karanja, N. and Kahindi, J. (2012). Impact of land use types and farming practices on occurrence of arbuscular mycorrhizal fungi (AMF) Taita-Taveta district in Kenya. *Agriculture, Ecosystem and Environment*. 157: 32-39.

35. Joner, E.J and Leyval, C. (1997). Uptake of [109]Cd by roots and hyphae of a *Glomus mosseae/Trifolium subterraneum* mycorrhiza from soil amended with high and low concentrations of cadmium. *New Phytologist.*135: 353-360.

36. Joner, E.J., Leyval, C. and Briones, R. (2000). Metal binding capacity of arbuscular mycorrhizal mycelium. *Plant Soil, (in press).*

37. Kabata-Pendias, A. and Mukherjee, A.B. (2007). Trace elements from soil to human. *Springer-Verlag, Berlin, New York.*

38. Kaldorf, M., Kuhn, A.J., Schroder, W.H., Hildebrandt, U. and Bothe, H. (1999). Selective element deposits in maize colonized by a heavy metal tolerance conferring arbuscular mycorrhizal fungus. *Journal of Plant Physiology.* 154: 718-728.

39. Keller, C., McGrath, S.P. and Dunham, S.J. (2002). Trace metal leaching through a soil–grassland system after sewage sludge application. *Journal of Environmental Quality.* 31: 1550-1560.

40. Kelly, C.N. Morton, J.B. and Cumming, J.R. (2005). Variation in aluminum resistance among arbuscular mycorrhizal fungi. *Mycorrhiza.* 15: 193-201.

41. Khade, S.W. and Adholeya, A. (2009). Arbuscular mycorrhizal association in plants growing on metal-contaminated and noncontaminated soils adjoining Kanpur tanneries, Uttar Pradesh, India. *Water Air and Soil Pollution.* 202: 45-56.

42. Khan, A.G., Kuek, C., Chaudhry, T.M., Khoo, C.S. and Hayes, W.J. (2000). Plants, mycorrhizae and phytochelators in heavy metal contaminated land remediation. *Chemosphere.* 41: 197-207.

43. Klauberg-son, O., Siqueira, O.J. and Moreira, F.M.S. (2002). Arbuscular mycorrhizal fungi in polluted area of soils with heavy metals. Journal of Soil Science. 26: 125-134.

44. Koomen, I., McGrath, S.P. and Giller, K. (1990). Mycorrhizal infection of clover is delayed in soils contaminated with heavy metals from past sewage sludge applications. *Soil Biology Biochemistry.* 22: 871-873.

45. Lambert, D.H. and Code (Jr), H. (1980). Effect of mycorrhizae on establishment and performance of forage species in mine spoil. *Agronomy Journal.* 72: 257-260.

46. Lester, J. (1987). Heavy metals in wastewater and sludge treatment process. *CRC Press, Inc., Boca Raton, Floraida, USA.* 1-40

47. Leyval, C., Turnau, K. and Haselwandter, K. (1997). Effect of heavy metal pollution on mycorrhizal colonization and function: physiological, ecological and applied aspects. *Mycorrhiza.* 7: 139-153.

48. Li, X.L. and Christie, P. (2000). Changes in soil solution Zn and pH and uptake of Zn by arbuscular mycorrhizal red clover in Zn-contaminated soil. *Chemosphere.* 42: 201-207.

49. Liang, C.C., Li, T, Xiao, Y.P., Liu, M.J., Zhang, H.B and Zhao, Z.W. (2009). Effects of inoculation with arbuscular mycorrhizal fungi on maize grown in multi-metal contaminated soils. *International Journal of Phytoremediation.* 11: 692-703.

50. Malcova, R., Vosátka, M. and Gryndler, M. (2003). Effects of inoculation with *Glomus intraradices* on lead uptake by *Zea mays* L. and *Agrostis capillaris* L. *Applied Soil Ecology*. 23: 55-67.

51. McGrath, S.P., Chaudri, A.M. and Giller, K.E. (1995). Long-term effects of metals in sewage sludge on soils, microorganisms and plants. *Journal of Indian Microbiology*. 14: 94-104.

52. Mehrotra, V.S and Prakash, A.P. (2006). Mycorrhizal fungi of the revegetated coal mine spoil of the northern India. In Prakash A. & Mehrotra V.S. (Eds.), *Mycorrhiza*. Scientific publisher (India), Jodhpur. 11-19.

53. Mertz, W. (1981). The essential trace elements. *Science*. 213: 1332-1338.

54. Nogales, A., Cortes A., Velianos, K., Camprubi, A., Eutaun, V. and Calvet, C. (2012). *Plantago lanceolata* Growth and Cr uptake after of mycorrhizal inoculation in Cr amended substrate. *Agricultural and Food Science*. 21: 72-79

55. Oehl, F., Sieverding, E., Ineichen, K., Mäder, P. and Boller, T. (2003). Impact of land use intensity on the species diversity of arbuscular mycorrhizal fungi in agrosystems of central Europe. *Applied Environmental Microbiology*. 69: 2816-2824.

56. Orlowska, E., Zubek, S., Jurkiewicz, A., Szarek-ukaszewska, G. and Turnau, K. (2002). Influence of restoration on arbuscular mycorrhiza of *Biscutella laevigata* L. (Brassicaceae) and *Plantago lanceolata* L. (Plantaginaceae) from calamine spoil mounds. *Mycorrhiza*. 12: 153-160.

57. Ortega-Larrocea, M.P., Siebe, C., Estrada, A. and Webster, R. (2007). Mycorrhizal inoculum potential of arbuscular mycorrhizal fungi in soils irrigated with wastewater for various lengths of time, as affected by heavy metals and available P. *Applied Soil Ecology*. 37: 129-138.

58. Pandolfini, T., Gremigni, P. and Gabbrielli, R. (1997). Biomonitoring of soil health by plants. In C.E. Pankhurst, B.M. Doube, V.V.S.R. Gupta (Eds.), *Biological indicators of soil health*. CAB International, New York. 325-347.

59. Parádi, I., Bratek, Z. and Lang, F. (2003). Influence of arbuscular mycorrhiza and phosphorus supply on polyamine content, growth and photosynthesis of *Plantago lanceolata*. *Biologia Plantarum*. 46: 563-569.

60. Pawlowska, T., Chaney, R., Chin, M. and Charvat, I. (2000). Effects of metal phytoextraction practices on the indigenous community of arbuscular mycorrhizal fungi at a metal-contaminated landfill. *Applied Environmental Microbiology*. 66: 2526-2530.

61. Rattan, R.K., Datta, S.P., Chandra S. and Sahran, N. (2002). Heavy metals and environmental quality: Indian scenario. *Fertiliser News*. 47(11): 21-40

62. Redon, P.O., Béguiristain, T. and Leyval, C. (2009). Differential effects of AM fungal isolates on *Medicago truncatula* growth and metal uptake in a multimetallic (Cd, Zn, Pb) contaminated agricultural soil. *Mycorrhiza*. 19: 187-195.

63. Regvar, M., Vogel, K., Irgel, N., Wraber, T., Hildebrandt, U., Wilde, P. and Bothe, H. (2003). Colonization of pennycresses Thlaspi sp. of the Brassicaceae by arbuscular mycorrhizal fungi. *J. Plant Physiology*. 160: 615-626.

64. Rillig, M.C., Wright, S.F., Nichols, K.A., Schmidt, W.F. and Torn, M.S. (2001). Large contribution of arbuscular mycorrhizal fungi to soil carbon pools in tropical forest soils. *Plant and Soil*. 23: 167-177.

65. Sambandan, K., Kannan, K. and Raman, N. (1992). Distribution of vesicular-arbuscular mycorrhizal fungi in heavy metal polluted soils of Tamil Nadu. *Journal of Environmental Biology*. 13: 159-167.

66. Sauve, S., Cook, N., Hendershot, H.W. and McBridge, B.M. (1996). Linking tissue copper concentrations and soil copper pools in urban contaminated soils. *Environ. Pollut.* 94: 153-157.

67. Shen, Z.G., Li, X.D., Wang, C.C., Chen, H.M. and Chua H. (2002). Lead Phytoextraction from contaminated soil with high-biomass plant species. *Journal of Environmental quality*. 31: 1893-1900.

68. Silva, I.R., Smyth, T.J., Raper, C.D., Carter, T.E. and Rufty, T.W. (2001). Differential aluminum tolerance in soybean: An evaluation of the role of organic acids. *Physiology Plant*. 112: 200-210.

69. Smith, S.E. and Read, D.J. (1997). *Mycorrhizal Symbiosis*, Academic Press, San Diego, USA.

70. Sonjak, S., Beguiristain, T., Leyval, C. and Regvar, M. (2009). Temporal temperature gradient gel electrophoresis (TTGE) analysis of arbuscular mycorrhizal fungi associated with selected plants from saline and metal polluted environments. *Plant Soil*. 314(1-2): 25-34.

71. Steinberg, P.D. and Rillig, M.C. (2003). Differential decomposition of arbuscular mycorrhizal fungal hyphae and glomalin. *Soil Biology Biochem*istry. 35: 191-194.

72. Sylvia, D. M. and Williams, S. E. (1992). Vesicular-arbuscular mycorrhizae and environmental stresses. In Bethlenfalvay, G.J. and Linderman, R.G. (eds.), *Mycorrhizae in Sustainable Agriculture*. Madison, USA. 101-124.

73. Takacs, T., Biro, B. and Voros, I. (2001). Influence of Cd, Zn and NI on the diversity of AM fungi. 49(3/4): 456-478.

74. Tao, H.Q. (1997). Effect of arbuscular mycorrhiza on resistance of red clover to heavy metal Zn and Cd pollution, MS thesis, China Agricultural University, Beijing.

75. Tonin, C., Vandenkoornhuyse, P., Joner, E.J., Straczek, J. and Leyval, C. (2001). Assessment of arbuscular mycorrhizal fungi diversity in the rhizosphere of *Viola calaminaria* and effect of these fungi on heavy metal uptake by clover. *Mycorrhiza*. 10: 161–168.

76. Turnau, K. and Mesjasz-Przybylowicz, J. (2003). Arbuscular mycorrhiza of *Berkheya coddii* and other Ni-hyperaccumulating members of Asteraceae from ultramafic soils in South Africa. *Mycorrhiza*. 13: 185-190.

77. Turnau, K., Kottke, I. and Oberwinkler, F. (1993). Element localization in mycorrhizal roots of *Pteridium aquilinum* L. Kuhn collected from experimental plots treated with cadmium dust. *New Phytologist*. 123: 313-324.

78. Turnau, K., Ryszka, P., Gianinazzi-Pearson, V. and van Tuinen, D. (2001). Identification of arbuscular mycorrhizal fungi in soils and roots of plants colonizing zinc wastes in southern Poland. *Mycorrhiza*. 10: 169-174

79. Vallino, M., Massa, N., Lumini, E., Bianciotto, V., Berta, G. and Bonfante, P. (2006). Assessment of arbuscular mycorrhizal fungal diversity in roots of Solidago gigantea growing in a polluted soil in Northern Italy. *Environmental Microbiology*. 8 (6): 971-983.

80. Van der Heijden, M.G.A. and Scheublin, T.R. (2007). Functional traits in mycorrhizal ecology: their use for predicting the impact of arbuscular mycorrhizal fungal communities on plant growth and ecosystem functioning. *New Phytologist*. 174: 244-250.

81. Van der Heijden, M.G.A., Klironomos, J.N., Ursic, M. and Moutoglis, P. (1998). Mycorrhizal fungal diversity determines plant biodiversity, ecosystem variability and productivity. *Nature*. 396: 69-72.

82. Vivas, A., Marulanda, A., Gómez, M., Barea, J.M. and Azcón, R. (2003). Physiological characteristics (SDH and ALP activities) of arbuscular mycorrhizal colonization as affected by *Bacillus thuringiensis* inoculation under two phosphorus levels. *Soil Biology Biochemistry*. 35: 987-996.

83. Voegelin, A., Barmettler, K. and Kretzschmar, R. (2003). Heavy metal release from contaminated soils: Comparison of column leaching and batch extraction results. *Journal Environmental Quality*. 32: 865-875.

84. Vosatka M. and Dodd, J.C. (2002). Ecological considerations for successful application of arbuscular mycorrhizal fungi inoculum. In Gianinazzi S, Schuepp H, Barea J.M & Haselwandter K (eds.), *Mycorrhizal technology in agriculture*. Birkhauser, Basel. 235-247.

85. Weast, R.C. (1984). CRC Handbook of chemistry and physics, 64th edn. Boca Raton, CRC Press

86. Weissenhorn, I. and Leyval, C. (1994). Differential tolerance to Cd and Zn of arbuscular mycorrhizal fungal spores isolated from heavy metal-polluted and unpolluted soils. *Plant Soil*. 167: 189-196.

87. Weissenhorn, I., Leyval, C. and Berthelin, J. (1993). Cd-tolerant arbuscular mycorrhizal (AM) fungi from heavy-metal polluted soils. *Plant Soil*. 157: 247-256

88. Weissenhorn, I., Leyval, C., Belgy, G. and Berthelin, J. (1995). Arbuscular mycorrhizal contribution to heavy metal uptake by maize (*Zea mays* L.) in pot culture with contaminated soil. *Mycorrhiza*. 5: 245-251.

89. Wu, Q.S., Zou, Y.N. and He, X.H. (2010). Contributions of arbuscular mycorrhizal fungi to growth, photosynthesis, root morphology and ionic balance of citrus seedlings under salt stress. *Acta Physiologiae Plantarum*. 32: 297-304.

90. Yang, R.Y., Zan, S.T., Tang, J.J., Chen, X. and Zhang, Q. (2010). Variation in community structure of arbuscular mycorrhizal fungi associated with a Cu tolerant plant-*Elsholtzia splendens*. *Applied Soil Ecology*. 44 (3): 19-197.

91. Yao, Q., Li, X., Weidang, A. and Christie, P. (2003). Bi-directional transfer of phosphorus between red clover and perennial ryegrass via arbuscular mycorrhizal hyphal links. *EuropeanJournal of Soil Biology*. 39: 47-54.

92. Yuan, Z.L., Zhang, C.L. and Lin, F.C. (2010). Role of diverse non-systemic fungal endophytes in plant performance and response to stress: progress and approaches. *Journal of Plant Growth Regulation*. 29: 116-126.

93. Zarei, M., Hempel, S., Wubet, T., Schfer, T., Savaghebi, G. and Jouzani, G.S. (2010). Molecular diversity of arbuscular mycorrhizal fungi in relation to soil chemical properties and heavy metal contamination. *Environmental Pollution*. 158(8): 2757–2765.

94. Zarei, M., Saleh-Rastin, N., Jouzani, G.S., Savaghebi, G. and Buscot, F. (2008). Arbuscular mycorrhizal abundance in contaminated soils around a zinc and lead deposit. *Europian Journal of Soil Biology*. 44: 381-391.

95. Zenk, M.H. (1996). Heavy metal detoxification in higher plants: a review. *Gene*. 179: 21-30.

96. Zhang, Y.L. Dai, J.L. Wang, R.Q. and Zhang J. (2008). Effects of long-term sewage irrigation on agricultural soil microbial structural and functional characterizations in Shandong China. *European Journal of Soil Biology*. 44: 84-91.

97. Zhu, Y.G., ChristFie, P. and Laidlaw, A.S. (2001). Uptake of Zn by arbuscular mycorrhizal white clover from Zn-contaminated soil. *Chemosphere*. 42: 193-199.

10

Global Biodiversity Assessment: Current Status and Threats

Rinku, A. Sharma, S.K. Patel and G.S. Singh*

*Institute of Environment and Sustainable Development,
Banaras Hindu University, Varanasi-221005, Uttar Pradesh, India
E-mail: gopalsingh.bhu@gmail.com*

Abstract

Environment, society and economy are the three pillars whose successful integration paves the way for sustainable development. Variation in one pillar affects the other and consequently disturbs the natural ecosystem of an organism. Earth has unique place in universe due to the presence of life that exhibits immense diversity. Various species and diverse ecosystems like deserts, ocean, mountains, and rivers comprise the biological diversity or biodiversity. Loss of biodiversity, reduce the capacity of ecosystem to provide immense products and services to human life. Various anthropogenic activities affect the earth systems as whole and cause change in the biotic structure of the planet. Various threats to biodiversity caused the decline in species richness in particular and ecosystem services in general. Habitat loss is the single largest threat to biodiversity. As the population is increasing at an alarming rate, their demands related to shelter, food, and fuel are also increasing subsequently. These increasing demands of human are destroying the natural habitat of living organisms. Overexploitation, biological invasion, climate change, pollution, modernization, infrastructure and disease to organisms are also some major threats to biological diversity. Besides these threats some other threats includes loss of keystone species, extinction, and change in species dynamics, accidental mortality and natural disaster. Sustainable development practices to conserve biodiversity and natural resources are the need of hour.

Keywords: Biodiversity, Extinction, Overexploitation, Sustainable Development, Threat

Introduction

Earth has unique place in the universe where life exists and variety of living things is one of the major characteristics of the earth. Biodiversity is variety of life living organisms found on earth. Biodiversity is formally defined by the Convention on Biological Diversity (CBD) as: "the variability among living organisms from all sources including, among others, terrestrial, marine and other aquatic ecosystems and the ecological complexes of which they are part; this includes diversity within species, between species and of ecosystems" (UN 1992 Article 2).Different types of plants, animals and microorganisms found on earth comprise biodiversity.

Earth is very rich in terms of its biological diversity which makes the blue planet beautiful. But this beauty of earth is declining as the consequence of various threats most of which are anthropogenic. Various human activities, directly or indirectly have become the main cause for biodiversity loss on the earth (Ramakrishnan 2014; Singh et al., 2014). Preindustrial era, more or less was much diverse and ecologically balanced than the modern era. Anthropogenic pressure on biodiversity is increased at much faster rate by Industrial revolution in the eighteenth century.

Vertebrate species populations across the globe decreased by fifty percent in last 40 years (Living Planet Report, 2014).

For the name of development and just only for making delightful life, human brutally exploited the nature overpopulation is threatening the biodiversity at an alarming rate. In the year 2011, the global population has passed more than 7 billion and for making everybody's need of food and shelters, more and more natural resources are exploiting.

Man is busy in construction of their home and high rising buildings and at the same time destroying the habitat of other organisms. Habitat loss is now days become the single largest threat to biodiversity. Life is found in all the spheres of earth viz; lithosphere, hydrosphere, and atmosphere but none of them is isolated from the anthropogenic threats. Environmental pollution and climate change affecting all the spheres of earth and converting the natural habitat of the species into an inhabitable place.

Overexploitation of natural resources is linearly related with increasing population. Overpopulation and overconsumption is the root cause of the environmental degradation and ecological imbalance. If this process of exploitation continues, the time is not far when there would be only concrete buildings but not a single tree on the earth. Mahatma Gandhi, the great Indian moral leader quoted that "there is enough on earth for everybody's need, but not for everybody's, greed". This quote has great relevance in this modern time. Biological invasion also become a major threat in which a non-native invasive species invade the foreign habitat and compete out the native species. Various pathogenic microorganisms by hind erthe bio-capability of organisms' to live and made them susceptible to death.

Loss of keystone species, extinction, and change in native species dynamics, accidental mortality, and natural disasters are some other cause of reducing biodiversity from the planet earth. The sustainable practices are the solution to secure our present and future for better life in harmony with nature.

Threats to Biodiversity

Any direct or indirect human activity that threatens the planet's biological diversity in the form of genes, populations, species, ecosystems, or other levels of biological organization is considered a threat to survival (Sechrest *et al.*, 2002). Biodiversity is under serious threat due to anthropogenic activities.

Major threats to Biodiversity

Major threats to biodiversity are habitat loss/degradation/fragmentation, overexploitation, climate change, invasive species, pollution, disease and other threats; the contribution of various threats to biodiversity loss is described in the Fig.1.

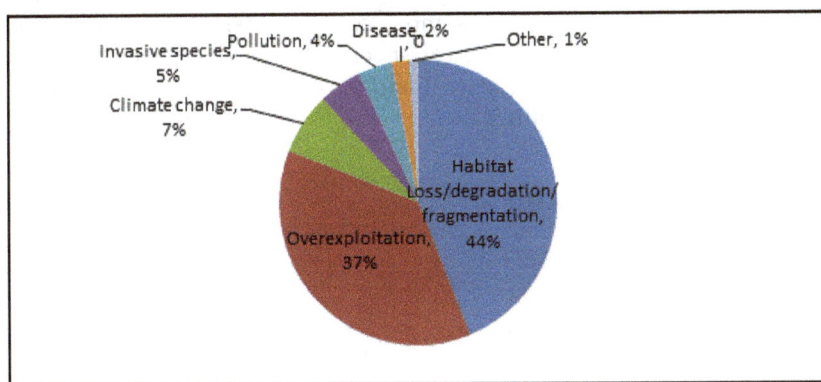

Figure 1: Major Threats to Biodiversity *(Source: WWF Living Planet Report, 2014)*

Major threats and their causes:

Various anthropogenic activities are main causes of most of the threats and are enlisted in the following Table 1.

Table 1: Major threats and their causes.

Threats	Causes
Habitat loss, degradation, fragmentation	Urbanization, industrialization, agriculture, wood cutting for fuel and furniture, mining, road and highway construction, dam construction, human settlements, tourism, natural disaster, power lines.
Overexploitation	Harvesting of various resources for food, fuel, medicines, fodder and fiber, cultural, scientific and leisure activities
Climate change	Global warming due to GHG emission by burning of fossil fuels, deforestation and natural variability and anthropogenic activities.

Threats	*Causes*
Invasive species	International trade and transport, gardening practices, exotic trees in forestry, climate change and exotic pests released in the wild.
Pollution	Domestic and industrial waste, water, agricultural runoff, transportation, burning of fossil fuels, energy production, mining activities, heavy metal works, oil spills, forest fires, etc.
Disease	Pollution, biological invasion, habitat destruction, pest infestation, climate variability etc.

IUCN Red List of Threatened species

It is the catalogue of taxa that are facing the risk of extinction. It was founded in 1963 and this list is the world's most comprehensive inventory of global conservation status of all the biological species. Various red list categories and red list criteria are given in the Figure 2 and various red list categories and their meaning is described in the Table 2.

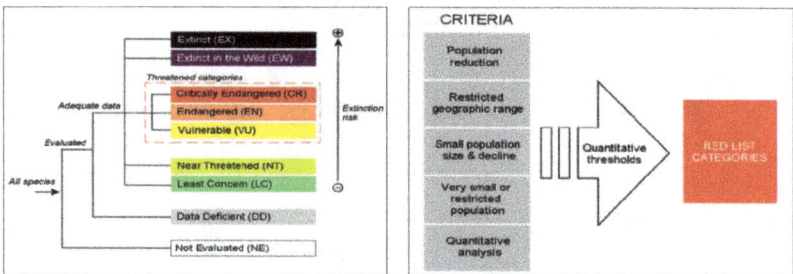

Figure 2: Red list categories and red list criteria

(Source: IUCN, 2010, Guidelines for Using the IUCN Red List Categories and Criteria)

Table 2: Red List categories and their meaning

Categories	*Meaning*
Extinct(EX)	A taxon is Extinct when there is no reasonable doubt that the last individual has died.
Extinct in the wild(EW)	A taxon is Extinct in the Wild when it is known only to survive in cultivation, in captivity or as a naturalized population (or populations) well outside the past range.
Critically Endangered(CR)	A taxon is critically endangered when it is facing extremely high risk of extinction in the wild.
Endangered(EN)	A taxon is Endangered when it is facing very high risk of extinction in the wild.

Categories	Meaning
Vulnerable(VU)	A taxon is considered vulnerable when it is facing high risk of extinction in the wild.
Near Threatened(NT)	A taxon is near threatened when it has been evaluated against the criteria but does not qualify for critically endangered, endangered and vulnerable.
Least Concern(LC)	A taxon is near threatened when it has been evaluated against the criteria but does not qualify for critically endangered, endangered, vulnerable, and near threatened.
Data Deficient(DD)	A taxon is data deficient when there is inadequate information to make a direct, or indirect, assessment of its risk of extinction based on its distribution and /or population status.
Not Evaluated(NE)	A taxon is Not Evaluated when it is has not yet been evaluated against the criteria.

(Source: IUCN, 2010, Guidelines for Using the IUCN Red List Categories and Criteria)

Current status of global biodiversity

Global biodiversity is under serious risk and many rare species have been extinct in Anthropogenic (human dominated era). According to the Living planet index (LPI), there has been about 52 percent decline in the vertebrate population between 1970 to 2010 as shown in Figure 3.

Global Living planet index (LPI) tracks trend in large number of populations of species and it is based on the trends of 10380 population of 3038 mammal, bird, reptile, amphibian, and fish species around the globe (Living Planet Report, 2014).

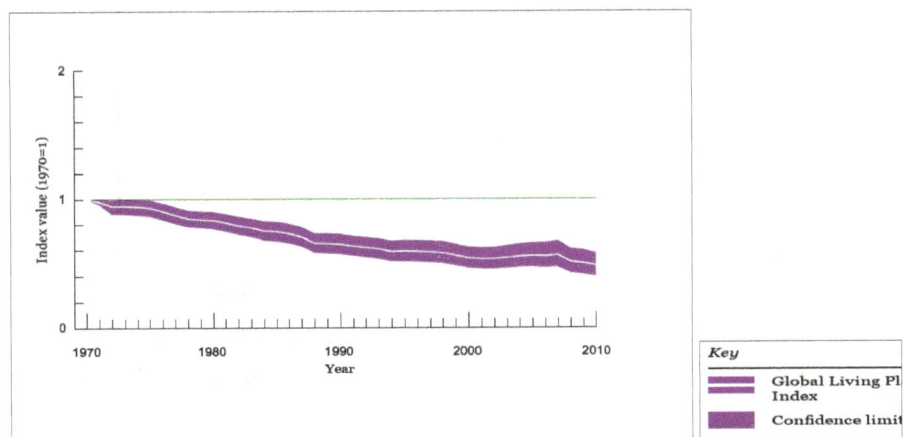

Figure 3: Global Living planet Index (Source: Living planet report, 2014)

Status of the threatened species in the World

In July 2013, a total of 70294species globally was assessed by the International Union for the Conservation of Nature and Natural resources. Nearly 799 species have extinct from the earth and moreover 61 species have extinct in wild. The proportion of various species in different threat categories is given the Figure 4.

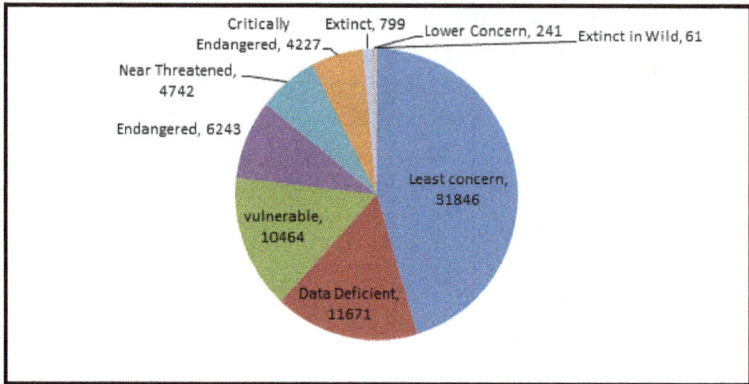

Figure 4: Proportion of species in different threat categories as per the source of IUCN 2013red list

Status of the threatened species in India

Indian fauna under IUCN (2013) threat categories

In the year 2013, a total of 4681 animal species in India were assessed by the IUCN and out of 4681 animal species one species have been extinct and 73 are critically endangered. Indian fauna under IUCN (2013) threat categories are represented in the Figure 5.

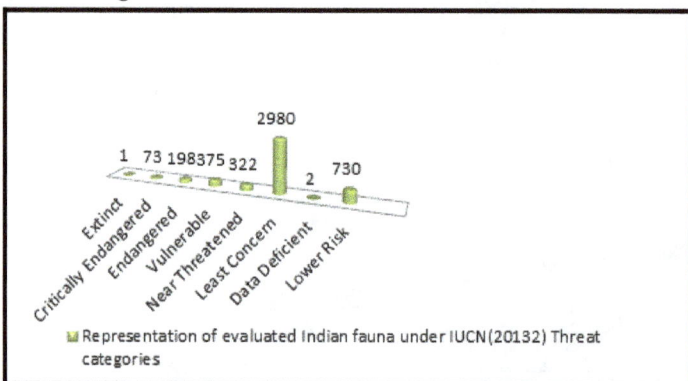

Figure 5: Representation of evaluated Indian fauna under IUCN (2013) Threat categories

(Source: India's Fifth national Report to CBD 2014)

Indian flora under IUCN (2013) threat categories

In the year 2013, a total of 1218 plant species were assessed by IUCN and found that 6 species have been extinct and 60 are critically endangered. Indian flora under IUCN (2013) threat categories are represented in the Figure 6. According to IUCN (2013), 6 plant species in India have extinct and 2 plant species have also extinct in the wild.

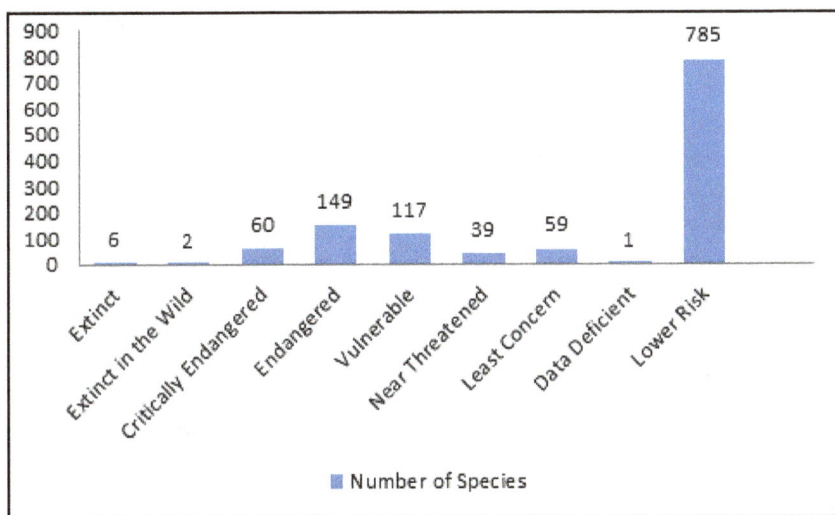

Figure 6: Numbers of evaluated Indian Flora in various IUCN (2013) threat Categories

(Source: India's Fifth National Report to CBD 2014).

HOTSPOTS: Areas on earth facing extreme Threats

The idea of hotspots was first given in 1988 by ecologist Norman Myers, who defined a hotspot as an area of exceptional plant, animal and microbe wealth that is under threat. Endemism (restricted to particular location and found nowhere else) and threat levels are the main criteria for determining a hotspot.

A hotspot ideally contains 0.5% of the plant species and should have lost 70% or more of its primary vegetation (Myers *et al.*, 2000). Globally there are thirty four terrestrial biodiversity hotspots and eleven marine biodiversity hotspots (Conservation International, 2013). Out of the thirty four global biodiversity hotspots, four (viz., Himalaya, Western Ghats, North-East and Nicobar Islands) are present in India (Conservation International, 2013). The biodiversity and major threats of Indian hotspots are described in Table 3.

Table 3: Characteristics and threats of Biodiversity hotspots in India

Hotspots	Location in Global Biodiversity Hotspots	Major Biodiversity	Major Threats
The Himalaya	Western and Eastern Himalaya form part of Himalayan global Biodiversity hotspot.	10000 plant species and out of which3160 endemic.	Tourism, global warming, agriculture, overexploitation
Western Ghats	Part of Western Ghats-Sri Lanka global biodiversity Hotspot.	7388 species of flowering plants. UNESCO World Heritage site	Overexploitation, climate change, pollution, biological invasion
North-East	Part of Indo-Burma global biodiversity Hotspot.	13,500 vascular plant species, of which about 7000 (52%) are endemic.	Climate change, overexploitation, pollution
Nicobar Islands	Part of the Sundaland global biodiversity Hotspot.	3500 plant species, 648 species (13.11% endemic).	Global warming, tourism, marine pollution

(Sources: Conservation International, 2013; UNESCO, 2012; and Planning Commission, 2007)

Major Threats to Biodiversity

Habitat Loss, Degradation and Fragmentation:

Habitat means place or site where an organism or population naturally occurs (CBD, UN, 1992 Article-2). Habitat loss, degradation and fragmentation are the most important drivers of biodiversity loss.

Habitat loss is in general elaborated as change in the structure of the habitat. It is the most devastating threat to biodiversity. All species have a specific food pattern and particular habitat. It leads to disturbance of the natural ecosystem in which particular species have their home the species becomes threatened.

Habitat degradation is commonly defined as the change in the structure and function of the habitat. For instance, the cutting of trees will lead to loss of canopy, this is the degradation.

Habitat fragmentation occurs when parts of a habitat become separated from one another because of changes in a landscape, such as the construction of roads. For instance, a road is constructed through the forest (as shown in Figure 7), the forest is fragmented into different parts and the size of the original forest is reduced. Here each part is a separate forest ecosystem. Any species that needs a large home range, such as a grizzly bear, will not survive if the area is too small. The small ecosystems are strongly affected by their surroundings, in terms of climate, and dispersing species. As a consequence, the ecology of a small ecosystem may differ from that of a similar ecosystem on a larger scale.

(a) Undisturbed/Intact Forest Ecosystems

(b) Fragmentation due to road construction indicating Two distinct forest ecosystems

(c) Fragmentation due to additional road construction leading three distinct forest ecosystems

Figure 7: Diagrammatic representation of Fragmentation of Forest Ecosystems due to road construction

Edge effect is the consequence of the habitat fragmentation. Edge effects refer to the changes in population or community structures that occur at the boundary of two habitats (Levin and Simon, 2009). For example, meadow and paved street. The edges of natural ecosystems are more susceptible to light, wind, and weather than interior areas, so they are less suitable habitat for species that live in inner forest areas. Edges are also vulnerable to invasive species. **Species Area Relationshipis the simplest empirical model to describe the effects of habitat loss on biodiversity.** In simple words, this model describe that large area will company the large number of species and vice versa. McArther and Wilson (1967) proposed a Law of Species Area relationship. According to this law Number of species encountered is proportional to the power of area sampled. Mathematical representation is, $S\alpha A^z$, which is derived as $S=CA^z$. This equation is commonly called **Power function equation**. Where S is the number of species; A is area and C and z are the regression defined coefficient.

All types of habitats, including forests, grasslands, wetlands and river systems, continue to be fragmented and degraded. Forests are one of the most important natural habitat on earth.

Deforestation is increasing at an alarming rate and consequently biodiversity is reducing in terrestrial ecosystems. Current status forest worldwide is represented in the Fig.8.It is shown in map that the intact forests are mostly present only in amazon basin, central Africa, north-east Himalaya and some Asia-pacific countries.

Intact Forest: These are the unbroken expanses of natural ecosystem greater than 50000 hectares.

Managed Forest: These are those forests that are fragmented by roads and/or managed for wood production.

Degraded (partially degraded) forest: These are the landscapes where there has been a significant decrease in the tree canopy density.

Deforested: Previously forested landscapes which have been converted into non-forest.

In the last five years during 2010 to 2015 Brazil has lost maximum forest in the world. Countries reporting the greatest annual forest area reduction (2010–2015) are described in the Table 4.

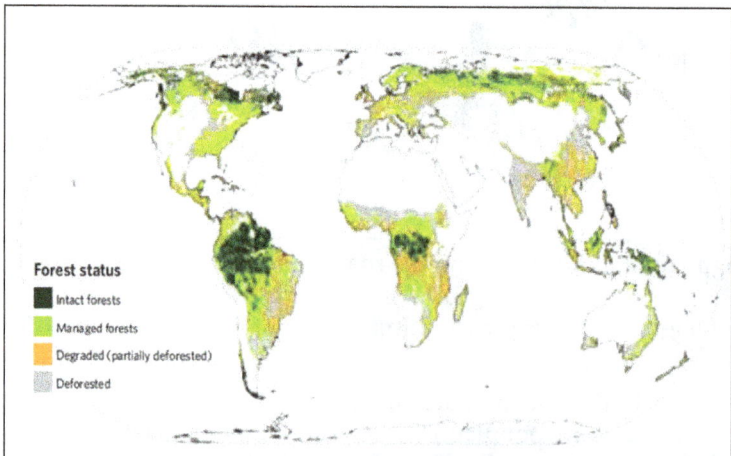

Figure 8: The extent of deforestation and forest degradation worldwide (*Source:* Global Biodiversity Outlook 4)

Table 4: Country wise significant annual forest area reduction (2010-2015)

Country	Forest Loss (000 ha)	% of the 2010 forest area
Brazil	984	0.2
Indonesia	684	0.7
Myanmar	546	1.7
Nigeria	410	4.5
United Republic of Tanzania	372	0.8
Paraguay	325	1.9
Zimbabwe	312	2.0

(Source: Global Biodiversity Outlook 4)

Factors responsible for Habitat Loss/Degradation/Fragmentation

Agricultural activities, demand of fuel, various extraction activities like mining, fisheries, wood collection, infrastructure development lead to habitat loss. Large-scale commercial agriculture has adversely affected biodiversity; particularly agro-biodiversity (Belfrage, 2006; Rosset, 1999). Direct habitat loss is major threat to coastal ecosystems through aquaculture (Valiela *et al.*, 2004).

Overexploitation

Overexploitation, or unsustainable use, happens when biodiversity is removed faster than it can be replenished and, over the long term, can result in the extinction of species. Overexploitation of wild species to meet consumer demand threatens biodiversity, with unregulated overconsumption contributing to declines in terrestrial, marine and freshwater ecosystems (Peres, 2010; Vorosmarty *et al.*, 2010; Kura *et al.*, 2004; Dulvy *et al.*, 2003). Overexploitation of anything in nature leads to an imbalance in the system. Similarly, overexploitations of various natural resources of biodiversity become a challenging threat presently. In wake of increasing human population at an alarming rate, the demands of natural resources are correspondingly increasing which shows a linear correlation.

Natural resources such as portable water, fodder, food, fuel, shelter, minor forest products and many more are receding substantially as per demand of the ever-rising human population. Although overexploitation is often difficult to quantify in terrestrial systems, major exploited groups include plants for timber, food and medicine; mammals for wild meat and recreational hunting; birds for food and the pet trade; and amphibians for traditional medicine and food (Vie et al. 2009). Threat to traditional crops and associated indigenous knowledge are equally vulnerable. The threat to vertebrates from overexploitation is particularly severe, driven, in particular, by demand for wildlife and wildlife products from East Asia. All this is done at the cost of precious biodiversity of the earth. Human is most successful consumer of natural resources on earth.

Ecological footprint

It is the simplest way to reflect our pressure on nature. It is a measure of how much biologically productive land and water an individual, population or activity requires to produce all the resources it consume and to absorb the waste it generates, using prevailing technology and management practices (Global Footprint network, 2012).It is measured in global hectares (gha).Global hectare is the hectare of biologically productive land and water with world average bioproductivity in a given year.

Biocapacity

It is the capacity of ecosystems to produce useful biological materials and to absorb waste materials (specially, CO_2) generated by humans, using current management schemes and technologies (Global Footprint Network, 2014). It is measured in global hectares (gha).

Our demands are unsustainable and increasing

We are consuming more than our actual needs. Thus we are creating the hardship for the sustenance of future generations. In the Fig.9 it is shown that 1.5 Earth would be required to meet the demands if we consume like our present style (Living Planet Report, 2014). But presently we have only one planet Earth which supports the more than 7 billion human populations. From where half more earth would come; for making situation balanced we have to cut our demands and use the natural resources in a sustainable manner. Economic developed countries are consuming more and thus they have challenge to make a balance between three pillars of the Sustainability. Thus if we want conserve our precious biodiversity we should limit our demand of natural resources like forest products which come from the very natural home of the species.

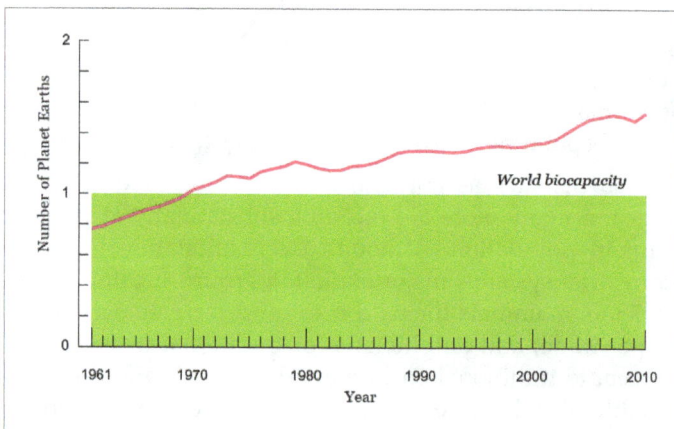

Figure 9: Humanity's Ecological Footprint

(source: Global Footprint Network 2014)

Overpopulation is one of the major causes for declining world natural resources.In the year 2011 the global population has crossed the mark of 7 billion. Thus for fulfilling the demands of food and shelter for every person, it is obvious more land for agriculture and settlements will be required and consequently more and more forest would be removed.

The trends in ecological footprint, biocapacity and population from 1961 to 2010 are shown in the Figure 10.

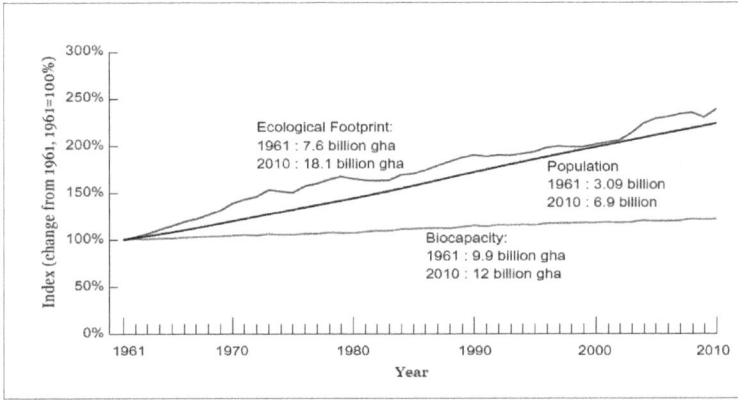

Figure 10: Trends in total biocapacity, Ecological footprint and world population from 1961 to 2010.

(Source: Living Planet Report 2014)

Climate Change

Climate change refers to a change in the state of the climate that can be identified (e.g. using statistical tests) by changes in the mean and/or the variability of its properties and that persists for an extended period, typically decades or longer. It refers to any change in climate over time, whether due to natural variability or as a result of human activity (IPCC, 2007)

For example, it could show up as a change in climate normal (expected average values for temperature and precipitation) for a given place and time of year, from one decade to the next. The changing climate is a threat to species and ecosystem (Tripathi and Singh, 2013; Tripathi *et al.*, 2015).

Every species is adapted to particular climate and the distribution of living things on earth is largely determined by the climate. Since mid-1800s the average global temperature increased by 0.6⁰C. Eleven of the last twelve years (1995-2006) rank among the twelve warmest years in the instrumental record of global surface temperature since 1850 (IPCC *Fourth Assessment Report*).

Climate change is the result of both internal variability within climate system and external factors including both natural and anthropogenic. The Fourth Assessment Report of the Intergovernmental Panel on Climate Change (IPCC) concludes, "that most of the observed increase in the globally averaged temperature since the mid-20th century is very likely due to the observed increase in anthropogenic greenhouse gas concentrations.

Among all the Greenhouse gases largest contribution to global warming is of Carbon Dioxide. Burning of fossil fuels like coal, petrol, natural gases, etc. are the major source for anthropogenic emission of Carbon dioxide. Current Carbondioxide concentration in the atmosphere is above 380 ppm as shown in the Figure 11 which is increased in linearly since 1958.

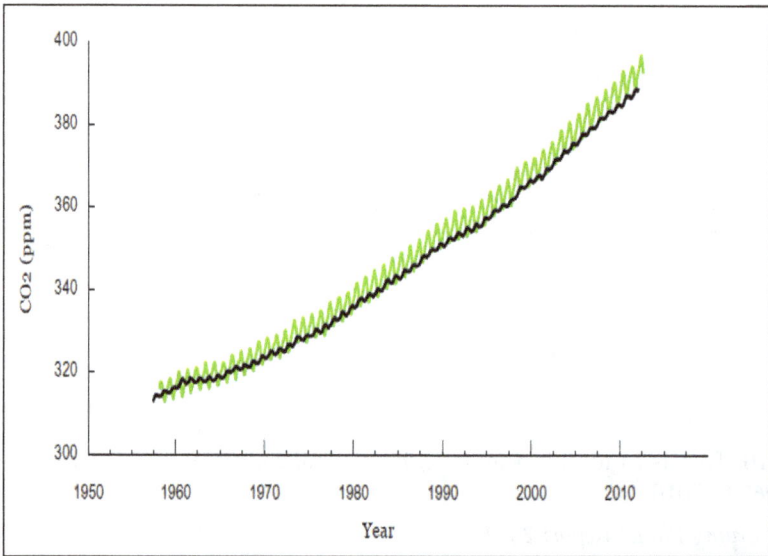

Figure 11: Atmospheric CO_2 concentrations since 1958measured at Mauna Loa Observatory, Hawaii.

(Source: Living planet report, 2014; *IPCC, 2013)*

The atmospheric Carbon dioxide concentration in November 2015 has reached a level of 400.16 ppm (OAA, 2015). The mean annual temperature of earth is 15°C. But in the absence of Greenhouse gases this temperature would drop to about-20° C. Thus greenhouse effect help in the warming of earth to temperature at which life can survive. But this global temperature is increasing due to enhanced greenhouse effect which is known as **Global warming.**

Impacts of Climate Change on Biodiversity:

People and natural environment have become vulnerable to the impacts of climate change. Climate change affects all the ecosystems found on earth. The impacts of climate change on the biodiversity in various ecosystems are described in Table 5.

Table 5: Impacts of Climate Change on Biodiversity of various ecosystems

Ecosystems	Characteristics	Impacts of Climate Change on Ecosystem and Biodiversity
Polar ecosystems	High latitude regions on earth with low temperature.	Melting of ice, reduction in the population of Polar bear
Agricultural ecosystems	Cover about one third of the land area on earth and about 7000 species of plants are cultivated for food.	Dryness of soil, drought, loss of agro-biodiversity.
Dry and humid land ecosystems	Cover about 40% of the terrestrial surface on earth and about 2 billion people live in these ecosystems.	Wildfires and change in rainfall patterns are reducing the biodiversity.
Forest Ecosystems	About one third of the earth surface is covered with forests and approximately two thirds of all known terrestrial species are found in forests	An increase of 1 degree C in the temperature can modify the functioning and composition of forests.
Inland Water Ecosystems	May be fresh or saline water system within continental and island boundaries and make up only 0.01% of world's water but they are the house of about 100000 species of living organisms.	Vulnerable to climate change because freshwater species are experiencing declines in biodiversity far greater than those in most terrestrial ecosystems.
Islands Ecosystems	Very rich biodiversity and Islands ecosystem are very fragile	Rise in sea level and the potential increase in the frequency of storms are major threats to island biodiversity.
Marine and Coastal ecosystems	Largest ecosystems of the earth covering more than 70% of the earth's surface. Contain world's most diverse ecosystems including mangroves, coral reefs and sea grass beds.	Coral bleaching due to increase in sea temperature and change in water chemistry. Climate fluctuations in North America reduce plankton populations, the main source of food of the North Atlantic right whale
Mountain Ecosystems	About 27% of earth's land is covered with mountains. Mountain ecosystems are the house of about 22% of the total human population	Retreat and sometimes disappearance of alpine species. Shrinking of glaciers modifies the water-holding capacities of mountains, affecting the quantity of freshwater available to both humans and biodiversity

(Source: Based on CBD, UNEP 2007)

Invasion of Non-native Species (Alien species)

The movement of organisms around the globe represents one of the greatest threats to biodiversity (Global Biodiversity Outlook4). Invasive species are those that occur outside their natural range, spread rapidly and cause harm to other species, communities or entire ecosystems and to human well-being (Singh *et al.*, 2014).Invasive species are also known as Non-natives, introduced, non-indigenous, exotic and foreign species. But, all the non-natives are not invasive. Invasive Alien Species (IAS) is that subset of alien species of plants, animal, fungi, viruses and bacteria whose establishment and spread threatens ecosystems, habitats, or species with economic or environmental significance (McNeely *et al.*, 2001).Alien species are distributed throughout the globe. Biological invasion is the best example of biological pollution. Biological invasion may be intentional or accidental.

The increased fragmentation, degradation and destruction of habitats, along with other threats, will certainly open more niches for non-native species introductions.

Introduction pathways of invasive species:

Introduction pathways of known cases of introduction of over 500 invasive alien species profiled in the Global Invasive Species Database (GISD) are given in the Table 6.

Table 6: Major introduction pathways of Invasive Alien Species

Major Pathways		*Sub –pathways*
Transport	(1)	Cargo shipments by air, land, or sea,
	(2)	Air crafts
	(3)	Ship
	(4)	Boats
Release	(1)	Aquaculture
	(2)	Horticulture
	(3)	Aquarium or pet trade
	(4)	Sea food trade
	(5)	Botanical gardens
	(6)	Zoos
Other Pathways	(1)	Research activities
	(2)	Biological control
	(3)	Medicines
	(4)	Religious
	(5)	Smuggling

(Source: Global Biodiversity outlook 4, ISSG-GISP, Invasive Pathways team Final report, 2003 National invasive species council and US department of Agriculture.)

Process of biological invasion:

The process which an invasive species follow to establish a non-native area and affecting the native biota is described in Figure 12.For any invasion process the invasive species must transported to a foreign place either intentionally or accidentally. After this, it is released to a new habitat and established to non-native habitats. As the establishments are complete, invasive species increase their population and tried to spread and expansion to a larger geographical range. The next step is to affect the native flora and fauna and the impacts of invasion are become responsible to threatening the native biodiversity.

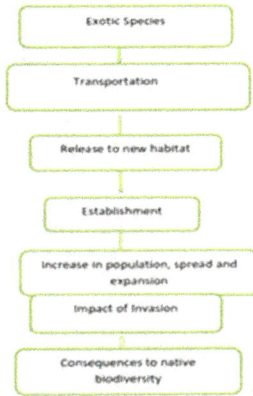

Figure 12: Diagrammatic representation of various steps involved in Biological invasion

Characteristics of invasive species

For a very successful invasion, the invasive species have some attributes that make their path easy to invade a non-native habitat and establish there. Some of the general characteristics of the invasive species are shown in the Figure 13.

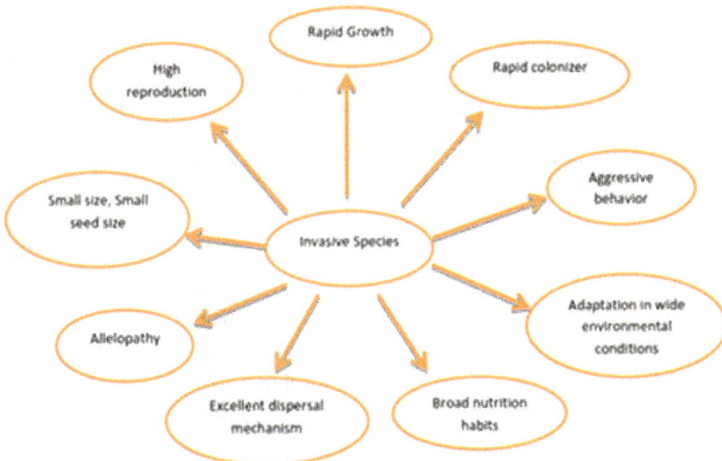

Figure 13: General characteristics of Invasive species.

World's worst invasive species:

Out of the 100 world's worst invasive species, some species are given in Table 7.

Table 7: World's worst Invasive species. (Source: ISSG, IUCN, SSC, Global Invasive Species Programme, GISP)

Name	Nativity	Ecological Impacts
Water Hyacinth (*Eichhorniacrassipes*)	Amazon basin	Cover lakes and ponds entirely; and affect water flow, blocks sunlight from reaching native aquatic plants, and starves the water of oxygen, often killing fish (or turtles).
Rinderpest virus (*Rinderpest virus*)	Asia	Cause rinderpest also known as cattle plague, a viral infectious disease in cattle.
Chestnut blight (*Cryphonectriaparasitica*)	Japan	The fungus enters through wounds and grows in and beneath the bark, eventually killing the cambium all the way round the twig, branch or trunk.
Crazy ant (*Anoploepisgracilipes*)	Africa and Asia	Easily established and dominant in new habitat due aggression toward other ant species
Cane toad(*Bufomarimus*)	Central and South America	When the cane toad is introduced to a new ecosystem, it poses a serious threat to native species.
Indian myna bird (*Acridotherestristis*)	Asia	Threats to native birds

Invasive Indian flora

Some important invasive plant species in India are described below in the Table 8.

Table 8: Some Invasive Indian Flora and its ecological characteristics.

Invasive species	Common name	Family	Nativity	Ecological Impacts
Partheniumhysterophorus	Gazargrass,-Congress grass	Asteraceae	Tropical America	The presence of *Parthenium* pollen grains inhibits fruit set in tomato, brinjal, beans, and a number of other crop plants.
Daturainnoxia	Datura, Mad Plant	Solanaceae	Tropical America	Aggressive colonizer, occasional weed on disturbed ground.

Invasive species	Common name	Family	Nativity	Ecological Impacts
Lantaanca-mara	Lantana, Wild Sage	Verbena-ceae	Central and south America	In disturbed native forests it can become the dominant understory species, disrupting succession and decreasing biodiversity
Xanthium strumari-um	Rough Cocklebur	Asteraceae	North America	Readily invades overgrazed pastures and destroy native species. Seedlings are toxic to domestic livestock.
Cuscutare-flexa	Dodder, Sky Creeper	Cuscuta-ceae	Mediter-ranean	Aggressive colonizer. Occasional stem parasite on garden shrubs, trees and over hedges in scrub lands.
Celosia argentea	Wool flower	Amaran-thaceae	Tropical Africa	Aggressive colonizer. Common weed of cultivated fields and scrub lands.

(Source: Reddy, 2008;Walton, 2006, Global Invasive Species Database)

Pollution

Pollution in simple words is the undesirable change in the environment. These undesirable changes affect the natural habitat of species or directly the species. Environmental pollution is mainly of three types namely; Air pollution, water pollution, and Soil Pollution. The effect of pollution on biodiversity is described below as:

Impacts of Air pollution on the biodiversity

(i)Effects of Air pollution on Plant biodiversity

In general air pollutants have drastic impacts on plant physiology through decrease in photosynthesis, chlorosis, necrosis, change in root-shoot ratio, abscission, increase in leaf senescence, decline in stomata activity, and consequently some plants adapted to polluted environment and many are susceptible to death. The impacts of air pollution on various plant groups are given in the Table 9.

Table 9: Impacts of air pollution on various groups of plants

Group	Impacts of air Pollution
Algae	Particularly blue green algae are susceptible to air pollutants and species of blue green algae are at the verge of extinction
Phytoplankton	In many acidified Swedish lakes, numbers of species of phytoplankton have fallen by over 50%
Lichens	Diversity declines dramatically due to SO_2 and wet acid deposition.

Group	Impacts of air Pollution
Bryophytes	Similarly highly sensitive to many air pollutants, particularly in the case of tree living or bog mosses.
Pteriodyophytes	Evidence for decline in some fern and many club moss species in polluted air.
Herbaceous flowering plants.	Increasing body of evidence for decline, through SO_2 pollution.
Broadleaved trees.	Sensitive to acute damage by ozone and other air Pollutants
	Many trees decline in polluted environments, due to the impacts of both air pollution and other stress factors: the multiple stress problems.
Coniferous trees	Sensitive to acute damage by ozone and other air Pollutant.
Aquatic flowering plants	Declined due to acidification in freshwater

(*Source: Air pollution and biodiversity: Areview: Nigel Dudleyand Sue Stolton, 1996*)

(ii)Effects of Air Pollution on Animal biodiversity

Generally many animal species have decline due to reproductive failure and food chain affected due air pollution. The impacts of air pollution on various animal groups are given in Table 10.

Table 10: Impacts of air pollution on various animal groups

Group	Impacts of air pollution
Inverte-brates	Almost all lower invertebrates decline in acid waters.
Arthropods	Many crustaceans and insects decline in acid waters.
Fish	Some disappearing in slightly acid waters and others able to with-stand even fairly severe acidification. Decline has occurred widely in Europe and North America.
Amphibi-ans	Many species decline in acidified waters, primarily due to repro-ductive failure.
Birds	A minority of species decline due to food chain effects from loss-es, particularly in acidified water. Others are apparently directly affected by SO_2
Mammals	Main effects noted for mammals come from food chain effects in species.

(*Source: Air pollution and biodiversity: a review: Nigel Dudley and Sue Stolton, 1996*)

Impacts of water pollution on Biodiversity

(i)Eutrophication:

The excessive growth of algae due nutrient enrichment is called as eutrophication. The process and impacts of eutrophication is described below in the Figure 14

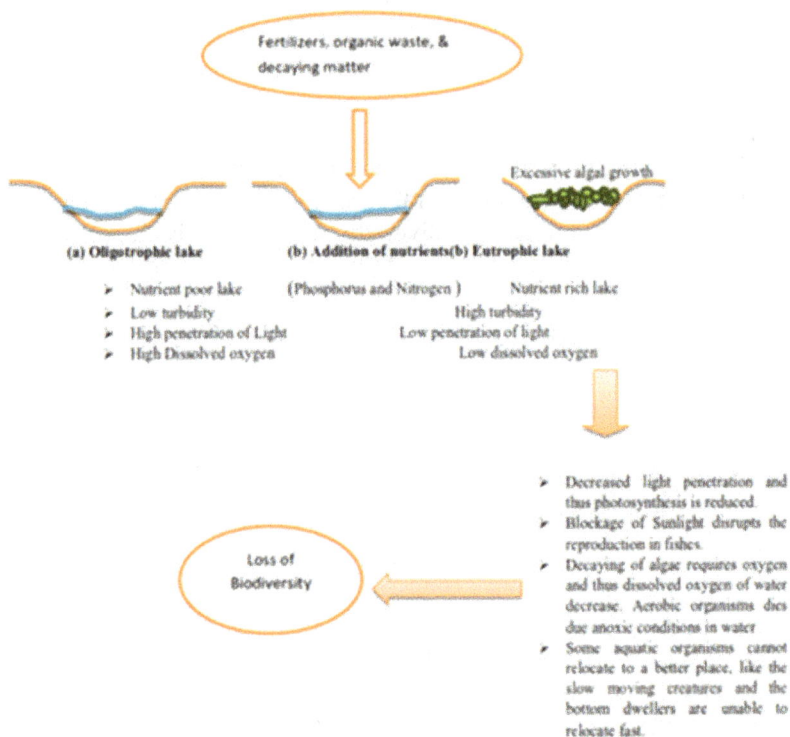

Figure 14: Diagrammatic representation of Eutrophication and its impacts on Biodiversity

(ii)Biomagnification

It is theincrease in the concentration of a toxicantfrom one link in a food chain to another (Mader and Sylvia, 1996). It starts after the bioaccumulation that is a process in which toxic substance accumulates in the organism's body. A very good example of biomagnification of DDT in aquatic food chain is described in the Figure 15. DDT is persistent organic pollutants which after entering into water body magnify successively in a food chain and cause lethal effects in species at higher trophic level.

Thus many bird species are under threat due to biomagnification of such harmful pesticides and other chemicals.

(iii)Oil spills and biodiversity

Oil spill is the release of a liquid petroleum product into the environment especially marine areas, due to anthropogenic activities and is harmful to marine life. Population of various phytoplankton and zooplankton is decreased after incidence of oil spill in marine ecosystem. Phytoplankton is the base of marine food chain and thus other organisms depended on phytoplankton also reduced. Copepods are most sensitive zooplankton and die after 3-4 days of oil immersion in sea. Many lobsters, oysters, scallops and calms are very sensitive to oil pollution. The slimy mucus on the gills of fish is somewhat resistant to oil what at a large concentration it is affected and gills get laden with oil and cause fish mortality. Heavy coatings of petroleum hydrocarbon disrupt the flying and swimming capacity of birds and subsequently cause death. In otters oil destroys the ability of fur to insulate, causing them to die hypothermia (lowering of body temperature). Oil also causes cancer in sea otters. Coral reefs are the tropical rainforest in ocean and support enormous biological diversity. They are highly complexes and productive. Corals are sensitive to oil and thus organisms living in corals become threatened.

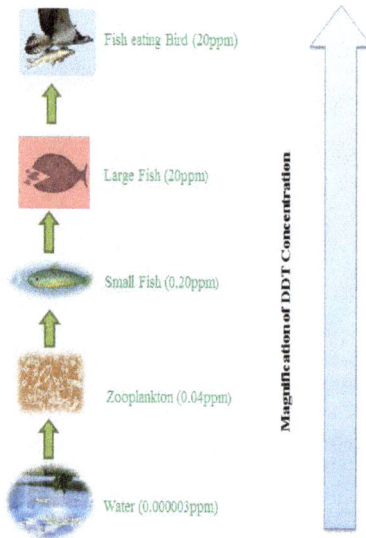

Fig.15: Biomagnification of DDT in a aquatic food chain (based on Energy in the ecosystem: Review by Slichter 1997)

Figure 15: biuomagnification of DDT in a aquatic food chain (based on Energy in the ecosystems: Review by Slichter 1997).

Impacts of soil pollution on Biodiversity

Soil provides foundations on which any plant grows and makes a contribution to the biodiversity. It has been noticed that polluted soil system and unhealthy soil pool would hamper plants growth. Soil is important not only for plants but also for animals.

Many organisms live in soil like earthworm, bacteria and various other small organisms which make the soil biodiversity unique and maintain the biochemical properties of soil and ultimately support the biodiversity in soil system. Pesticides like DDT may persist for about ten years or more in soil. Organisms do not have enzymes to degrade these pesticides so they persist for long time in system. Persistent Organic pollutants (POPs) through soil enter into the terrestrial food chain and causes death of organisms at higher trophic level as a result of biomagnification. Thus over use of pesticides in agriculture is threatening the precious biodiversity on the blue planet.

Decomposition of organic matter in soil and rainfall patterns affects the soil pH level. Areas of high rainfall generally have low pH because various nutrients like calcium, and potassium which are responsible for high pH are leached out and reduce the pH. Plants, microbes and other small organisms found in the soil live in a particular range of pH and altering pH level threat their survival in the soil. Acid deposition due to Air pollution also acidifies the soil and affects the soil flora and fauna. Acid rain dissolves the mineral calcite (calcium carbonate) that is necessary for the formation of shells in the snails. Various sensitive plants do not grow in the soil with altered chemistry.

Disease

Disease to wildlife is become a threat to biodiversity. Disease in plants and animals are caused by various pathogenic bacteria which hinders the immune system of the organism. Every organism is imbibed with an immune system for defense mechanism to fight against disease causing agent. Sometimes pathogens are so powerful and overshadow the defense mechanism and cause the disease in the living beings.

Diseases to organisms are sensitive to environmental disturbance. Some processes affecting infectious disease and their transmission are deforestation, climate change, urbanization, land–use change, and intentional or accidental biological invasion (WHO: Climate change and Human health).Flood affected area becomes sensitive to disease causing microbes. After flood, epidemics shadow the area under their attack and many local flora and fauna are extinct. Expansion of disease to the flora and fauna is become a sensitive threat to global biodiversity. Pollution in marine ecosystems in terms of acidification reduces the immune response of fishes. Climate change may expand the range of vector borne disease from the biodiversity rich tropical zone to the less bio diverse temperate zone (Dobson and Carper, 1992; Harvell *et al.*, 2002).

Amphibians declines and extinct worldwide, due to a disease Chytridiomycosisthe infectious disease caused by *Batrachochytrium dendrobatidis*. *Chytridiomycosis* was first detected in Spain in 1999, and it have caused to nearly complete extinction of mid wife toad in that area. (Biodiversa.2013 Policy brief, Europe).Crayfish plague (*Aphanomyces astaci*), a water mold, is considered responsible for declines in native white clawed crayfish numbers in United kingdom (Cunningham *et al.*, 2003).

In some cases it is observed that disease distribution is controlled by the geographical factors like altitude as shown in Figure 16. (U.S. Geological Survey, 2012).In Hawaiian Islands, Climate warming leads to the transmission of mosquito borne avian malaria at higher elevations and threatens native bird species of Hawaiian Island. Avian malaria is caused by the *Plasmodium relictum* which is protist and whose real vector is mosquito *Culex quinquefasciatus*. On the Hawaiian Island avian malaria is restricted to warmer altitudes below1500 meters (Van Riper, III and others 1986). As per some prediction, if higher elevations will become warmer, parasite development would also increase subsequently.

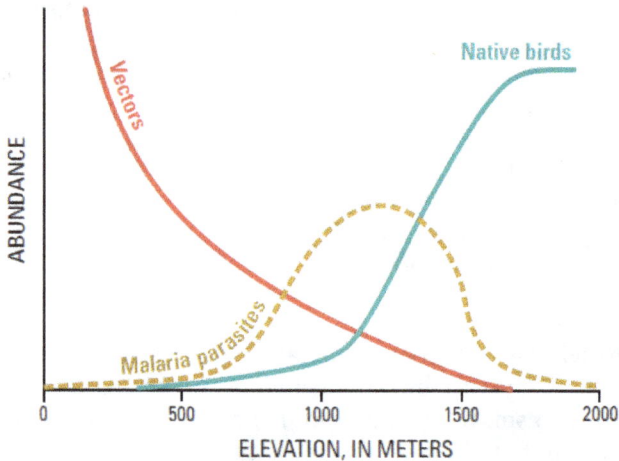

Figure 16: Graph showing the abundance of avian malaria parasite in Hawaiian birds, disease carrying mosquitoes and native Hawaiian birds with respect to elevation (Riper,van Charles III *et al.*, 1986; U.S. geological Survey, 2012).

Other Threats

Loss of Keystone Species

A keystone species is a species which play an important role in the ecosystem functioning and without which ecosystem would collapse.Keystone species have disproportionate effect on its environment relative to its abundance.A small number of keystone species can have a large impact on the environment.

Disappearance of keystone species from an ecosystem pave the way for biological invasion in which native species are out compete by non–native species.

Extinction

The complete loss of a species from the earth is known as extinction. Extinction occurs when no more individuals of a taxonomic group survive, either within a specified part of their range or forever lost across their entire range.Extinction and speciation are two fundamental processes in the geologic history of earth, species extinct and new species come to earth. There are three kinds of extinction like Natural extinction, Mass extinction and anthropogenic extinction.

Natural extinction

It is also known as background extinction. With the change in environmental conditions, some species disappear and others which more adapted to changed conditions take their place. This loss of species which occurred in the geological past at very slow rate is called Natural extinction.

Mass extinction

There have been several periods in the earth's geological history when large number of species became extinct because of catastrophe. Mass extinction occurred over millions of years. In geological past 5 big mass extinction has occurred namely,(a) Late Ordovician, (b) Late Devonian, (c) Late Permian, (d) Late Triassic, and (e) End Cretaceous.Average durations of species were generally less than 10 million years and biological composition of earth at species level changed drastically several times. The most serious extinction occurred near the end of the Permian period (250million years ago), removing 60.9% of all life (Benton 1995). Human activities are now paving the way for sixth mass extinction.

Anthropogenic extinction

It is human caused extinction of species from the earth. It is occurred within the short period of time.Rate of extinction is very high.Current rate of extinction is 1000-10000 times higher than the background rate. About 1000 species out of one million species extinct per decade.

Factors which make some of species susceptible to extinction

Large body size, small population size, feeding at higher trophic level in the food chain, fixed migratory routes and habit, localized and narrow range of distribution are some factors which make some species susceptible to extinction.

Change in native species dynamics

There are various factors which cause the change in native species dynamics and consequently loss to biodiversity. Competition, Hybridization among species,predation,mutualism, pathogen and parasite attacks are some factors which make change in native species dynamics.

Accidental Mortality

Several species are injured or even dead due to various accidental incidents. Fishery related practices like hooking and netting etc. cause injury orsometimes death to fishes.Collision of species with vehicle, tower and building make them injured or cause mortality which subsequently lost the biodiversity.

Natural Disasters

A natural disaster is a major adverse event resulting from natural processes of earth. Natural disasters have always made man feel helpless in spite of technological advances. The recent tsunami of December 2004 in the South Asian region is the worst ever in recorded history.

Apart from the trail of death and destruction of human property that it left behind, it also caused widespread destruction of ecological habitats with lasting effect on the populations of several species.

Natural disaster may destroy particular species completely or alter the species composition. Earthquake, volcanic eruptions, flood, drought, wildfires, landslides and storms are come under the natural disasters.

Various natural disaster and their impacts on biodiversity

(i)   *Earthquake:* It is a form of energy of wave motion transmitted through the surface layer of the earth in widening circles from a point of sudden energy release, the focus (Singh, 2015). Many plants are uprooted in highly seismic prone ecosystems such has mountain and many rare species are killed during major earthquakes.Earthquake can change the structure and composition of some fragile ecosystem and thus lead to species extinction.

(ii) *Volcanic eruptions:* Volcanic eruptions are the outflow of molten rock material in the form of lava from the volcanoes. Fallout of immense volcanic material destroys the nearby vegetation and animals. Thick covers of green forest on Mount St. Helens were completely destroyed due to forest fires produced by hot lava. Aquatic organisms are threatened due to increase in acidity, turbidity and temperature after volcanic eruptions (Volcano world, Department of Geoscience, Oregon State University, 2015).

(iii) *Floods:* It is caused by rain, heavy thunderstorms, and thawing of snow. Many rare species are washed out in heavy flood in downstream. After flood epidemics spread their wings which is harmful to living organism of that area.

(iv) *Drought:* A drought is a period of below-average precipitation in a given region, resulting in prolonged shortages in its waters supply, whether atmospheric, surface or ground water. Many lakes and wetlands dried due severe drought and organisms of these habitat died. Drought is also lethal to crops and thus decreases the agro-biodiversity of a particular area.

(v)   *Wildfires:* Forest fires remove many threatened plant species and burnt many animal species which are precious for biodiversity. Many small herbs, shrubs are burnt due to surface fires and small organisms living in them are also injured and many die. Invasive species like *Lygodiummicrophyllum*and *Bromustectorum* can grow rapidly in areas that were damaged by forest fires and thus further threat to new biologicalsuccession (Harris, 2013)

(vi) *Landslides:* It is a geological phenomenon that includes ground movements such as rock falls,deep failure of slopes,and shallow failure of slopes and shallow debris flow.Cause death to many animals due to dumping by huge rock material. Many plant species are uprooted.

(vi) *Storms:* Storm is a disturbance of the normal condition of the atmosphere, manifesting itself by winds of unusual force or direction, often accompanied by rain, snow, hail, thunder, and lightning, or flying sand or dust. High velocity and energy of wind cause injury or death to plants, animals, as well as human.

Conclusion

Biodiversity is precious gift of nature to mankind. On the blue planet there are different varieties of living things which in combination of physical factors of environment make the earth unique planet in the universe.Decrease in the numbers of various species of animal,birds, and plants have drawn world attention to the problem of species at risk.Habitat loss is become the greatest threat to biodiversity. It is the fragmentation, degradation, and loss of forests, wetlands, coral reefs, and other ecosystems that poses the threat to biodiversity. Every species needs a specific natural habitat for their sustenance. Species richness is decrease with decrease in area. Human population is increasing at an exponential rate and threatening biodiversity due to overexploitation of natural resources. Forestsare destroying at an alarming rateand thus habitat of wildlife is on the verge of extinction. Various anthropogenic activities like intensive agriculture and infrastructure development have become the major cause of habitat loss. Tropical rain forests aredeclining with a higher rate as compared to other type of forest. Great pace industrialization, transportation, mining, construction, domestic sewage and agricultural runoff have become the main cause of pollution and climate change. Anthropogenic emission of greenhouse gases particularly carbon dioxide is increasing the average temperature of earth's surface. Due to this global warming the biodiversity in all ecosystems, particularly in marine and polar ecosystems have been declining. Thinning of ozone is lethal to living things on the earth because of ultraviolet (UV) radiation hazards. Other threats to biodiversity include invasion by non-native species, extinction, natural disaster etc. If natural habitat is degraded all other environmental factors like invasion by alien species, climate change, pollution, and natural disaster predominate in that area and ultimately reduced the biotic component of the area. To mitigate these threats we need to limit our exploitation rate of natural resources and apply sustainable development for better future of life on planetearth. Traditional knowledge based biodiversity conservation and management would reflect a better option for long-term technical know-how and do-how management.

References

1. Baillie, J.E.M., Hilton-Taylor, C. and Stauart, S.N. (2004). (eds.).*IUCN Red List of Threatened Species. A Global Species Assessment.IUCN, Gland, Switzerlandand Cambridge, UK.*

2. Belfrage, K. (2006). The effects of farm size and organic farming on diversity of birds, pollinators and plants in Swedish landscape. *Ambio.* 34(8): 582-588.

3. Biodiversa. (2013).Wildlife Diseases on the Increase: a Serious Threat for Europe's Biodiversity.

4. Butchart, S.H.M., Walpole, M., Collen, B., van Strien, A., Scharlemann, J.P.W., Almond, R.E.A., Baillie, J.E.M., Bomhard, B., Brown, C., Bruno, J., Carpenter, K.E., Carr, G.M., Chanson, J., Chenery, A.M., Csirke, J., Davidson, N.C., Dentener, F., Foster, M., Galli, A., Galloway, J.N., Genovesi, P., Gregory, R.D., Hockings, M., Kapos, V., Lamarque, J.-F., Leverington,

F., Loh, J., McGeoch, M.A., McRae, L., Minasyan, A., Hernández Morcillo, M., Oldfield, T.E.E., Pauly, D., Quader, S., Revenga, C., Sauer, J.R., Skolnik, B., Spear, D., Stanwell-Smith, D., Stuart, S.N., Symes, A., Tierney, M., Tyrrell, T.D., Vié, J.-C. and Watson, R. (2010). Global biodiversity: indicators of recent declines. *Science.* 328(5892): 1164-1168.

5. CBD. (1992). *United Nations 1992, Article 2.*

6. CBD. (2000). *Secretariat of the Convention on Biological Diversity.*

7. CBD. (2007). *Biodiversity and climate change.*

8. CBD. (2010). *Global Biodiversity Outlook 3.* Secretariat of the Convention on Biological Diversity, Montreal.

9. *CO_2.Now.org, National* Oceanic and Atmospheric Administration (NOAA), Earth System research laboratory.

10. *Conservation International 2013; UNESCO 2012; and Planning Commission, 2007.*

11. Cunningham, A.A., Daszak, P. and Rodriguez, J.P. (2003). Pathogen pollution: Defining a parasitological threat to biodiversity conservation. *Journal of Parasitology* 89: S78-S83.

12. Danielsen, F., Beukema, H., Burgess, N.D., Parish, F., Brühl, C.A., Donald, P.F., Murdiyarso, D., Phalan, B., Reihnders, L., Struebig, M. and Fitzherbert, E.B. (2009). Biofuel plantations on forested lands: double jeopardy for biodiversity and climate,*Conservation Biology* 23: 348-358.

13. Dobson, A. and Carper, R. (1992). Global Warming and Potential Changes in Host–Parasite and Disease–Vector Relationship. Yale University Press, Connecticut. 201-217.

14. Draft on National Biodiversity Action Plan (2007, *Ministry of Environment and Forest, Government of India.*

15. Driscoll, C.T., Lawrence, G.B., Bulger, A.J., Butler, T.J., Craonan, C.S., Christopher, E., Lambert, K.F., Likens, G.E., stoddaed, J.L. and Weathers, K.C. (2001). Acidic deposition in the northeastern united states: sources and inputs, ecosystem effects, and management strategies. *BioScience.* 51: 180-198.

16. Fifth Assessment report. (2013). IPCC.

17. Global Biodiversity Outlook. (2006). Convention on Biodiversity.

18. Global Footprint Network. (2012). A report of Global Footprint Network, International Environment House 2, Geneva Switzerland.

19. Global Footprint Network. (2014). A report of Global Footprint Network,International Environment House 2, Geneva Switzerland.

20. Harris, A. (2013). Environmental Biodiversity. *Random Exports, New Delhi.*

21. Harvell, C.D., Mitchell, C.E., Ward, J.R., Altizer, S. and Dobson, A.P. (2002). Ecology – climate warming and disease risks for terrestrial and marine biota. *Science.* 296: 2158-2162.

22. India's fifth national Report toCBD. (2014). Ministry of Environment and Forest, Govt. of India.

23. *Invasive Pathways team Final report. (2013)*: National invasive species council and US department of Agriculture.

24. IPCC Fourth Assessment Report. *(2007)*.

25. IUCN. (2010). *Guidelines for Using the IUCN Red List Categories and Criteria.*

26. Kuniyal J.C., Vishvakarma S.C.R. and Singh, G.S. (2004). Changing crop biodiversity and resource use efficiency of traditional versus introduced crops in the cold desert of the northwester Indian Himalaya: a case of the Lahaul Valley. *Biodervisity and conservation*. 13: 1271-1304.

27. Kura, Y., Revenga, C., Hoshino, E. and Mock, G. (2004*)*. Fishing for Answers. *World Resources Institute, Washington, DC.*

28. Living planet report. (2014). WWF International.

29. MA. (2005a). Ecosystems and Human Well-being: Synthesis. Millennium Ecosystem Assessment. *World Resources Institute. Island Press, Washington, DC.*

30. MA. (2005b). Ecosystems and Human Well-being: Wetlands and Water Synthesis. Millennium Ecosystem Assessment. *World Resources Institute. Island Press, Washington, DC.*

31. Mader, S.S. (1996). Biology-5th Ed.

32. Mc Arther and Wilson. (1967). *The Theory of Island Biogeography.*

33. McNeely, J.A., Mooney, H.A.,Neville, L.E., Schei, P. and Waage J.K. (eds.). (2001). A Global Strategy on Invasive Alien Species. IUCN Gland, Switzerland, and Cambridge, U.K., in collaboration with the Global Invasive Species Programme.

34. Myers, N., Mittermeier, R.A., Mittermeier, C.G., da Fonseca, G.A.B. and Kent, J. (2000). Biodiversity hotspots for conservation priorities. *Nature.* 403: 853-856.

35. Ramakrishnan, P.S. (2014). Ecology and Sustainable Development. National Book Trust, Ministry of Human Resource Development, Govt. of India.

36. Red list of Threatened Species. *(2013)*. IUCN, Gland, Switzerland.

37. Reddy, C.S. (2008). Catalogue of invasive alien flora of India, Forestry and Ecology Division, *National Remote Sensing Agency*, Balanagar, Hyderabad.

38. Riper, V.C., Riper, V.S., Golf, M.L. and Laird, M. (1986). Ecological monograph. 56(4): 327-344.

39. Sechrest, W.W. and Brooks, T.M. (2002). Biodiversity-Threats,*Encyclopedia of life science.* Macmillan Publishers Ltd, Nature Publishing Group.

40. Secretariat of the *Convention on Biological Diversity. (2014).* History of the Convention on Biological Diversity.

41. Singh, G.S. (1999). Utility of non-timber forest products in a small watershed in the Indian Himalaya: The threat of its degradation. *Natural Resource Forum.* 23: 65-77.

42. Singh, G.S., Ram S.C. and Kuniyal, J.C. (1997b). Changing traditional land use patterns in the great Himalayas: A case study from Lahaul Valley. *J. Environmental Systems.* 25: 195-211.

43. Singh, G.S., Rao, K.S. and Saxena, K.G. (1997a). Energy and economic efficiency of mountain farming system: A case study in the north western Himalaya. *J. Sustainable Agriculture.* 9: 25-49.

44. Singh, J.S., Gupta, S.R., and Singh, S.P. (2014). Ecology, Environmental science and Conservation (revised). *Delhi: Chand Publication.*

45. Singh, S. (2015). Earthquakes. *Physical Geography* Allahabad: Pravalika Publications.129-144.

46. Synthesis report IPCC. (2007).

47. U.S. Geological Survey. (2012). Climate change and wildlife Health: direct and indirect effect.

48. UNEP. (2007). Biodiversity and Climate Change.

49. UNEP. (2007). Global Environment Outlook 4: Environment for Development, *United Nations Environment Programme.* Progress Press, Valletta.

50. Valiela, I., Rutecki, D. and Fox, S. (2004). Saltmarshes: biological controls of food websin a diminishing environment. *Journal of Experimental Marine Biology and Ecology.* 300(1–2): 131-159.

51. Van-Riper, C., Van-Riper, S.G., Goff, M.L. and Laird, M. (1986). The epizootiology and ecological significance of malaria in Hawaiian land birds. *Ecological Monographs.* 56(4): 327-344.

52. Vie, J.C., Hilton-Taylor, C. and Stuart, S.N. (eds.) (2009). Wildlife in a Changing World. An Analysis of the 2008 IUCN Red List of Threatened Species. *International Union for Conservation of Nature, Gland.*

53. Volcano world, Department of Geoscience, *Oregon State University.* (2015).

54. WHO: Climate change and Human health.

11

Biodiversity Assessment at Larsemann Hills, East Antarctica

Pawan Kumar Bharti[1,2]

[1]*Bharti Station, Larsemann Hills,*
Ingrid Christenson Coast, East Antarctica
[2]*Antarctica Laboratory, R & D Division,*
Shriram Institute for Industrial Research, Delhi-110007, India
E-mail: gurupawanbharti@rediffmail.com

Abstract

India has conducted scientific investigations at Dakshin Gangotri and Maitri stations in Antarctica in different disciplines. It now intends to broaden the scope of its scientific research by complementing the existing studies from an additional location. The new location for the new Indian research base is at Larsemann Hills, in East Antarctica. The Larsemann Hills (69°20′–69°30′S lat: 75°55′–76°30′E long.), named after Larsemann Christensen, is an ice-free coastal oasis with exposed rock and low rolling hills. However, it is mandatory to have some background ecological information prior to the initiation of station activity in the proposed area. Hence, the lichens-one of the major biological elements of Antarctica and highly privileged environmental indicators in addition to Moss communities, are studied to generate baseline information for future biomonitoring studies in the area to assess anthropogenic activities in the area after the construction of the third Indian research station in Larsemann Hills. Rock is the major substratum in the island accommodating many lichen species followed by moss species. True soil is virtually absent in the studied area, but a thin soil may be accumulated in rocks crevices, base of the rocks or in moss beds. The closely packed soil grains form a hard crust, a suitable habitat for lichens, as different species have been collected from such habitats. Organic matter comprising dead birds was frequent in the island and *Caloplaca citrina* was found growing luxuriantly in such habitats. Most of the lichens are substrate-specific, while some were found growing on all available substrate. *Buellia frigida, Candeleriella flava* and *Rhizoplaca melanophthalma* were found the most abundant and dominant lichen species in various islands/peninsulas of Larsemann Hills especially at Bharti Promontory and Fisher Island.

Key words: Biodiversity, Lichens, Moss community, Antarctica

Introduction

Antarctica is the most precious asset on the earth and is the last heritage of human kind. Antarctica is the only area on earth planet which is strictly devoted to scientific research and the continents of extremes come to be known as the 'Continent of Science'. It is the nature biggest laboratory on earth where no outside anthropogenic interference has taken place over the centuries till recent times. Being at a unique geographic location, it offers unique opportunities for Scientists to conduct number scientific research experiments. Antarctica is attracting world attention because of the tremendous biological species in surrounding seas and likelihood of vast hydrocarbons. Even though it is difficult to survive at Antarctica, still Scientists all around the worlds have been engaged in pursing the exciting scientific research investigations. The investigations are essential not for the exploitation of natural resources buried under the region but for the preservation of environment and ecology on earth; especially in the light of climate change. The north and south poles of Arctic and Antarctic polar region maintain the heat budget of the world in balance. The heat transported through the atmosphere and the oceans to the poles is dissipated in space in the form of long wave radiation. The cold air from Antarctica, when meets the warm air in the atmosphere of the lower latitude, charges into moisture bearing clouds. Thus, Antarctica polar region regulates the global climate and more particularly southern atmosphere.

Antarctica provides a unique, unpolluted and stable pure environment for carrying out scientific observation. It is far away from all sources of environmental contamination and thus remains an unpolluted datum point from which global changes due to pollution could be monitored and is suitable for a wide range of scientific research.

Antarctic Ocean supports biological communities of few species with large population and short food chain magnification. It is among the richest biological provinces on the earth. It is far away from main earth. Further, life forms in the Antarctica are also concentrated primarily in coastal lakes, which are poor in species and low in productivity, or in the surrounding seas, which are relatively rich in species. Thus, this is an ideal location for study of new species undiscovered so far.

Antarctica polar region contains almost 90 % of the world ice which constitutes about 70 % of the world's store of fresh water reserves. Further, huge amounts of water from all the regions of the world are constantly mixing together making Antarctic Ocean absolutely essential in the earth climate and heat balance (Hodgson, 2001). Since Antarctica is virtually untouched by human civilization and has no higher plant and animal life.

Because of the influence of world weather, and climate change, Antarctica lies at the heart of the debate on climate change and has become the premier location to study the effects of global warming and climate change.

The Larsemann Hills (69°20'–69°30'S lat: 75°55'–76°30'E long.), named after Larsemann Christensen, is an ice-free coastal oasis with exposed rock and low rolling hills. The Larsemann Hills contain hundreds of freshwater lakes of varying sizes, depth and biology (Hodgson *et al.*, 2001).

Important information about Antarctica:

Parameters	Value
• Total area, million sq km	14.2
• 98% of Ice Sheets	
• 2% Ice free areas	
• Antarctica possess	
• 90% of world ice	
• 70% of world fresh water resources (Stored)	
• Highest peak (Vinson Massif), m	4,897
• Height of surface of South Pole, m	2,635
• Maximum Known Thickness of ice, m	4, 776
• Shortest Distance across Antarctica, km (via South Pole)	3250

Antarctica is the worlds coldest and the most isolated continent on the earth. The Indian Ocean, Pacific Ocean and Atlantic Ocean surround the continent. Antarctic continent covers 10% of the earth surface and has a surface area of nearly 14 million square kilometres. It also has 70% of the world's fresh water resources in the form of ice sheets. Antarctica possesses 20 % of the world seas and ocean. Thick ice sheets cover the whole continent (almost) 98% of the surface. As a result of the environmental conditions, the remaining (~ 2%) portion without ice cover is basically the barren soil and rocks. The Antarctica is 990 km away from Cape horn- the southernmost tip of Argentina, which is the nearest land area to Antarctica.

Antarctica is the most inaccessible to human beings in comparison to the other continent on the earth. Antarctica continent is endowed with the harshest climate on earth. The coldest temperature recorded in the Antarctica was –90°C at Vostok Station (in the year 1983). The mean temperature in Antarctica lies between –20°C to –30°C on the coast of Antarctica and –40°C to –90°C in the interior of Antarctica (Bharti and Gajananda, 2013). The Antarctica region in addition to extreme cold is also full of ultraviolet rays during the summer period. It is also the stormiest, driest and windiest place on the earth. There are frequent blizzards, snow storms and snow drifting in Antarctica. These factors coupled with others make Antarctica conditions harsh stressful, which makes human life very difficult in the region and survival in such conditions for the mankind is a real struggle. It is also the remotest continent. Antarctica has very few living organisms such as lichens, mosses, algae, yeast etc. However, the Antarctica lacks higher forms of plants and wild lives except Penguins, Skua, Snow Petrel, Tern, etc.

Study area

Larsemann Hills is at the Ingrid Christensen Coast of east Antarctica (69° 21' 00" S to 69^0 27' 00" S and $75^0$57' 00" E to 76° 27' 00" E). It is an ice-free Antarctic coastal oasis, located between the Amery Ice Shelf and the Vestfold Hills. Isolated islands, peninsulas and nunataks occur along the coastal continental ice. The areas are mainly exposed hilly and rocky terrains. There are two main peninsulas in the Larsemann Hills, of east Antarctica (i) Broknes Peninsula and (ii) Stornes Peninsula. In between these two peninsulas, there are numbers of promontory, islands of varying shapes, sizes, heights and water bodies.

Bharti Island (69^0 24' to 69^0 26' S; 76^0 10' to 76^0 15' E), is situated over the Larsemann Hills of east Antarctica. The ice patches over the lake and the rocky regions are typical characteristics of the oases regions (Hall and Walton, 1992). The poplar ice is also seen in the background of the Bharti Island, which is one of the main feeders of fresh water to various lakes in this oasis region. This promontory is sandwiched in the northern direction by McLeod Island; eastern direction comprises the Quilty bay, north-eastern direction by Fisher Island, western direction is the Stornes peninsula and the southern direction by the Antarctic continent and polar ice caps.

The ecosystem of this area is simple and in the primary stage of ecological succession. The area is devoid of any higher organisms and plants except for some sea birds, seals, penguins, skuas, algae, lichen and mosses. Thick mosses, algae and lichens crusts were observed mainly near the lakes banks. Most of the areas are ice/snow free.

A major feature of the climate of the Larsemann Hills is the existence of persistent, strong katabatic winds that blow from the northeast in summer days. Daytime air temperatures from December to February frequently may exceed 4°C, with the mean monthly temperature a little above 0°C (SIIR, 2012). Mean monthly winter temperatures are between –15°C and –40°C. Pack ice is extensive inshore throughout summer and the fjords and bays are rarely ice-free. Precipitation occurs as snow and is unlikely to exceed 250 mm water equivalent annually (Bharti and Gajananda, 2013). Snow cover is generally deeper and more persistent on Stornes than Broknes, due to northeasterly prevailing winds and the perennial sea ice held in by the islands offshore from Stornes.

Biological study

Weak development of the weathering forms on the rock surfaces and fresh traces of glacial impact, indicate recent ice disappearance. The weathering of rocks releases minerals, an essential component for the survival and development of the ecosystem (Hall and Walton, 1992). It may be noted that the weathering of rocks leads to the formation of sandy soil, which cannot support normal forms of plant and animal lives. At the same time, brown rocks absorb solar energy, leading to a much higher temperature of the rock surfaces, thus providing a better habitat for the unique micro flora and fauna of Antarctica.

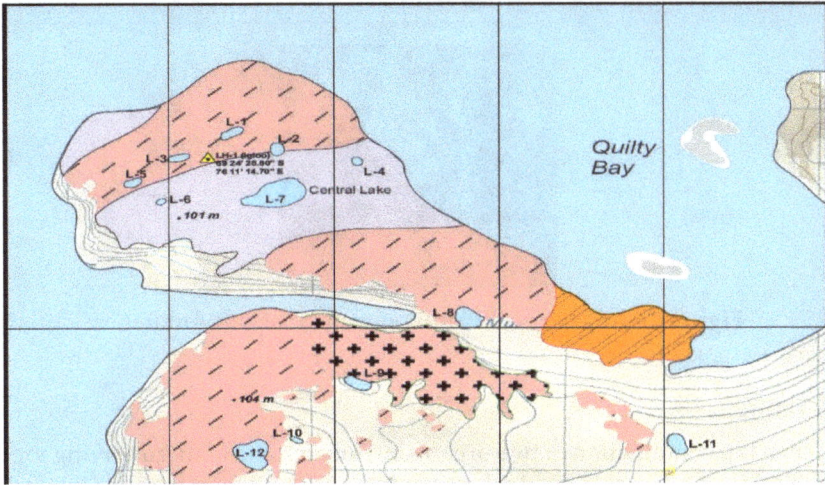

Figure 1: Bharti Island, Larsemann hills, east Antarctica

a

b

c

Figure 2: (A-C): Lichen, Moss and Algae observed at various Islands of Larsemann Hills

Figure 3: Collection of Plankton from a lake in Antarctica

Lichens

Lichen is a combination of two organisms, an alga and a fungus, living together in symbiotic association. The algal component in the lichen is called phycobiont or photobiont while fungus as mycobiont. The phycobiont and the mycobiont loose their original identity during the association and the resulting entity (lichen) behave as a single organism, both morphologically and physiologically. Hence the lichen is called as a composite organism. In lichen thallus (body) the mycobiont predominates with 90% of the thallus volume and provides shape, structure and colour to the lichen with partial contribution from algae. Whatever visible from out side in a lichen thallus is fungal part, that holds algal cell inside. Hence the lichens are placed in the Kingdom – Mycota (Fungi). The fungi present in lichens are called as lichenized fungi. Among the 20,000 lichen species known in the world 95% belongs to the Ascomycetes group of fungi while Bacidiomycetes and Deuteriomycetes groups are represented by only 3% and 2% of species respectively (Bharti, 2012).

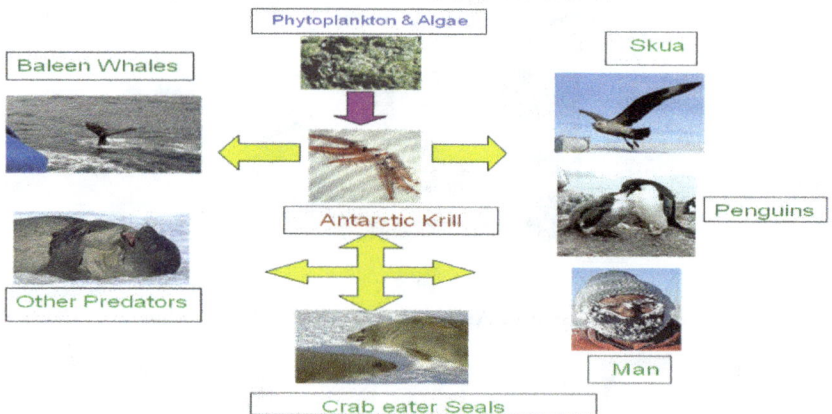

Figure 4: Antarctica Ecosystem

Lichens can grow in diverse climatic conditions and on diverse substrates. The lichens that are growing on tree trunk and bark are called corticolous lichens; twig inhabiting ones are ramicolous, on wood-legnicolous, on rocks and boulders – saxicolous (epilithic), on moss-muscicolous, on soil-terricolous and on evergreen leaves – foliicolous (epiphyllous). In general any lichen growing on other plant is called as epiphytic. The lichens can grow on underwater rocks, but not freely in water or on ice. The lichens are widely distributed in almost all the phytogeographical regions of the world. Sufficient moisture, light and altitude, unpolluted air and undisturbed, perennial substratum often favour the growth and abundance of lichens (SIIR, 2012).

By their appearance the lichens can be grouped into three main categories of growth forms:

- **Crustose lichens:** The thallus in crustose lichen is closely attached to the substratum without leaving any free margin. The thallus usually lacks lower cortex and rhizines (root like structure). Such lichens are collected along with their substratum for the detailed study.

- **Foliose lichens:** They are also called as leafy lichens. The thallus in this case is loosely attached to the substratum at least at the margin. Lichens are collected by scraping them from the substratum.

- **Fruticose lichens:** Here the lichen thallus is attached to the substratum at one point and remaining major portion is either growing erect or hanging. The lichen usually appears as small shrub or bush and easy to collect with hand.

Table 1: List of lichens from McLeod Island, Larsemann Hills, East Antarctica (Singh *et al.*, 2007).

Lichen taxa	Substratum	Growth form
Acarospora gwynnii C. W. Dodge & E. D. Rudolph	Rock	Crustose
Arthonia lapidicola (Taylor) Branth & Rostr.*	Rock	Crustose
Buellia frigida Darb.	Rock	Crustose
Buellia grimmiae Filson# Moss,	soil	Crustose
Caloplaca athallina Darb. #	Moss	Crustose
Caloplaca sp. A	Moss	Crustose
Caloplaca citrina (Hoffm.) Th. Fr.	Rock	Crustose
C. lewis-smithii Søchting & Øvstedal*	Moss	Crustose
C. saxicola (Hoffm.) Nordin* Rock,	soil	Crustose
Candelariella flava (C. W. Dodge & Baker) Castello & Nimis Rock,	soil, moss	Crustose
Carbonea vorticosa (Flörke) Hertel*	Rock	Crustose
Huea coralligera (Hue) C. W. Dodge & G. E. Baker*	Moss	Crustose

Lichen taxa	Substratum	Growth form
Lecanora expectans Darb. Soil,	moss	Crustose
L. geophila (Th. Fr.) Poelt*	Moss	Crustose
Lecidea cancriformis C. W. Dodge & G. E. Baker Rock,	soil	Crustose
Lecidella patavina (A. Massal.) Knoph & Leuckert*	Rock, soil	Crustose
L. siplei (C. W. Dodge & G. E. Baker) May. Inoue*	Rock	Crustose
Lepraria sp.# Rock, soil,	moss	Crustose
Physcia sp.* Rock,	soil	Foliose
P. caesia (Hoffm.) Furnr.	Rock, moss	Foliose
P. dubia (Hoffm.) Lettau*	Rock	Foliose
Pseudophebe minuscula (Nyl. ex Arnold) Brodo & D. Hawksw. Rock,	soil	Fruticose
Pseudephebe minuscule (Nyl. ex Arnold)#	Rock	Fruticose
Rhizoplaca melanophthalma (Ram.) Leuckert & Poelt Rock, soil,	moss	Crustose
Rinodina olivaceobrunea C. W. Dodge & G. E. Baker Soil,	moss	Crustose
R. peloleuca (Nyl.) Mull. Arg.*	Rock	Crustose
Sarcogyne privigna (Ach.) A. Massal.*	Rock, soil	Crustose
Umbilicaria decussata (Vill.) Zahlbr.#	Rock	Foliose
Usnea antarctica Du Rietz#	Rock	Foliose
Xanthoria elegans (Link) Th. Fr.	Rock	Foliose
Xanthoria mawsonii C. W. Dodge	Rock	Foliose

Moss

Mosses are small, soft plants that are typically 1–10 cm (0.4–4 in) tall, though some species are much larger. They commonly grow close together in clumps or mats in damp or shady locations. They do not have flowers or seeds, and their simple leaves cover the thin wiry stems. At certain times mosses produce spore capsules which may appear as beak-like capsules borne aloft on thin stalks. There are approximately 12,000 species of moss classified in the Bryophyta. The division Bryophyta formerly included not only mosses, but also liverworts and hornworts. These other two groups of bryophytes now are often placed in their own divisions. Botanically, mosses are bryophytes, or non-vascular plants. They can be distinguished from the apparently similar liverworts (Marchantiophyta or Hepaticae) by their multi-cellular rhizoids. Other differences are not universal for all mosses and all liverworts, but the presence of clearly differentiated 'stem' and 'leaves', the lack of deeply lobed or segmented leaves, and the absence of leaves arranged in three ranks, all point to the plant being a moss (Bharti, 2012).

Benthos

Benthic macroinvertibrate are the animals inhabiting the sediment, or living on or in other available bottom substrates of freshwater, estuarine and marine ecosystems. During all or part of their life cycles, these organisms may construct attached cases, tubes or nets that they live on or in; roam free over rocks, organic debris and other substrates; or burrow freely in substrates. Although they vary in size from small form, difficult to see without magnification, to other individuals large enough to see without difficulty, macroinvertibrate are considered historically by definition to be visible to the unaided eye and retained on a US standard No. 30 sieve (0.595 mm or 0.600 mm openings) (Ellis-Evans *et al.*, 1998).

The major macroinvertibrate found in freshwater are flatworms, annelids, mollusks, crustaceans and insects. The major macroinvertibrate groups included in estuarine and marine waters are bryozoans, sponges, annelids, mollusks, roundworms, cnidarians (coelenterates) crustaceans, insects and echinoderms.

Periphyton

Biodiversity play a significant role in the structure and balance of an ecosystem. The biodiversity rich ecosystem is generally called as wealthy ecosystem. Primary producers, primary consumers, secondary consumers, ultimate consumers and decomposers all are the part of biodiversity and all have the specific role to sustain the ecosystem.

Microorganisms growing on stones, sticks, aquatic macrophytes and other submerged surfaces are useful in assessing the effect of pollutants on lakes, streams and estuaries.

Included in this group of organisms, here designated periphyton are the zoogleal and filamentous bacteria, attached protozoa, rotifers, algae and the free living microorganisms that swim, creep and lodge among the attached forms.

Plankton

The term Plankton refers to those aquatic forms having little or no resistance to currents and living free floating and suspended in natural waters. Planktonic plants (Phytoplankton) and planktonic animals (Zooplankton) are covered in this section.

Plankton, particularly phytoplankton have long been used as indicators of water quality. Some species flourish in highly eutrophic waters while other is very sensitive to organic and chemical waste. The species assemblage of phytoplankton and zooplankton may be useful in assessing water quality.

Many researchers have already discovered many species of bacteria in Antarctica. There are a lot of opportunities to find out the new species of plankton and bacteria in Larsemann Hills area in east Antarctica (Bharti and Niyogi, 2015a; Bharti and Niyogi, 2015b).

Microbes

Microbiological study was also conducted for the Lake water ecosystem at various Islands of Larsemann Hills in east Antarctica (Bharti and Niyogi, 2015c).

These few microbial parameters were selected for study-

1. Total Bacterial Count/ml,
2. Pschrophillic count/ml,
3. MPN coliform/100 ml,
4. Yeast and Mold counts/ml,
5. Salmonnela/ 25ml
6. Staphylococcus aureus/25 ml,
7. Pseudomonas spp./10 ml, and

The results of microbiological studies in Antarctic freshwater lakes may be very beneficial and excited due to the potential possibilities of occurrence of new microbe species in the study area. Few species of Pseudomonas can be observed in few lakes (Chauhan *et al.,* 2015).

Biodiversity in Antarctica

Many species of Antarctic wildlife are unique to the southern region. Each of the major species shares a variety of adaptations that enables them to survive in the harsh environment. They all take their food from the sea that surrounds the continent; indeed, most live at the shore, although some breed on land.

The flightless birds-Penguins are able to withstand the extreme cold because of insulation provided by their short, densely-packed feathers forms a waterproof coat. A thick layer of fat or blubber also serves as an energy store. These adaptations, among others, enable them to minimize heat loss in icy cold waters so they can cope with the harsh Antarctic conditions.

Most of the seal pups can swim on the first day of their birth. Among the 35 varieties of the seals, 6 types live in Antarctica. They are Leopard seals, Crabeater seals, Ross seals, Weddell seals, Southern elephant seals & Antarctic fur seals.

Along with seals & penguins, whales also share the Antarctic waters. They belong to the family Cetacea, which includes about 75 species of huge whales, smaller dolphins & porpoises. The blue, fin, sei, humpback & minke whales migrate regularly in summers feeding on krill & small fish. These are the most dangerous of all the Antarctic animals (SIIR, 2012).

At the center of the Antarctic marine ecosystem is the krill. These are underwater shrimp like creatures up to 8 cm in length. They live on phytoplanktons, which grow in the warm sunlight in summer. Obviously these are the days when krill are found in unimaginable numbers. And why not? Imagine the female laying 2000 to 3000 eggs twice a year, which grow into adults in 2 or 3 years. One also may be surprised to notice that nearly 250 million tonnes of krill are consumed every year.

Antarctica also has a distinct variety of birds, which are distinguished by a thick layer of insulating fat under their skins & extensive fat deposits throughout their bodies. Their feathers provide excellent protection from the cold. There are 43 species of birds that breed within the limits of Antarctic convergence. The giant

Albatrosses, the colorful shags, ferocious skuas & the small, white, pigeon like sheathbills are some of them.

A variety of fresh water & saline lakes in Antarctica contain a limited range of aquatic life. The only vegetation observed here are mosses & lichens. They inhabit an exposed ground where moisture is available. Mosses grow rarely more than 100 mm deep even in most favorable conditions. Lichens are best adapted to survive at lower temperatures, & with less light & water. More than 300 species of non-marine algae are found under stones. They may form spectacular red, yellow or green patches on the areas of permanent snow. A few species of mites, insects & invertebrates are also rarely found in the outer environment of Antarctica.

One must appreciate that even after the advent of man on this continent, the life here has remained as unspoilt & innocent as it was. And the credit for this mainly goes to the Antarctic treaty, which has played and is playing a pivotal role in the explorations of this continent.

References

1. Bharti, P.K. (2012). Assessment of Lichen patches and Moss communities in the vicinity of Larsemann hills, east Antarctica, In: Biodiversity Conservation and Environmental Management (Eds.-Khanna *et al.*). *Biotech Books, Delhi*. 221-235.

2. Bharti, P.K. and Gajananda, K.H. (2013). Environmental monitoring and assessment in Antarctica, In: Environmental Health and Problems (Eds.-Bharti, P.K. and Gajananda, K.H.). *Discovery Publishing House, Delhi*. 178-186.

3. Bharti, P.K. and Niyogi, U.K. (2015a). Assessment of Pollution in a freshwater lake at Fisher Island, Larsemann Hills over east Antarctica. *Science International*. 3(1): 25-30.

4. Bharti, P.K. and Niyogi, U.K. (2015b). Plankton diversity and aquatic ecology of a freshwater lake (L3) at Bharti Island, Larsemann Hills, east Antarctica. *Global Journal of Environmental Science and Management*. 1(2): 137-144.

5. Bharti, P.K. and Niyogi, U.K. (2015c). Environmental conditions of psychrophilic Pseudomonas spp. in Antarctic Lake at Stornes Peninsula, Larsemann Hills, east Antarctica. *International Journal of Environmental Monitoring and Protection*. 2(3): 27-30.

6. Chauhan, A., Bharti, P.K., Goyal, P., Varma, A. and Jindal, T. (2015). Psychrophilic pseudomonas in Antarctic freshwater lake at Stornes peninsula, Larsemann hills over east Antarctica. *SpringerPlus*. 4(582): 1-6.

7. Ellis-Evans, J.C., Laybourn-Parry, J., Bayliss, P.R. and Perriss, S.J. (1998). *Arch. Hydrobiol*. 141: 209-230.

8. Hall, K.J. and Walton, D.W.H. (1992). 'Rock weathering, soil development and colonization under a changing climate'. Philosophical Transactions of the Royal Society of London. Series B, Biological Sciences. 338(14): 269-277.

9. Hodgson, D.A., Noon, P.E., Vyverman, W., Bryant, C.L., Gore, D.B., Appleby, P., Gilmour, M., Verleyen, E., Sabbe, K., Jones, V.J., Ellis-Evans,

J.C. and Wood, P.B. (2001). Were The Larsemann Hills Ice-Free Through The Last Glacial Maximum? *Antarctic Science.* 13(4): 440-454.

10. SIIR, (2012). Long term environmental monitoring study at new scientific base Bharti at Larsemann Hills, 27th, 28th, 29th, 30th ISEA combined report to NCAOR, submitted by Shriram Institute of Industrial Research, Delhi, India.

11. Singh, S.M., Nayaka, S. and Upreti, D.K. (2007). Lichen communities in Larsemann Hills, East Antarctica. *Current Science.* 93(12): 1670-72.

12

Measuring and Monitoring Biodiversity

Anjana Pant

Wild Life Fund, SATS Integrated Conservation Solutions Pvt. Ltd.,
716 Imperial, Supertech Estate,
Vaishali 9, Ghaziabad-201010, Uttar Pradesh, India
E-mail: anjana.pant@gmail.com

Abstract

Knowledge of status and trends of species and ecosystems is critical to the conservation of biodiversity which by nature is a dynamic. To that end, biological assessment and monitoring programmes play a key role. Biodiversity measurement and monitoring is the process of determining the status and tracking changes in living organisms and the ecological complexes of which they are a part. The challenge of quantifying patterns of diversity at the species level, even when the organisms are known to science, is complicated by the problem of detecting rare species and the underlying complexity of the environmental template. Biodiversity monitoring is important because it provides a basis for evaluating the integrity of ecosystems, their responses to threats, and the success of actions taken to conserve or recover biodiversity. Research addresses questions and tests hypotheses about how these ecosystems function and change and how they interact with stressors. Ecological research provides the context for interpreting these monitoring results. Policy and management needs guide the development of monitoring. Monitoring is necessary for an adaptive management approach and the successful implementation of ecosystem management. This chapter aims to present a basic approach towards '*measuring and monitoring biodiversity*' particularly with reference to tropical ecosystems.

Keywords: Biodiversity, Monitoring, Assessment, Species Diversity. Ecosystem Diversity.

Introduction

Life on Earth is diverse at many levels, beginning with genes and extending to the wealth and complexity of species, life forms, and functional roles, organized in spatial patterns from biological communities to ecosystems, regions, landscapes and beyond. The study of biodiversity encompasses the discovery, description, and analysis of the elements that underlie these patterns as well as the patterns themselves. *'Biodiversity'* provides the basis for ecosystems and the services they provide, upon which all living organisms including human beings depend. We rely on biodiversity in our daily lives, often without realizing it. It contributes towards people's well-being in terms of provisioning products such as food and fibres, whose values are widely recognized. Biodiversity underpins a very wide range of services, which are currently highly undervalued. Ranging from the tiny microscopic bacteria and microbes that transform waste into usable products, insects that pollinate crops and flowers, coral reefs and mangroves that protect coastlines to the biologically-rich landscapes and seascapes that provide aesthetic values are just a few and while much still remains to be understood regarding the relationships between biodiversity and ecosystem services, it is well established that if the products and services that are provided by biodiversity are not managed effectively, future options will become ever more restricted, for rich and poor people alike. It is actually the poor people who tend to be the most directly affected by the deterioration or loss of ecosystem services, as they often live in places most vulnerable to ecosystem change. Current rapid loss of biodiversity is also restricting the future development options. Ecosystems are being transformed by man, and, in some cases, irreversibly degraded. A huge number of species have either gone extinct in recent history or are at the verge of extinction. Biodiversity losses continue also because the current policies and economic systems do not incorporate the values of biodiversity effectively in either the political or the market systems, and many current policies are not fully implemented.

Functioning ecosystems are crucial as buffers against extreme climate events, as carbon sinks, and as filters for waterborne and airborne pollutants. It is now well established that the currently ongoing changes to biodiversity on terrestrial as well as wetland (freshwater as well as marine) ecosystems are more rapid at this time than at any time in human history and have led to a serious degradation in many of the world's ecosystem services and it is a 'now or never situation' at this point in time. In order to develop and implement strategies to conserve biodiversity in tropical moist forests, it is important to measure the key external threats and inherent weaknesses and their impacts of forest disturbance as well as the different ongoing forest management practices.

A global overview of the status of Biodiversity: Ecosystems vary greatly in size and composition, ranging from a small community of microbes in a drop of water, to the entire Tropical Forest. The very existence of people, and that of the millions of species with which the planet is shared, is dependent on the health of our ecosystems. Humans are putting increasing strain on the world's terrestrial and aquatic ecosystems. Despite the importance of ecosystems, they are being modified in extent and composition by people at an unprecedented rate, with little understanding of the implications this will have in terms of their ability to function and provide services in the future (MA, 2005). For more than half of the world's 14 biomes, 20–50 per cent of their surface areas have already been converted to croplands (Olson and others 2001). Tropical dry broadleaf forests have undergone the most rapid conversion since 1950, followed by temperate grasslands, flooded grasslands and savannahs. Approximately 50 per cent of inland water habitats are speculated to have been transformed for human use during the twentieth century (Finlayson and D'Cruz 2005). Some 60 per cent of the world's major rivers have been fragmented by dams and diversions (Revenga and others 2000), reducing biodiversity as a result of flooding of habitats, disruption of flow patterns, isolation of animal populations and blocking of migration routes. River systems are also being significantly affected by water withdrawals, leaving some major rivers nearly or completely dry. In the marine realm, particularly threatened ecosystems include coral reefs and seamounts.

Need for Biodiversity Assessment and Monitoring: Natural Resources are always limited and thus to maximise the benefits of any action. In terms of global trends, measuring progress towards the global target of reducing the rate of biodiversity loss by 2010 relies on monitoring species abundance, threat of extinction, extent and condition of habitats, and ecosystem goods and services (Dobson, 2005). Much of the work of practicing conservationists entails making judgements about the relative importance of different areas in a range of different scales (Sutherland, 2000).

Assessment of biodiversity can be best made by responding to three key questions:

    i.   What are the key pressures on our biodiversity? There could be many like Habitat loss/degradation, flow modification (in terms of rivers), Pollutions (in terms of wetland ecosystems), Harvesting of resource – timber/non-timber, wildlife, fishes or any other, impact of climate change, Invasive alien species/weeds, Genetically modified organisms or even hybridization (see Figure 1).

    ii.   The status of biodiversity (how it is doing) – be it in terms of species or even ecosystem or both.

   iii. Evaluate our response both in terms of remedial action as well as institutional capacity building to ensure that it is having the desired effect.

   iv. Most monitoring systems in tropical forests include the establishment of permanent vegetation plots where all plants above a certain size are identified to species and measured, for instance every 5 years. Such monitoring can generate data for rigorous hypothesis testing and provide important scientific evidence. However, since the frequency of data collection is low and limited data is collected on the use of resources, such exercise rarely provides any input to management (Danielson et al, 2007). Most monitoring systems in tropical forests include the establishment of permanent vegetation plots where all plants above a certain size are identified to species and measured, for instance every 5 years. Such monitoring has the potential to generate data for rigorous hypothesis testing and provide important scientific evidence. However, since the frequency of data collection is low and limited data is collected on the use of resources, such exercise rarely provides any input to management. For conservation purposes protected areas in developing countries need monitoring that is realistic and at the same time useful for guidance rather than what is ideally required for in depth studies of how community structure and species richness are affected by different environmental changes.

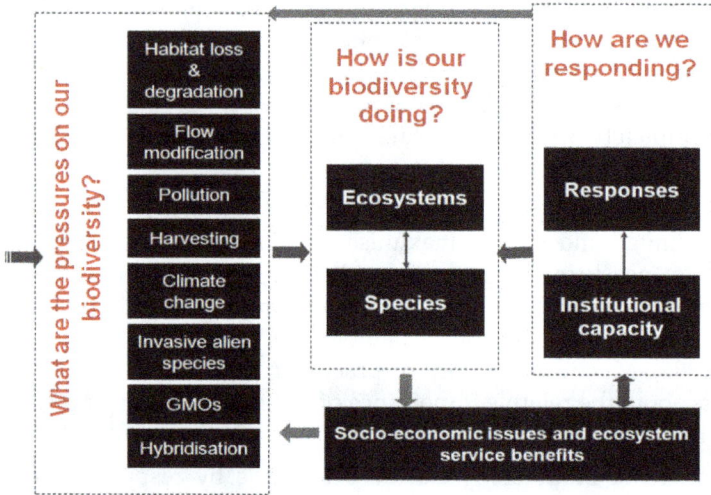

Figure 1: Biodiversity Assessment Needs

The conservation importance of an area is typically determined by assessing its biodiversity and as the basic unit of biodiversity may be considered to be species, this is mainly done by species presence and their abundance at the primary level (Sutherland, 2000). This helps us set conservation priorities through prioritisation of species by classifying then into IUCN defined classes like Extinct, Extinct in the wild, Critically Endangered, Endangered, Vulnerable, Conservation dependent, Low Risk, Data deficient and Not evaluated. Similarly, a habitat with high priority species may be considered as high priority sites for conservation.

Biodiversity Measurement and Monitoring

The primary step is for any biodiversity project is to set priorities as it is impossible to conserve everything. Conservation priorities are usually seen in the form of the conservation status of

(a) individual species like vulnerability to extinction, evolutionary distinctiveness, popular appeal, likelihood of recovery and local appeal;

(b) habitats like habitat preferences of high priority species, local and global distribution of the habitat and threats to the same *e.g.* global biodiversity hotspots. In practice, the four criteria most frequently used in site evaluation include naturalness, diversity, rarity/uniqueness and size (Smith and Theberge, 1986; Usher, 1986).

(c) A combination of both.

Whatever technique is used, it is important to quantify the amount of effort put into sampling as otherwise it is difficult to determine the extent to which a long species list reflects the importance of the site or the enthusiasm and skill of the observers (Block et al, 1987; Gaston, 1996). Many biodiversity assessment methods involve sampling by selecting sample areas or transect routes. Most of these methods can be used to gain information on the abundance of each species as well as biodiversity. Time-based observation methods such as time-restricted search and timed species counts are only practical if field identification is rapid and reliable. These are thus appropriate only for experienced naturalists. A measure of biodiversity on its own may be pretty meaningless. The objective of assessing biodiversity is usually to compare sites or to provide the data that can be used to characterise and compare sites. It therefore, irrespective of the method, important that the methodology is consistent and clearly-states so that it can be repeated. It is also important to note that for each method, the results will vary with weather, time of day, season and habitat structure. Biodiversity encompasses numerous levels of natural organization. Species, however, are the focus of biodiversity, because they are the most easily defined (Noss, 1990; Huston, 1994). Although other levels of biodiversity organization, such as genetic diversity (Watson-Jones et al., 2006) and landscape configuration (Roy and Tomar, 2000; Lindenmayer et al., 2006), are important, we focus the development of our biodiversity index on species.

Some of the key methods include:

i. Total species list, Genus and/or family list: This includes listing of all the species/genera/families and addition to the existing list as and when detected. A number of searching techniques like searching for direct sighting, indirect signs (droppings, footprints), and different life stages (eggs larvae and pupae). Talking to local people is also useful here and documenting and including pictures of the species will be a useful tool for identification.

ii. Parallel-line searches: In order to assess the presence of visible and fairly sedentary species in reasonably small areas is parallel-line search Nelson, 1987). In this method the area is divided into blocks not exceeding 10 hectares and each block crossed systematically by parallel paths across shortest width recording all species and marking location of rarities. It is important to note that this method will provide more detailed data but will also be more time consuming.

iii. Habitat sub-sampling: Some species occupy microhabitats and it may be worthwhile to conduct a more detailed sub-sampling of these sites even while using rapid survey methods especially for getting information on smaller species. Condit et al (1986) found rectangular quadrats contained 10% more species than square ones and very narrow quadrats (100mx1m) contained 18% more species than square quadrats in the same area.

iv. Time-restricted search: This is also known as rapid inventory or rapid biodiversity assessments (Crump & Scott, 1994) and here the habitat is searched for a set period of time (an hour or a day) and species recorded using expert field naturalists. A range of techniques are available but they need to be consistent so that the data is comparable. This method is a good one to compare sites that have little information. There is likely to be a variation in data between observers or teams here.

v.   Encounter rates: The most basic way of incorporating effort into abundance estimates is encounter rates. The total number of records are divided by the time surveyed (sometimes the reciprocal or minutes per individuals, is also used). Eg. If 300 person hours are used for 60 sightings, we get a value of 0.2 observations per hour. Measuring encounter rates is not considered to be a very good method but is certainly an improvement over the simple species list. The site is often classified into types so that the encounter rate can be calculated into each.

vi. Species discovery curves: This is essentially to know for how long the number of species will go on increasing with an increase in the survey time. A simple method is to record the total number of species versus time spent in the field. This is called the species discovery curve. The species discovery curve shows the point at which further effort is unlikely to reveal further species. Goff et al (1982) describe a similar approach for survey of the plants where they stop surveying once 30 minutes have been spent without finding a new species.

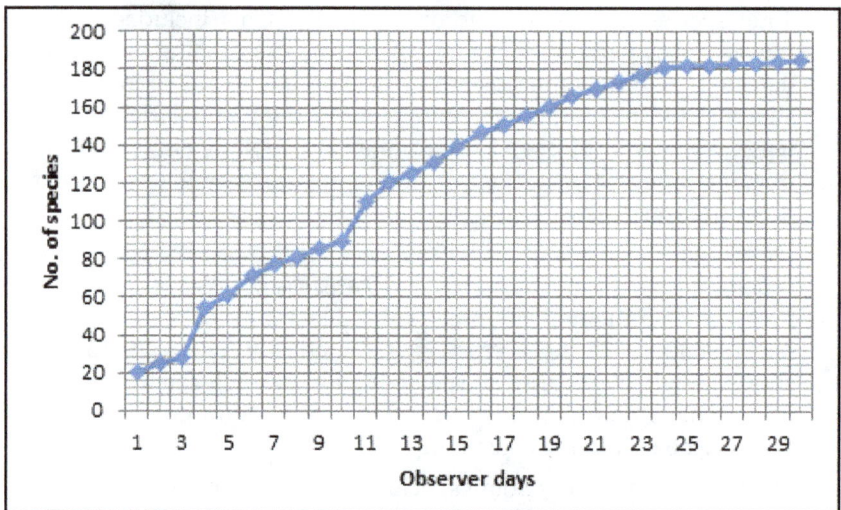

Figure 2: Species area curve for birds in a 5 sq km wetland

vii. MacKinnon Lists: MacKinnon lists (MacKinnon and Phillips, 1993) includes a series of surveys, each of the individual surveys stopping until a fixed number of species (say 20) are found. The next survey finds a fresh new list. The cumulative number of species observed from all counts combined is then plotted against the counts. Sites with high biodiversity will have a higher cumulative species total after a given number of counts. Again this does not indicate the maximum number of species that would be recorded but will show when further counts are unlikely to produce further species. This method too underestimates inconspicuous and gregarious species because a group counts only once (Robertson and Liley, 1998). More rigorous analysis of these curves is complex (Soberon and Llorente, 1993).

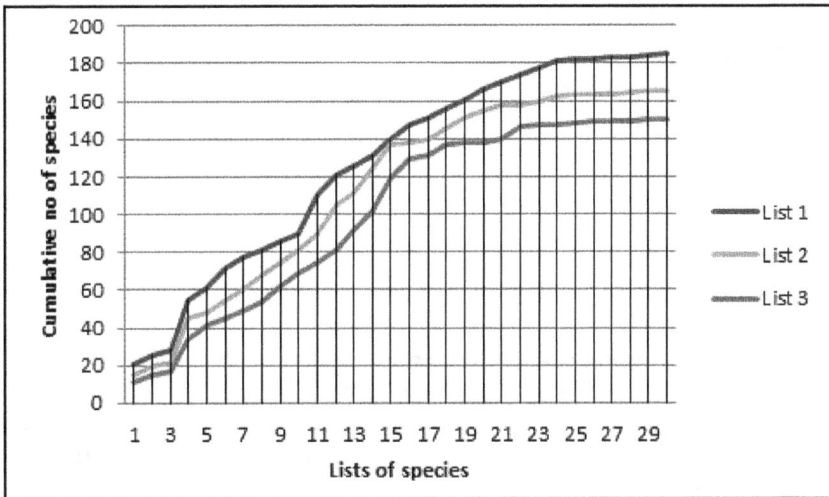

Figure 3: MacKinnon curves plotting the total number of bird species against the number of 30 bird species for three sites.

viii. Timed Species counts: This method takes advantage of the fact that common species are likely to be seen first soon after starting a survey while rare species, if seen, are likely to be seen in the end. The method involves dividing a one hour distribution period into six 10 minute blocks (using the alarm function of the watch/mobile phone). A list of species is made for every 10 minute slot. Once a species is recorded for the first instance, it is ignored in later time slots during the hour. Those which are recorded in the first 10 minutes are given a score of 6, those in the next 5 and so on. These values are averaged over a number of census periods (usually 10-15) (Pomeroy and Tengecho, 1986). Just as with MacKinnon lists, the scores underestimate inconspicuous and gregarious species. One disadvantage for this method is that this method has been rarely used and therefore there is little data for comparison (Sutherland, 2000).

ix. Recording absence: Absence is ironically harder to record than presence and needs more detailed surveys. Reed (1996) used the concept of statistical power to calculate in theory the necessary number of visits, N

$$N = \frac{\ln (\alpha \text{ level})}{\ln (1\text{-}P)}$$

where, P is the probability of detecting and species on a given visit and α level is the acceptable risk that the species is present but recorded as extinct. Of course, the ideal is to be absolutely positive that the species is extinct (α level = 0) but this is impossible. As is obvious more number of species is needed for elusive species.

x. Habitat feature assessment: This method involves bringing together a range of experts to get information on important considerations for site quality. Eg. Altitude, sub-surface water and soil type for grasslands or presence of caves, den sites and fruiting trees for bears. It is to be noted that this method is for an initial survey to identify areas which warrant a more detailed survey e.g. for studying impacts of planned linear structures (rail, roads, electric lines) on an forest tract.

One of the key tools that can be is Remote Sensing and GIS. utilized on larger tracts The amount of change that is occurring in tropical parts of the World has been of considerable interest in the past ten years. Remote sensing offers perhaps the only practical method of analyzing large areas over time. Green and Sussmann (1990) used a combination of aerial photography, forest maps, and satellite images to estimate deforestation rates in Madagascar from 1950 to 1985, spanning a total of 35 years. With the advent of availability of satellite remote-sensing data, several countries have recently launched temporal monitoring of forest cover, which facilitates analyzing biodiversity losses. Prediction of the spatial distribution and relative abundance of wildlife on the basis of multi-temporal satellite data and simulation models is also a recent development, Coops and Catling (2002) extensively reviewed such approaches.

Discussion

In terms of global trends, measuring progress towards the global target of reducing the rate of biodiversity loss by 2010 relies on monitoring species abundance, threat of extinction, extent and condition of habitats, and ecosystem goods and services (Dobson, 2005). Biodiversity is threatened by modern human activities (Hooper et al., 2005). The current extinction crisis is one of the most significant in earth's history, with habitat loss, spread of non-native species, and global climate change the greatest threats (Wilcove et al., 1998; Chapin et al., 2000). Maintenance of biodiversity is important as its erosion will result in less stable ecosystems with reduced function (Naeem et al., 1994, 1995; Tilman et al., 1996; Stachowicz et al., 1999). Reduced function and stability eventually lead to greater uncertainty in ecosystem services, including a number critical for human welfare (Costanza et al., 1997; Millennium Ecosystem Assessment, 2005). The value of such services is substantial, with global natural capital estimated at \$33 trillion (US) per year in 1997; nearly double the global gross national product (Costanza et al., 1997).

Given economic values and social-ethical concerns, governments, organizations, and scientists have attempted to quantify the 'state' of biodiversity by assessing status and trends, setting targets for mitigating biodiversity loss, and/or identifying hot spots for biodiversity protection (Dobson et al., 1996, 2001; Myers et al., 2000; Weber et al., 2004; Scholes and Biggs, 2005). Despite the need for consistency in monitoring programmes, no single method of measuring or reporting biodiversity has emerged (Purvis and Hector, 2000). When biodiversity is measured and reported, it is not always evident what benchmark to use for comparison and indexing (Allen et al., 2003). Three general approaches have been used: (1) desired goal or target; (2) time-zero; and (3) protected areas. In desired goal or target, expert opinion or social values determine reference (benchmark) conditions. Floristic quality assessments, for instance, have been used to assess ecological integrity of the Midwestern USA (Herman et al., 1997; Taft et al., 1997) using prior assignment of coefficients of conservation for each species. Such assignments are impractical when dealing with hundreds to thousands of species necessary to inform biodiversity and for taxonomic groups about which little knowledge exists. Moreover, additional quantitative information, such as relative abundance (density, percent cover, etc.), is not fully considered. As an alternative to desired states, time zero referencing has been suggested. Here, a point in time is selected (normally the start of the monitoring programme) to compare and index against current conditions. The Living Planet Index uses 1970 as a benchmark to report on the state of the planet's ecosystems and species (Loh et al., 2005). Without a sufficiently distant past, time zero references fail to fully inform conservation-based boundaries for restoration and status assessments. Local areas within many ecosystems were already highly degraded in the year 1970.

Furthermore, comparisons between monitoring programmes are compromised unless year of time zero and level of degradation are similar. Protected areas have also been used as comparison benchmarks. Sites of interest are compared against 'natural' or 'intact' reference sites, such as national parks (Mayer and Galatowitsch, 2001; Sinclair et al., 2002; Scholes and Biggs, 2005). Existing protected areas do not always contain a representative sample of biodiversity (Scott et al., 2001; Hansen and Rotella, 2002), since they often occur in remote high elevation areas lacking the potential for cultivation (Margules and Pressey, 2000; Scott et al., 2001). Without controlling for environmental gradients, differences among target and control areas can be solely due to natural patterns in species distributions, rather than anthropogenic influence. Furthermore, protected areas are being degraded over time by human activity resulting in sliding benchmarks. We propose a fourth alternative for calculating benchmarks and biodiversity intactness. By estimating empirical relationships between species occurrence/abundance and human footprint we are able to estimate reference conditions under a pristine situation. These statistically-derived reference conditions are then compared to current species occurrence and abundance to index intactness. Deviation from reference (decreasing sensitive species or increasing non-native species) results in loss of intactness. With species as the basic unit of measure, numerous levels of organization can be reported (i.e., guilds, taxonomic group, or overall biodiversity). We demonstrate the utility of the approach using winter mammal monitoring data collected from the boreal forest of Alberta, Canada.

Conclusion

Biodiversity assessment and monitoring are critical to any natural resource conservation and management programme. This is not a static but a dynamic one and therefore, it is essential that the data is collected using appropriately selected based on the requirements and clearly stated. It is also essential to understand the pros and cons of each method for instance if there is a time constraint and the data to be collected is humungous then rapid survey methods are very useful although the precision in this case is not too high. However, where detailed data is required e.g. survey of small species that are sensitive and inhabit special microhabitats. In this context 10 assessment techniques have been described in the chapter including Total species list/genus/family list, Parallel line searches, habitat sub-sampling, Time-restricted search, Encounter rates, Species discovery curves, MacKinnon lists, Timed species counts, recording absences and Habitat feature assessments.

References

1. Block, W.M., Brennan, L.A. and Gutierrez, R.J. (1986). Evaluation of guild indicator species resource management for use in single. *Environmental Management*. 11: 265-9.

2. Chapin III., F.S., Zavaleta, E.S., Eviner, V.T., Naylor, R.L., Vitousek, P.M., Reynolds, H.L., Hooper, D.U., Lavorel, S., Sala, O.E., Hobbie, S.E., Mack, M.C., Dı́az, S. (2000). Consequences of changing biodiversity. *Nature*. 405: 234–242.

3. Costanza, R., d'Arge, R., de Groot, R., Farber, S., Grasso, M., Hannon, B., Limburg, K., Naeem, S., O'Niell, R.V., Paruelo, J., Raskin, R.G., Sutton, P., van den Belt, M. (1997). The value of the world's ecosystem services and natural capital. *Nature*. 387: 253–260.

4. Crump, M.L and Scott., N.J. 1994. Visual Encounter Surveys. In: Heyer, W.R. Donnelly, M.A. McDiarmid, R.W., Hayek, L.C., and Foster. M. S. (sds. Measuring and Monitoring Biological Diversity. Standard method for amphibians. Washington: Smithsonian Institution Press. 84-92.

5. Condit, R., Hubbel, S. P. and La Frankie, J.V. (1996). Species-area and species-individual relationship for a tropical tree. *Journal of Ecology*. 84: 549-62.

6. Danielson, F., Danilo S. B., Michael, K. P., Enghoff, E., Nozawa, C.M. and Jensen, A. E. (2007). A simple system for monitoring biodiversity in protected areas of a developing country.

7. Dobson, A.P. (1996). Conservation and Biodiversity. New York. *Scientific Publication*.

8. Dobson, A.P. (2005). Monitoring global rates of biodiversity change: challenges that arise in meeting the Convention on Biological Diversity (CBD) 2010 goals. Philosophical Transactions of the Royal Society B-Biological Sciences. 360: 229-241.

9. Finlayson, C.M. and D'Cruz, R. (CLAs). (2005). Inland Water Systems. Chapter 20. In *Ecosystems and Human Well-being: Current Status and Trends*. Millennium Ecosystem Assessment. Island Press, Washington, DC.

10. Gaston, K.J. (1986). Biodiversity-congruence. *Process in physical Geography*. 20: 105-12.

11. Gaston, K.J. (1986). Species: measure and measurement. In: Gaston K.J. (ed.) Biodiversity: biology of numbers and difference. Oxford: Blackwells Science Ltd.

12. Goff, F.G., Dawson, G. A. and Rochow, J.J. (1982). Site examination for threatened and endangered plant species. *Environmental Management*. 6(307-16): 76-98.

13. Hansen, A.J. and Rotella, J.J. (2002). Biophysical factors, land use, and species viability in and around nature reserves. *Conservation Biology*. 16: 1112–1122.

14. Herman, K.D., Masters, L.A., Penskar, M.R., Reznicek, A.A., Wilhelm, G.S., Brodowicz, W.W. (1997). Floristic quality assessment: Development and application in the state of Michigan (USA). *Natural Areas Journal*. 17: 265–279.

15. Hooper, D.U., Chapin, F.S., Ewel, J.J., Hector, A., Inchausti, P., Lavorel, S., Lawton, J.H., Lodge, D.M., Loreau, M., Naeem, S., Schmid, B., Setala, H., Symstad, A.J., Vandermeer, J., Wardle, D.A., 2005. Effects of biodiversity on ecosystem functioning: A consensus of current knowledge. *Ecological Monographs*. 75: 3–35.

16. Huston, M.A. (1994). Biological diversity: The coexistence of species on changing landscapes. Cambridge University Press, Cambridge, United Kingdom.

17. Lindenmayer, D.B., Franklin, J.F. and Fisher, J. (2006). General management principles and a checklist of strategies to guide forest biodiversity conservation. *Biological Conservation*. 131: 433–445.

18. Loh, J., Green, R.E., Ricketts, T., Lamoreux, J., Jenkins, M., Kapos, V. and Randers, J. (2005). The Living Planet Index: using species population time series to track trends in biodiversity. Philosophical Transactions of the Royal Society B: *Biological Sciences*. 360: 289–295.

19. Margules, C.R. and Pressey, R.L. (2000). Systematic conservation planning. *Nature*. 405: 243–253.

20. MacKinnon, J. and Phillips, K. (1993). A field guide to the birds of Sumatra, Java and Bali. Oxford: Oxford University Press.

21. Mayer, P.M. and Galatowitsch, S.M. (2001). Assessing ecosystem integrity of restored prairie wetlands from species production diversity relationships. *Hydrobiologia*. 443: 177–185.

22. Millennium Ecosystem Assessment. (2005). Ecosystems and Human Well-being: Biodiversity Synthesis. World Resources Institute, Washington, DC.

23. Myers, N., Mittermeier, R.A., Mittermeier, C.G., da Fonseca, G.A.B. and Kent, J. (2000). Biodiversity hotspots for conservation priorities. *Nature*. 403: 853–858.

24. Naeem, S., Thompson, L.J., Lawler, S.P., Lawton, J.H. and Woodfin, R.M. (1994). Declining biodiversity can alter the performance of ecosystems. *Nature*. 368: 734–737.

25. Naeem, S., Thompson, L.J., Lawler, S.P., Lawton, J.H. and Woodfin, R.M. (1995). Empirical evidence that declining species diversity may alter the performance of terrestrial ecosystems. *Philosophical Transactions of the Royal Society of London B.* 347: 249–262.

26. Nelson, J. R. (1987). Rare plant surveys: Techniques for impact assessment. In: Elias, T.S. (ed.) Conservation and management of rare and endangered plants. *California: California Native Plant Society.* 155-66.

27. Nielsen, S.E., E.M. Baynea, E.M., Schieckb, J. Herbersa, J. and Boutina, S. (2007). A new method to estimate species and biodiversity intactness using empirically derived reference conditions.

28. Noss, R.F. (1990). Indicators for monitoring biodiversity: A hierarchal approach. *Conservation Biology.* 4: 355–364.

29. Olson, D.M., Dinerstein, E., Wikramanayake, E.D., Burgess, N.D., Powell, G.V.N., Underwood, E.C., D'Amico, J.A., Itoua, I., Strand, H.E., Morrison, J.C., Loucks, C.J., Allnutt, T.F., Ricketts, T.H., Kura, Y., Lamoreux, J.F., Wettengel, W.W., Hedao, P. and Kassem, K.R. (2001). Terrestrial ecoregions of the world: a new map of life on earth. In *BioScience.* 51:933-8.

30. Pomeroy, D. and Tengecho, B. (1986). Studies of birds on semi-arid area of Kenya III – the use of 'time-species counts' for studying regional avifaunas. *Journal of Tropical Ecology.* 2: 231-47.

31. Purvis, A. and Hector, A. (2000). Getting the measure of biodiversity. *Nature.* 405: 212–219.

32. Reed, J.M. (1996). Using statistical probability to increase confidence of inferring species extinction. *Conservation biology.* 10: 1283-95.

33. Revenga, C., Brunner, J., Henninger, N., Kassem, K. and Payne, R. (2000). *Pilot Analysis of Global Ecosystems: Freshwater Systems.* World Resources Institute, Washington, DC.

34. Robertson, P. and Liley, D. (1998). Assessment of sites – measurement of species richness and diversity. In: Bibby. C., Jones. M, and marsden, S. (eds.) Bird surveying and conservation. London: Royal Geographical Society.

35. Roy, P.S. and Tomar, S. (2000). Biodiversity characterization at landscape level using geospatial modelling technique. *Biological Conservation.* 95: 95–109.

36. Scholes, R.J. and Biggs, R. (2005). A biodiversity intactness index. *Nature.* 434: 45–49.

37. Scott, J.M., Davis, F.W., McGhie, R.G., Wright, R.G., Groves, C. and Estes, J. (2001). Nature reserves: Do they capture the full range of America's biological diversity? *Ecological Applications.* 11: 999–1007.

38. Sinclair, A.R.E., Mduma, S.A.R. and Arcese, P. (2002). Protected areas as biodiversity benchmarks for human impact: agriculture and the Serengeti avifauna. Proceedings of the Royal Society of London Series B – *Biological Sciences.* 269: 2401–2405.

39. Smith, P.G.R. and Theberge, J.B. (1986). A review of criteria for evaluating natural areas. *Environmental management*. 10: 915-34.

40. Soberon, J.M. and Llorente, J.B. (1993). The use of species accumulation functions for the prediction of species richness. *Conservation Biology*. 7: 480-8.

41. Soule, M.E. (1986). Conservation Biology: The Science of Scarcity and Diversity. Sunderland, MA.: Sinauer Associates. A classic collection of papers focusing on the conservation of biodiversity.

42. Stachowicz, J.J., Whitlatch, R.B. and Osman, R.W. (1999). Species diversity and invasion resistance in a marine ecosystem. *Science*. 286: 1577–1579.

43. Sutherland, William J. The Conservation Handbook – Research, Management and Policy

44. Usher, M.B. (1986). Wildlife Conservation Evaluation: attributes, criteria and values. In: Usher, M.B. (ed.) *Wildlife Conservation Evaluation London: Chapman and Hall*. 3-44.

45. Taft, J.B., Wilhelm, G.S., Ladd, D.M. and Masters, L.A. (1997). Floristic Quality Assessment for Vegetation in Illinois, A Method for Assessing Vegetation Integrity. *Erigenia*. 15: 3–95.

46. Tilman, D., Wedin, D. and Knops, J. (1996). Productivity and sustainability influenced by biodiversity in grassland ecosystems. *Nature*. 379: 718–720.

47. Watson-Jones, S.J., Maxted, N. and Ford-Lloyd, B.V. (2006). Population baseline data for monitoring genetic diversity loss for 2010: A case study for Brassica species in the UK. *Biological Conservation*. 132: 490–499.

48. Weber, D., Hintermann, U. and Zangger, A. (2004). Scale and trends in species richness: Considerations for monitoring biological diversity for political purposes. *Global Ecology and Biogeography*. 13: 97–104.

49. Wilcove, D.S., Rothstein, D., Dubow, J., Phillips, A. and Losos, E. (1998). Quantifying threats to imperiled species in the United States. *BioScience*. 48: 607–615.

50. Wilhelm, G.S. and Masters, L.A. (1995). Floristic Quality Assessment in the Chicago Region and Application Computer Programs, Morton Arboretum, Lisle, IL. 17 Appendices.

51. Young, W.J., Chessman, B.C., Erskine, W.D., Raadik, T.A., Wimbush, D.J., Tilleard, J., Jakeman, A.J., Varley, I. and Verhoeven, T.J. (2004). Improving expert panel assessments through the use of a composite river condition index-The case of the rivers affected by the snowy mountains hydro-electric scheme, Australia. *River Research and Applications*. 20: 733–750.

52. Green, G.M. and Sussmann, R.W. (1990). Deforestation history of the Eastern rain forests of Madagascar from satellite images. *Science*. 248: 212-215.

53. Coops, N.C. and Catling, P.C. (2002). Prediction of the spatial distribution and relative abundance of ground-dwelling mammals using remote sensing imagery and simulation models. *Land. Ecol*. 17: 173188.

13

Threats to Biodivesrity

Disha Jaggi* and Bhumesh Kumar*

Directorate of Weed Research, Jabalpur-482004, Madhya Pradesh, India
E-mail: kumarbhumesh@yahoo.com, disha.jaggi@gmail.com

Abstract

Since the beginning of industrial era, anthropogenic activities and other natural causes had threatened the natural diversity worldwide. Major factors responsible for the decline of biodiversity can be listed as the destruction and fragmentation of natural habitats; climate change; increase in level of pollution caused by agricultural and industrial practices; excessive exploitation of natural resources; and introduction of invasive species due to liberal trade and tourism. Existence of some species can be easily threatened by a combination of several of factors. In addition, modern agricultural practices also contributed towards loss of biodiversity. In many parts of the world, lack of stringent quarantine and preventive laws aggravated the problem due to invasion by alien species leading to the loss of biodiversity. At the current rate of biodiversity and ecosystems loss has attracted the attention of many ecologists. If not all, at least losses to the biodiversity driven by human actions must be minimized. An urgent focus and co-ordination is required at local, regional, national and global level.

Key words: Alien species, Climate change, Habitat, Hotspots, Invasion

Introduction
Biodiversity

Biodiversity consists of two words "biological" and "diversity." It refers as the variety of the planet's living organisms and their interactions. Biodiversity is not only the sum of all ecosystems, species and genetic material. Rather, it represents all of life's variation, expressed in genes, individuals, populations, species, communities and ecosystems.

Quantitative measures of biodiversity most often focus on a taxonomic unit, typically the species, although aspects of ecological diversity can also be measured. Biodiversity is a dynamic entity, and has changed throughout the history of life on Earth. Thus, biodiversity is the variation of taxonomic life forms within a given ecosystem, biome or for the entire Earth. It is often used as a measure of the health of biological systems. The term Biodiversity/biological diversity was first coined by Walter Rosen in the 1986 Forum on Biodiversity (Wilson, 1988). The 1992 United Nations Earth Summit in Rio de Janeiro defined 'biodiversity' as "the variability among living organisms from all sources, including, 'inter alia', terrestrial, marine and other aquatic ecosystems, and the ecological complexes of which they are part: this includes diversity within species, between species and of ecosystems" (UNEP, 1992). This come closest thing to a single legally accepted definition of biodiversity and also the definition adopted by the United Nations Convention on Biological Diversity (CBD). The concept of biodiversity involves an "understanding that all organisms interact like a web of life with every other element in their local environment" (SCBD, 2010).

An estimated 1.7 million species have been described to date although estimates for the total number of species existing on earth at present vary from five million to nearly 100 million. Biodiversity is not distributed evenly on Earth and it is consistently richer in the tropics and in other localized regions. Forests are more biologically diverse than any other land-based ecosystem, and contain more than two-thirds of the world's terrestrial species (ibid). Figure 1 represents the extent of global biodiversity conducted by Botanical Survey of India.

Biodiversity Hotspots

Biodiversity hotspots are a method to identify those regions of the world where attention is needed to address biodiversity loss and to guide investments in conservation. The idea was first developed by Norman Myers in 1988 to identify tropical forest 'hotspots' characterized both by exceptional levels of plant endemism and serious habitat loss. Conservation International adopted Myers' hotspots as its institutional blueprint in 1989, and in 1999, the organization undertook an extensive global review which introduced quantitative thresholds for the designation of biodiversity hotspots. A re-working on the hotspots analysis in 2004 resulted a system in place today. First time, 25 global biodiversity hotspots were demarcated in different parts of the world in the year 2000 (Myers *et al.*, 2000), Later, in 2009, another 9 hotspots were added (http://www.conservation. org) based on the criteria of exceptional concentration of endemic plants and higher degree of anthropogenic pressure, most of which occur in tropical forests (Figure 2). These hotspots together represent just 2.3% of Earth's land surface, but contain around 50% of the world's endemic plant species and 42% of all terrestrial vertebrates. Overall, hotspots have lost around 86% of their original habitat and are considered to be significantly threatened by extinctions induced by climate change (Mittermeier *et al.*, 2011).

India is situated north of the equator between 66°E to 98°E and 8°N to 36°N. The varied edaphic, climatic and topographic conditions have resulted in a wide range of ecosystems and habitats such as forests, grasslands, wetlands, coastal and marine ecosystems, and deserts. India represents: (i) two realms-the Himalayan region represented by 'Palearctic Realm' and the rest of the sub-continent represented by 'Malayan Realm'; (ii) five biomes' e.g. Tropical Humid Forests, Tropical Dry Deciduous Forests (including Monsoon Forests), Warm Deserts and Semi-deserts, Coniferous Forests, Alpine Meadows; and (iii) Ten biogeographic zones and 27 biogeographic provinces (Khandekar and Srivastava, 2014).

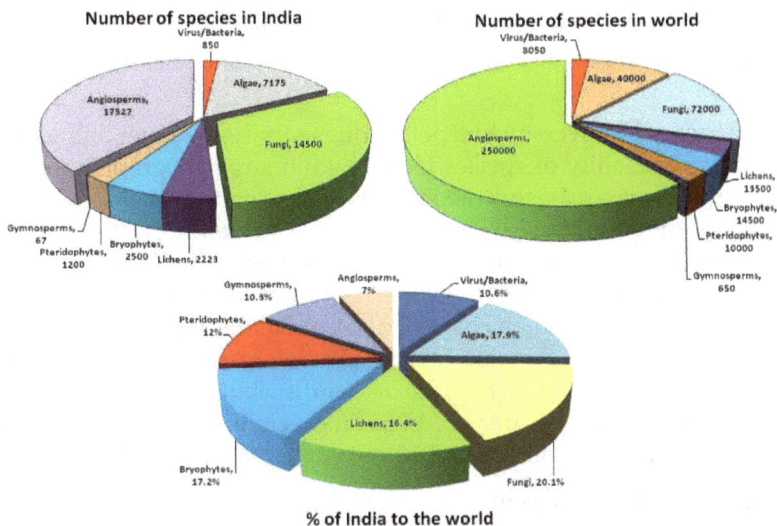

Figure 1: Extent of diversity in major groups of plants and microorganisms at global level (modified from BSI, 2009).

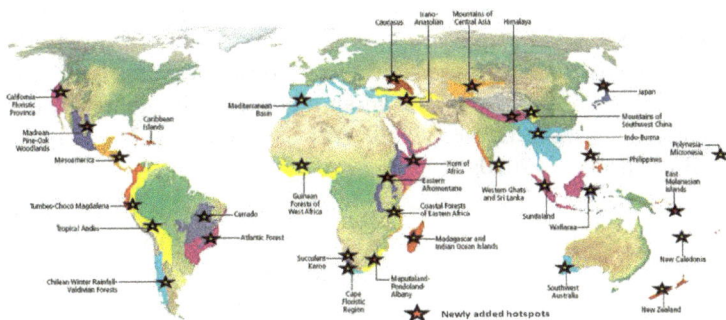

Figure 2: Global biodiversity hotspots (adopted and modified from source: www. nature.com)

Indian region has over 130,000 species of plants and animals which have been scientifically documented. The country has been referred to as one of the top mega diversity region of the globe with only 2.5% of the global land area. India accommodates parts of four global biodiversity hotspots – the Himalaya, the Western Ghats, Indo-Burma and Sundaland, which are facing challenges due to anthropogenic disturbance and climate change. The richness of the biodiversity of the regions is largely due to the occurrence of large diversity among species, genetic and ecological variability in different biogeographically defined zones. Nonetheless, there is no comprehensive study on these biodiversity hotspots regarding current status of vegetation cover and species richness (Chitale *et al.*, 2015).

Importance of biodiversity hotspots is due to the high vulnerability of habitats and high irreplaceability of species found within large geographic regions. This envisages that these areas and the species present within them are both under high levels of threat and are of immense global value due to their uniqueness. Therefore, rigorous and precise assessments of biodiversity of these hotspots is an urgent need to prevent further biodiversity loss within these areas. This is a global scale approach based on coarse scale eco-regions hence, has limited use for site-scale assessment and decision making. Therefore, more detailed assessments are needed to point out impact on the actual distribution of biodiversity within particular area(s) including areas of high biodiversity as well as degraded land and urban areas (Mittermeier *et al.*, 2011).

Major Threats to Biodiversity

Any direct or indirect human activity that adversely affects the natural diversity in the form of populations, species, gene pool, ecosystems, or other levels of biological organization is considered a threat to sustainability of biodiversity. Biodiversity is under serious threat as a result of human activities. Threats to species are principally due to human interferences with nature. At present, biodiversity is affected by multiple drivers and pressures that modify its ability to provide ecosystem services to people. The principal pressures on biodiversity include habitat loss, habitat fragmentation and degradation, extinction of species, overexploitation of natural resources. Impact of these pressures further aggravated to due to introduction of invasive alien species, climate change, pollution, population and diseases as depicted in Figure 3 (MA, 2005a; Vie *et al.*, 2009; Baillie *et al.*, 2010; Butchart *et al.*, 2010; CBD, 2010b).

Biodiversity losses can be attributed to the resource demands of our rapidly growing human population. The human population has increased from about 1 billion in 1900 to almost 6 billion today. Like other living beings, we use natural resources to survive, but we occupied much greater share of natural resources (food, water, space) had been destructive to other life-forms than any species previously known. As a result, there is less and less natural habitat remaining as land is developed for human habitation and activities.

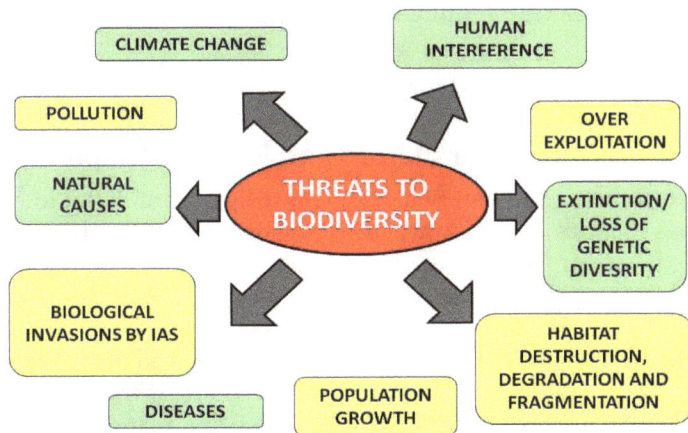

Figure 3: Principal threats to biodiversity

The negative impacts of our actions have become to the extent that we are losing biodiversity more quickly now than at any other time in Earth's recent history (Table 1). Scientists have assessed more than 47000 species and found that 36 percent of these are threatened with extinction. In addition, extinction rates are estimated to be between 50 and 500 times higher than those observed from fossil records or the so-called "background rate". The current rate of biodiversity loss has led many to suggest that the Earth is currently experiencing a sixth major extinction event, the one greater than that which resulted in the extinction of the dinosaurs. However, unlike past extinction events, which were caused by natural disasters and planetary changes, this one is being driven by human activities. In the last century, the Indian cheetah, lesser Indian rhino, pink-headed duck, and the Himalayan mountain quail are reported to have become extinct and several other species (39 mammals, 72 birds and 1,336 plants) have been identified as vulnerable or endangered (Khandekar and Srivastava, 2014).

The constraints and challenges to biodiversity conservation include: biodiversity information base; implementation of Biological Diversity Act and safeguarding traditional knowledge; new and emerging biotechnologies; economic valuation and natural resource accounting; policy, legal and administrative measures; and institutional support. The cold desert areas lie in the Trans-Himalayan zone and some species in the region those are endemic to Tibetan plateau and also include oasitic elements that comprises a variety of exotic as well as indigenous species. The area represents common herbaceous, shrubby and woody elements of temperate vegetation and alpine species and also harbours rare and endangered fauna and avifauna endemic to the region which contributes to its uniqueness. Livestock rearing, agricultural and horticultural practices and mode of agro-forestry are entirely different and the people living here have succeeded in developing their own distinguish culture (Khandekar and Srivastava, 2014). This region is unique and requires special attention and hence, a challenging avenue for biologists and conservationists. Similarly, impacts climate change on biodiversity have to be kept at low and potential mitigation strategies needed to be adopted which otherwise accelerating species extinctions, loss of natural habitat, and changes in the distribution and abundance of species and biomes over the 21st century (Pereira et al., 2010a).

Habitat Destruction, Degradation and Fragmentation

Habitat destruction is identified as the most significant threat to biodiversity. Habitat loss occurs when natural environments are transformed or modified to in order to cater human needs. Under diverse natural conditions, about 1.3 billion people in rural and urban areas live in harmony under a democratic system in India. Pressing needs for food, fibre, shelter, fuel and fodder for such a huge population combined with the growing need for infrastructure and economic development exert enormous pressure on natural resources. The loss and fragmentation of natural habitats affects all animal and plant species of a region. We need to not only stop any further habitat loss immediately but also to restore a substantial fraction of the wilderness that has been depleted in the past. Various species of plants and animals are on the decline due to habitat fragmentation and overexploitation of natural resources. Examples of such decline include habitats of Great Indian Bustard in Madhya Pradesh, Gujarat and Rajasthan, and of the Lion-tailed Macaque in Western Ghats (Khandekar and Srivastava, 2014).

Habitat loss in the plains has been caused largely due to efforts made for the expansion of agriculture. More than 30 per cent of land has been converted for agricultural production (Foley et al., 2011). Large-scale commercial agriculture involving cash crops and mono-cropping pattern has adversely affected biodiversity, particularly agro-biodiversity (Rosset, 1999). Moreover, the growing demand fuel energy has taken a toll at the expanses of forests. Natural lands in South East Asia being converted into mono-crop species for biofuels production (Danielsen et al., 2009; Fitzherbert et al., 2008). Direct habitat loss is also a major threat to coastal ecosystems through aquaculture. Wetlands in particular have faced a 50 per cent loss in the 20th century (MA, 2005b). Freshwater ecosystems are severely affected by fragmentation (Nilsson et al., 2005). In addition, floodplain ecosystems are also threatened (Tockner et al., 2008). Benthic habitats have been degraded as a consequence of bottom trawling and other destructive fishing methods. Though, some habitat loss is unavoidable to meet human needs, when natural habitats are changed or modified with little concern for biodiversity the results can be very negative and unfavorable.

One of the best examples is disappearance of common birds due to depleting green cover. Fruiting, flowering and nesting are all related events and due to excessive cutting of trees for human need and other commercial activities, birds have also started fleeing. Not only sparrows, it's difficult to spot hornbills, green pigeons and purple sunbirds which have migrated to the outskirts of the city and nearby villages where they can nest and feed more easily than in the urban areas. Traditional norms and practices for conservation of neighbourhood forest and common land are also diminishing. Habitat fragmentation is also one of the primary reasons leading to cases of human and wild life conflict. Common property resources like pastures and village forests, which served as a buffer between wildlife habitat and agriculture, have been gradually converted into agricultural fields, social entities like playgrounds and habitation. Due to this, the villagers are brought into a direct conflict with wild animals. Some examples of man-animals conflicts are frequent entries of leopards, elephants, tigers, monkeys, blue-bulls, wild boars and certain birds which sometime harm the local people to a great extent.

Overexploitation

Overexploitation of natural resources to meet growing demands of huge population threatens biodiversity and contributed to declines in terrestrial, marine and freshwater ecosystems (Vorosmarty *et al.*, 2010). India is rich with diverse forest types ranging from the tropical wet evergreen forests in North-East to the tropical thorn forests in the Central and Western India. Forests of the country can be divided into 5 major groups and further divided in subgroups and forest types based on climatic factors and species composition.

Forests face threats on account of land use conversion for agriculture, industry, human settlements, and other developmental plans. Growing needs for construction of roads and canals and other type of shifting cultivation and encroachments are other threats which contributing towards the shrinking forest areas. Degradation of forests also occurs from illegal cutting, excess removal of forest products, fodder, fuelwood, forest floor litter, overgrazing and forest fires. As a result, some of the funa and flora of the forest, sometime including many endemic forest species are now left below the critical threshold with a very with narrow populations, hence, warrant immediate attention of the conservationists and ecologists. The rich diversity of medicinal plants in India needs conservation with a stringent check over their utilization as their habitats are either degraded or the species are being overexploited for commercial purposes. About 90% of the medicinal plants being trade commercially are harvested from the forests areas posing a threat of extinction of many species.

Extinction and Gentetic Diversity Loss

Loss of habitats and over exploitation has led to depletion of genetic diversity in both faunal and floristic components. Narrowing genetic diversity leads to more vulnerability to diseases and pests and lesser adaptability to environmental changes. Such trend has emerged from the world-wide experience of drastically curtailed genetic diversity in agriculture after the farming revolutions (green and white) in agriculture-based economies, and more so in India. Need of conserving the large animal species (such as the lion, tiger, rhino and elephant) point out towards significant loss of biodiversity in India. Despite of partially success, such projects should also aim at broadening the genetic base (gene pool) for breeding purposes besides focusing on habitat protection which may be the decisive factor in saving critically endangered species and maintaining a minimum genetic base of inter-mating individuals rather than their total number.

Table 1: Various threats and their impacts on biodiversity

THREATS	MODE OF THREATNING/ EXAMPLES	IMPACTS
Human interference with Ecosystem	• Deforestation • Unlimited urbanization and industrialization • Alteration of vegetation • Habitat loss, habitat degradation and Fragmentation • Alteration of hydrology • Foreign exchange • Biotechnological methodologies • Population growth	• Change or Complete loss of ecosystem function or structure
Overexploitation, Extinction/ shrinking genetic diversity	• Excessive use of natural resources • decline of natural flora; leads to decrease in faunal diversity • loss of original genetic pool from ecosystem • Diseases • Pollution	• Increased heath risks or direct mortality of individuals which may result in decreases in population size and stability. • Loss of natural resources, vegetation and other related diversity.
Invasive Alien Species	• Competition • Allelopathy • Hybridization • Contamination of original genetic diversity • Introduced Pathogens • Blockage of water bodies/ways	• Change or loss of native diversity • Loss of productive lands • Reproductive impacts on populations • Proliferation of new diseases • Loss in crop yields • death of underwater species

THREATS	MODE OF THREATNING/ EXAMPLES	IMPACTS
Pollution	• Agricultural pollution • Air pollution • Water pollution • Chemical pollution • Soil pollution • Acid precipitation • Green house effect • non-renewable resources/ industrial waste • Contamination of ecosystem, habitats, and species from industrial by-products	• Change in ecological processes, green house effect. Change in weather patterns, health issues • Developmental, or reproductive impacts on individuals, populations, or species
Enhanced Climate Change	• Unpredicted weather patterns • Alteration of hydrological cycle and ice regimes • Extreme variation in temperature • Changes in precipitation (e.g. drought, flooding)	• Unexpected losses in agriculture • Change in some component of ecosystem function/ processes, structure, or composition • Shifting or loss of ecosystems
Natural Causes	• Natural outbreaks (e.g., wildfire, insect outbreaks, floods, geologic events)	• Change of some component of function, structure, or composition • Collapse of population

Another example of shrinking genetic base can be seen from the fact that out of 150 crops which provide food for most of the global population at present, just 12 of these provide 80% of food energy (with wheat, rice, maize and potato alone providing 60%). Out of about 30 mammalian and bird species which are used extensively for livestock products, just 15 of them account for about 90 per cent of global livestock products. The Indian scenario is not very different. Choice of crops and farm livestock in agricultural farming systems is now largely governed by market trends and changing lifestyles, affecting the variety, taste and nutrition value of our food basket (Khandekar and Srivastava, 2014), which sometime lead to the imbalanced dietary supplies creating situation of malnutrition.

Landraces grown traditionally by farming communities since long, locally adapted obsolete cultivars, and their wild relatives comprise crop genetic resources. These provide the building blocks used by frontline researchers and plant breeders as the raw material for breeding new plant varieties and also act as a reservoir of genes sought after for manipulation using latest gene technology tools for engineering of desirable plant types. Indigenous cultivars, adapted to local environments are, however, mostly low yielding (largely because of not receiving ample breeding effort) and are, hence, getting fast replaced by just a few high-yielding and pest-resistant superior varieties/hybrids under different crop. Narrowing genetic base has become a concern for plant breeders and alarm bells started ringing globally. Success of any breeding programme heavily relies on availability of a wide genetic base encompassing various traits which can be incorporated into plant types.

Many among the well known nearly 140 native breeds of farm livestock and poultry are also facing similar threat to their survival. Although the local breeds are more resilient to climatic stress, are more resistant to local parasites and diseases, and serve as a unique reservoir of genes for improving health and performance for modern commercial breeds. Conservation and proper utilization of locally adapted breeds will be most effective in achieving food and nutrition security objectives at global, regional as well as local level. Wild species, related closely to their domesticated forms, are valued by breeders as a source of genetic material for various abiotic and biotic stresses. Regular evolutionary development of these valuable species depends on adequate genetic diversity in their native populations. Increasing fragmentation, degradation and loss of their habitats over the years have seriously limited their availability and threatened their survival.

An assessment of necessity of plant genetic resources for food and agriculture clearly illustrates the need to conserve such resources as these contribute to people's livelihoods, food, medicine, feed for domestic animals, fibre, clothing, shelter, energy and a range of other products and services. India is remarkably rich in agriculturally important genetic resources. On the other hand, both the number of crops grown on commercial scale and the number of varieties being used under different agro-ecosystems have severely declined in last few decades. Such practices severely reduced the agricultural biodiversity at farm lands across the country.

Biological Invasions due to Alien Species

Among the major threats faced by native plant and animal species (and their habitats), the one posed by the invasive alien species (IAS) is truly alarming and is considered second only to that of the habitat loss. Invasive alien species are species that have spread outside of their native range and threaten biodiversity of new areas, are a major cause of biodiversity loss. These species are harmful to native biodiversity due to absence of predators, parasites, vectors (or carriers) of disease or direct competitors for habitat and food. In many cases predators are either not available in new environments at least at initial stages, or their population size is insufficient to check the growth and spread of such species (e.g. *Parthenium hysterophorus*). These may also cause severe economic, environmental damage, and may adversely affect human and animal health. The introduction of invasive alien species can be either intentional in the form of a new crop, ornamentals or livestock species; or accidental such as by transport means, contaminants of imported assignments and tourism activities. Unplanned economic introductions, surface, air and water transport, trade activities including trade in pets, garden plants and aquarium species some of the important pathways for the dispersal and spread of invasive species (Reise *et al.,* 2006). Invasive alien species affect native species mainly through predation, competition and habitat modification (McGeoch *et al.,* 2010; Vie *et al.,* 2009).

The major invasive alien species in India include *Parthenium hysterophorus, Lantana camara, Mikania micrantha, Eupatorium glandulosum, Ipomea carnea, Mimosa* species, *Eichhornia crassipes, Ulex enropaeus, Prosopis juliflora, Cytisus scoparius, Euphorbia royleana* etc. Alien aquatic weeds like water hyacinth, *Alternanthera* sp. and water lettuce are increasingly choking waterways and degrading freshwater ecosystems. *Lantana* and *Parthenium* cause major economic losses at global level and more so in Indian subcontinent. At present, *Parthenium* is distributed all over India and invading local diversity of native regions due to its highly competitive and allelopathic nature (Jaggi *et al.,* 2012, 2015). Climbers like *Chromolaena* and *Mikania* species are highly invasive and have shown their devastating impacts on the native vegetation in North-East Himalayan region and Western Ghats. Numerous pests and pathogens such as coffee berry borer, turnip stripe virus, banana bunchy top virus, potato wart and golden nematode have invaded agro-ecosystems becoming serious menace in certain areas. Himalayan Forest Research Institute, Shimla has identified some of the plant and insect species which though invasive have naturalized itself in the region thereby, posing a serious threat to the ecology of hilly regions. No ample attention to these invasive species had been paid over a period of time which had resulted into the present alarming situation. Several countries have now come over a single platform for fighting the menace caused by the invasive alien species and to devise potential strategies at global level. Accordingly special efforts are needed towards pest risk assessment of these invasive species. Surely, if we fail in this direction, it will lead to loss of endemic biodiversity on one hand and will expose the area to the exotic species on the other hand as suggested Khandekar and Srivastava (2014).

Diseases

Another threat to biodiversity caused by introduced species is the expansion of range of pathogens and parasites resulting from human activity. Habitat fragmentation can lead to high densities of a species build-up and cause a greater susceptibility to parasites and for disease to be transmitted. As wild species come into more contact with domestic animals there is a greater chance of spread of diseases. Disease can result from genetic disorders, pathogens such as viruses or bacteria, or parasites. Co-evolution of hosts and pathogens over evolutionary time results in co-existence of both host and pathogens. Diseases are often transmitted across different species, with the new host species often devastated by the new pathogen. For example, American chestnut (*Castanea dentata*) trees are wiped out due to the introduction of chestnut blight fungus (*Cryphonectria parasitica*) that had evolved in Asia with the closely related Chinese chestnut (*Castanea mollissima*). In addition, organisms that are affected by environmental contaminants, such as exposure to organochlorines, may play a role in lowering immune response and resistance to disease. Downing of immune functions resulting from contaminants or stress can potentially push populations or species at risk upto to the level of extinction. Wipe of wild black-footed ferrets (*Mustela nigripes*) due to canine distemper virus is an example from which lessons can be learnt. Introduced diseases are often more deadly as host–pathogen dynamics are usually the product of a long association which also involved the presence or absence of the inherent resistance against a given pathogens/parasite.

Pollution

Contamination of the natural environment is called pollution. Pollution can be in the form of liquids, solids, gases, or even forms of electromagnetic radiation input into air, water, or land. Since the beginning of industrial era, the input of organic and inorganic substances into the environment by humans has become a growing threat to biodiversity. In India, biodiversity is facing such threat from various sources of pollution. The major threats are from improper disposal of municipal solid waste, inadequate sewerage, changes life style, excessive use of agrochemicals and continuous use of hazardous chemicals even where non-hazardous alternatives are available. New industrial processes are generating a variety of toxic wastes, which cannot be dealt efficiently with by currently adopted approaches and technology in the country. Besides, economic constraints and problems related to the indigenization make the sustainable substitution of these technologies difficult and uneconomic to the end-users. Pollutants such as pesticides and fertilizer effluents from agriculture and forestry, industry including mining and oil or gas extraction, sewage plants, run-off from urban and suburban areas, and oil spills harm biodiversity directly either through mortality or by reduced reproductive success, and indirectly through habitat degradation/framentation (MA, 2005a).

Pollution can be acute, with a single incident, or chronic, with the gradual addition of substances to the environment over a continuous time period. Examples of acute environmental disasters include oil spills, refinery and shipping accidents, and nuclear accidents. Although the initial effects of these disasters can result in massive biodiversity loss, and often leave longer lasting repercussions as well. Prolonged ecological imbalance due to Chernobyl nuclear power plant explosion in Ukraine in 1986 is an example of an acute disaster with its long lasting effects through the region. Several other anthropogenic activities leading to industrial emissions, aerosol release, biomass burning, agricultural runoff, pesticides, erosion, and automobile emissions may cause chronic pollution. Although the immediate effects of chronic pollution may be small, sustained rates and accumulation of chronic pollutants can be more severe in impact than acute environmental disasters. There are several documented examples of pollution and its impact on biodiversity. Many of released toxic elements and compounds resist degradation and accumulate in the environment. Organic chemicals with these properties are called persistent organic pollutants (POPs); organisms incorporate toxins in their tissues and the pollutants are subsequently entered in food chain. Beluga whale (*Delphinapterus leucase*) has been reported with impaired reproductive or immunological function as a result of pollutants.

Emission of sulfur dioxide and nitrous oxides, mainly from industrial and automobile emissions, into the atmosphere cause acid deposition (either dry or wet) pollution. In this case other aerosol pollutants, soil and water far from the site of pollution emission are often affected. Damage to the upper atmosphere's ozone layer described as 'ozone hole' has been well documented and well ascribed to chlorofluorohydrocarbons (CFC's) and other ozone-depleting chemicals released into the atmosphere. Impact of widening of ozone is mainly due to more penetration of ultraviolet light, which can be harmful to many biological organisms. Pollution stress forced organisms towards adaptation at the cost of other activities. Species those cannot adapt to a desirable extent have to succumb to pollution pressure and may contribute to reduction in original diversity. On the other hand those species which can adapt well provide tough completion for the species having less adaptive potential. For example, insect resistance to pesticides can create hardy taxa that may have the potential threat to natural ecosystems balance. Long back, attention has been called by Rachel Carson's Silent Spring which paved the way to formulate and introduction of strict regulation of pesticide use in many countries. However, situation is far from its perfection especially in countries where restricted commodities (e.g. pesticides) are still being used despite of ban. Increasing dependency on agro-chemicals such as pesticides and herbicides in agriculture is also a matter of great concern with respect to their residual toxic effects on non-target organisms. Development of resistance due to continuous use of same pesticide resulted development of resistances in target organism. In such situation, higher dose applications of pesticides adopted by the farmers which led to the build-up of pesticide residues and entrance of residual molecules into the food chain. This may also a cause for shrinking genetic diversity in plants and other soil biota. Pollution either its environmental or agricultural often hit higher levels of organization, altering community and ecosystem structure and functions and services of biodiversity.

Accelrated Climate Change

Climate change is much talked threat to species and natural habitats. There is widespread evidence that changes in phenology, including the timing of reproduction and migration, physiology, behaviour, morphology, population density and distributions and dominance of different types of species are driven by climate change (Rosenzweig *et al.*, 2007). Climate change, which is caused by a build-up of greenhouse gases such as carbon dioxide and nitrous oxides, methane and CFCs in the atmosphere, is a serious concern of ecologists and biologists. As it can alter the climate patterns and ecosystems in which species have evolved, it is likely that such change would exert certain impacts in either direction depending on the species. By virtue of changing the temperature and rain patterns, shifts in ranges of many species is likely to occur. There are indications that the projected changes in temperature and CO_2 concentration may alter growth, reproduction and host-pathogen relationships in both plants and animals. In the light of documented evidences, it can also be believed that the ecosystems with undiminished species diversity, and species with their genetic diversity intact, are likely to be in a much better position to face the impact of climate change.

It is well known that atmospheric CO_2 concentration increases due to the human interference in nature like increased fossil fuel burning, industrialization and transformation of agricultural lands to non-agriculture purposes. A steady increase of the CO_2 concentration in atmospheric has been manifested since the industrial revolution and will continue in future. According to the Inter-governmental Panel on Climate Change (IPCC), there will be an expected rise in the CO_2 of ~2 ppm for the next decade, which can bring the CO_2 in 2100 near to 550 ppm if no effective mitigation strategies are adopted (IPCC, 2007). Studies conducted at ICAR-Directorate of Weed Research, Jabalpur, (India) on wheat and *Phalaris minor* (a dominant weed) at elevated CO_2 (550 ppm) and elevated temperature (ambient +2.5 °C) indicated significant difference in growth and other physiological parameters (Paraste *et al.*, 2015). Such trends point out towards dominance of weed species over crop in the future climate change regime.

The Intergovernmental Panel on Climate Change in its summary report released in February, 2007, has estimated huge loss of biodiversity for biodiversity-rich mega diverse countries like India, because of higher greenhouse gas emissions. Targeted research on impacts of climate change on forest types, eco-sensitive zones, crop yields and biodiversity is required under the changing climatic regime. Similarly, more scientific studies have brought out that strong inter linkages exist between desertification and biodiversity loss. Such trends warrant for more focused research on the impact of desertification, as also synergizing efforts to combat desertification and promote biodiversity conservation.

Population Growth

Unlimited increase in population coupled to current consumption and production patterns, is a strong driver of the above-mentioned threats to biodiversity. To meet the demand for food, water, medicine, clothes, shelter and fuel for over 7 billion people live on Earth, there will be definite over-exploitation of natural resources. Increase in global population and drivers which threat the biodiversity seems to be strongly inter-linked. And, if unchecked, the growing population can be seen a biggest pressure threatening the global biodiversity as well as destruction of an ecosystem.

Natural Causes

Natural threats like unpredicted weather patterns, sudden outbreaks of diseases and pests, irregular and excessive rainfalls, drought, earthquakes, cyclones, tsunami, floods leads to natural disasters etc. There many examples where natural extremes made a sizable loss to human life as well biodiversity. The recent earthquake in East Japan on 11 March, 2011 followed by the collapse of the Fukushima Atomic Power Plant not only destroyed local human and animal life and properties, but damaged biodiversity of the area significantly. Frequent cyclones in costal parts of India also impacted the biodiversity in terms of wild and domesticated life, range shifts in some areas and degradation of natural resources.

Conclusion

Developmental activities, liberal trades, tourism and other direct or indirect human interferences threaten the natural diversity. Threats that modify ecosystem services include habitat loss, extinction, overexploitation of natural resources, introduction of invasive alien species, climate change, pollution, and population pressure. Among the major threats, invasion by alien species is truly alarming and considered second only to that of the habitat loss. In recent past, invasive alien species (plants or animals) extended their range and spread. Climate change seems to be major driver responsible for such shifts. Unplanned or poorly planned developmental activity, transport, and tourism are important pathways for the introduction of invasive species. Adoption of modern agricultural practices mainly mono-cropping, excessive use of agrochemicals and mechanization also contributed towards loss of biodiversity at farming lands globally. Lack of stringent quarantine laws especially in developing countries prompted the easy introduction of invasive alien species posing a serious threat to the natural biodiversity. In current scenario, where a significant loss of biodiversity is already evident, an urgent attention is required at managerial and policy level to enforce stringent quarantine laws to check further introduction of accidental and intentional entries of alien species.

References

1. Baillie, J.E.M., Griffiths, J., Turvey, S.T., Loh, J. and Collen, B. (2010). Evolution Lost: Status and Trends of the World's Vertebrates. Zoological Society of London, London.

2. BSI. (2012). Botanical Survey of India. http://bsi.gov.in/floristics.shtm (Accessed on 11.01.2012).

3. Butchart, S.H.M., Walpole, M., Collen, B., Vanstrien, A., Scharlemann, J.P.W., Almond, R.E.A., Baillie, J.E.M., Bomhard, B., Brown, C., Bruno, J., Carpenter, K.E., Carr, G.M., Chanson, J., Chenery, A.M., Csirke, J., Davidson, N.C., Dentener, F., Foster, M., Galli, A., Galloway, J.N., Genovesi, P., Gregory, R.D., Hockings, M., Kapos, V., Lamarque, J.-F., Leverington, F., Loh, J., McGeoch, M.A., McRae, L., Minasyan, A., Hernandez Morcillo, M., Oldfield, T.E.E., Pauly, D., Quader, S., Revenga, C., Sauer, J.R., Skolnik, B., Spear, D., Stanwell-Smith, D., Stuart, S.N., Symes, A., Tierney, M., Tyrrell, T.D., Vie, J. C. and Watson, R. (2010). Global biodiversity: indicators of recent declines. *Science*. 328(5892): 1164-1168.

4. CBD. (2010b). Global Biodiversity Outlook 3. Secretariat of the Convention on Biological Diversity, Montreal.

5. Chitale, V. S., Behera, M.D. and Roy, P.S. (2015). *Current Science*. 108(2): 25. Global biodiversity hotspots in India: significant yet under studied.

6. ConservationInternational;http://www.conservation.org/where/priority_areas/hotspots (accessed on 5 February 2014).

7. Danielsen, F., Beukema, H., Burgess, N.D., Parish, F., Bruhl, C.A., Donald, P.F., Murdiyarso, D., Phalan, B., Reihnders, L., Struebig, M. and Fitzherbert,

E.B. (2009). Biofuel plantations on forested lands: double jeopardy for biodiversity and climate. *Conservation Biology*. 23: 348-358.

8. Fitzherbert, E.B., Struebig, M.J., Morel, A., Danielsen, F., Bruhl, C.A., Donald, P.F. and Phalan, B. (2008). How will oil palm expansion affect biodiversity. *Trends in Ecology and Evolution*. 23(10): 538-545.

9. Foley, J.A., Ramankutty, N., Brauman, K.A., Cassidy, E.S., Gerber, J.S., Johnston, M., Mueller, N.D., O'Connell, C., Ray, D.K., West, P.C., Balzer, C., Bennett, E.M., Carpenter, S.R., Hill, J., Monfreda, C., Polasky, S., Rockstrom, J., Sheehan, J., Siebert, S., Tilman, D. and Zaks, D.P.M. (2011). Solutions for a cultivated planet. *Nature*. 478: 337-342.

10. IPCC: Climate Change (2007). Synthesis Report. Summary for Policy-makers. Intergovernmental Panel on Climate Change, Cambridge University Press.

11. Jaggi, D., Knox, J. and Paul, M. S. (2012). *Parthenium hysterophorus*: A serious threat Biological Forum-*An International Journal*. 4(1): 132-138.

12. Jaggi, D., Varun, M. and Kumar, B. (2015). In vitro phytotoxic evaluation of wasteland weeds species. *Indian Journal of Weed Science*. (Accepted).

13. Khandekar, V. and Srivastava, A. (2014). Ecosystem Biodiversity of India, "Biodiversity-The Dynamic Balance of the Planet". ISBN 978-953-51-1315-7, Published: May 14.

14. MA, (2005a). Ecosystems and Human Well-being: Synthesis. Millennium Ecosystem Assessment. World Resources Institute. Island Press, Washington, DC.

15. McGeoch, M. A., Butchart, S. H. M., Spear, D., Marais, E., Kleynhans, E.J., Symes, A., Chanson, Mittermeier R. A., Turner, W.R., Larsen, F.W., Brooks, T. M. and Gascon, C. (2011). Global biodiversity conservation: the critical role of hotspots. In: Zachos FE, Habel JC (eds) Biodiversity hotspots: distribution and protection of conservation priority areas. Springer, Heidelberg.

16. Myers, N., Mittermeier, R. A., Mittermeier, C. G., Da Fonseca, G. A. and Kent, J. (2000). *Nature*. 403: 853-858.

17. Nilsson, C., Reidy, C.A., Dynesius, M. and Revenga, C. (2005). Fragmentation and flow regulation of the world's large river systems. *Science*. 308(5720): 405-408.

18. Paraste, K., Rathore, M., Chaudhary, P. P., Singh, R., Pagare, S., Jaggi, D., Varun, M., Tripathi, N., Chander, S., Sarathambal, C. and Kumar, B. (2015). Crop weed dynamics and weed management under the regime of climate change. Vol. IInd Pp. 575, In Proceedings of 25th APWSS Conference on Weed Science for Sustainable Agriculture, Environment and Biodiversity (13-16 Oct. 2015), Hyderabad, India.

19. Pereira, H. M., Belnap, J., Brummitt, N., Collen, B., Ding, H., Gonzalez-Espinosa, M., Gregory, R. D., Honrado, J., Jongman, R. H., Julliard, R., McRae, L., Proenca, V., Rodrigues, P., Opige, M., Rodriguez, J.P., Schmeller, D. S., Van Swaay, C. and Vieira, C. (2010a). Global biodiversity monitoring. *Frontiers in Ecology and the Environment*. 8: 459-460.

20. Reise, K., Olenin, S. and Thieltges, D.W. (2006). Are aliens threatening European aquatic coastal ecosystems? *Helgoland Marine Research*. 60: 77-83.

21. Rosenzweig, C., Casassa, G., Karoly, D. J., Imeson, A., Liu, C., Menzel, A., Rawlins, S., Root, T.L., Seguin, B. and Tryjanowski, P. (2007). Assessment of observed changes and responses in natural and managed systems. In Climate Change 2007: Impacts, Adaptation and Vulnerability. Contribution of Working Group II to the Fourth Assessment Report of the Intergovernmental Panel on Climate Change (eds. Parry, M.L., Canziani, O.F., Palutikof, J.P., van der Linden, P.J. and Hanson, C.E.). pp. 79–131. Cambridge University Press, Cambridge.

22. Rosset, P.M. (1999). The Multiple Functions and Benefits of Small Farm Agriculture. Policy Brief. Institute for Food and Development Policy, Oakland and Transnational Institute, Amsterdam.

23. SCBD (Secretariat of the Convention on Biological Diversity), (2010). Forest biodiversity-Earth's living treasure, Montreal.

24. Tockner, K., Bunn, S. E., Quinn, G., Naiman, R., Stanford, J. A. and Gordon, C. (2008). Floodplains: critically threatened ecosystems. In Aquatic Ecosystems (ed. Polunin, N.C.). pp. 45-61. Cambridge University Press, Cambridge.

25. UNEP (United Nationa Environment Programme), (1992). Convention on Biological Diversity, (NA92-7807), New York, UNEP.

26. Vie, J. C., Hilton-Taylor, C. and Stuart, S.N. (eds.) (2009). Wildlife in a Changing World. Analysis of the 2008 IUCN Red List of Threatened Species. International Union for Conservation of Nature, Gland.

27. Vorosmarty, C. J., McIntyre, P. B., Gessner, M. O., Dudgeon, D., Prusevich, A., Green, P., Glidden, S., Bunn, S. E., Sullivan, C. A., Reidy Liermann, C. and Davies, P. M. (2010). Global threats to human water security and river biodiversity. *Nature*. 467: 555-561.

28. Wilson, E. O. (1988). The current state of biological diversity, In Biodiversity edited by E O Wilson and F M peter, Washington D. C., National Academy Press.

14

Concept of Endangered Species

Surendra Singh*, Ankita Kachhwaha and Sameer Choudhary

Algal Biotechnology Laboratory, Department of Post Graduate
Studies and Research in Biological Science,
Rani Durgavati University, Jabalpur-482001, Madhya Pradesh, India
E-mail: singhbiosci@yahoo.co.in

Abstract

Biodiversity, or biological diversity, is a term used to describe the variety of living things in the natural environment: the different plants, animals and micro-organisms; the genes they contain; and the ecosystems in which they occur. The world's biodiversity is in trouble, species threatened with extinction are classified into categories such as Extent, Endangered, Threatened, Vulnerable or Imperilled on the International Union for the Conservation of Nature and Natural Resources (IUCN) Red List of Threatened Species. The number of plants and animals in these categories has increased in the past few decades. Disappearance or extinction of wild habitats, rampant trade and introduction of non-native species are the primary causes of threatening species. Diseases, pollution, pesticides, toxic chemicals, thinning of the ozone layer and other environmental problems contribute to their decline.

Keywords: Biodiversity, Endangered Species, Red Data Book, Extinction Causes.

Introduction

Flora and Fauna of the world is rich in diversity due to varied topography. According to the Census of Marine Life, the total number of species on Earth is about 8.7 million with 6.5 million species on land and 2.2 million in oceans. The forest wealth is getting depleted leading to degradation of land and changes in the climate to disturb the ecosystem of the area. As a consequence, various kind of environmental hazards such as floods, droughts, etc., appear in vulnerable places. Those species which are at a verge of extinction are called endangered species, meaning that there are so few left of their kind that they could disappear from the planet (Carr, 2005). Usually endangered species have a declining population or a very limited range. The current rate of extinction is thought to be far greater than the expected natural rate, with many species are going to be extinct before they have even been discovered. Current estimates suggest that a third of the world's amphibians, a quarter of all mammals and one in eight birds are endangered. To date more than 76,000 plant and animal species have been assessed with more than 22,000 at risk of extinction (Klemm and Shine, 1993; IUCN, 2015).

It has been estimated that there are 4 to 5×1030 prokaryotic cells on Earth (Whitman *et al.,* 1998). For oceanic protists, prokaryotes and viruses, respectively, abundances are related roughly as 1:1000:6250. A typical coastal water sample contains 107 viruses, 106 bacteria and 103 protists ml^{-1}. Microbes come in diverse forms; however, it is not known how many different species of microbes inhabit the planet. Most microbes probably do not suffer much from habitat fragmentation, since they need smaller habitats than larger species. Free-living microbes do not seem to be threatened by extinction, although the number of threatened species might be higher than anticipated. Associated microbes which are specific to the respective plant or animal host equally as threatened as their hosts can be considered as threatened (Weinbauer and Rassoulzadegan, 2007).

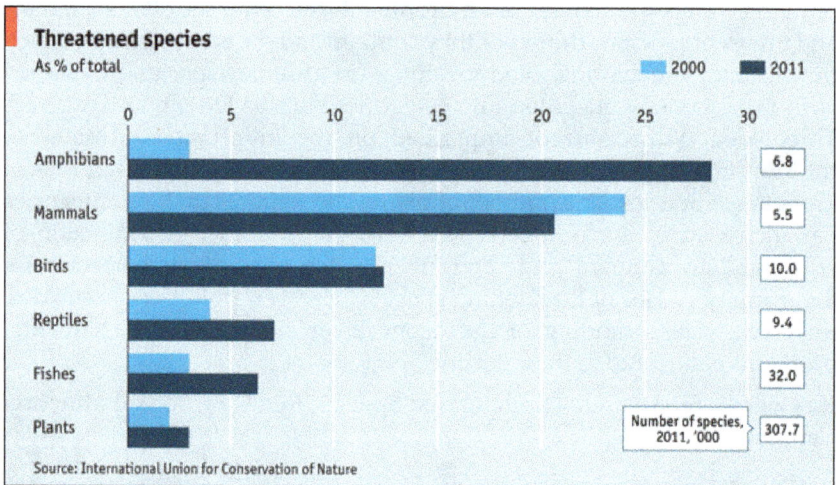

Figure 1: Comparative data of Threatened Species between 2000 and 2011 (IUCN, 2015).

World Forestry Day or International Day of Forests is celebrated worldwide every year on 21st of March at the international level in order to increase the public awareness among the communities about the values, significance and contributions of the forests to balance the life cycle on the earth. May 22 was proclaimed as the International Day for Biological Diversity (IDB) to increase understanding and awareness of biodiversity issues. Wildlife Week is celebrated every year in India between October 1 and 8 to promote the preservation of fauna and to encourage animal welfare (Udgata, 2011).

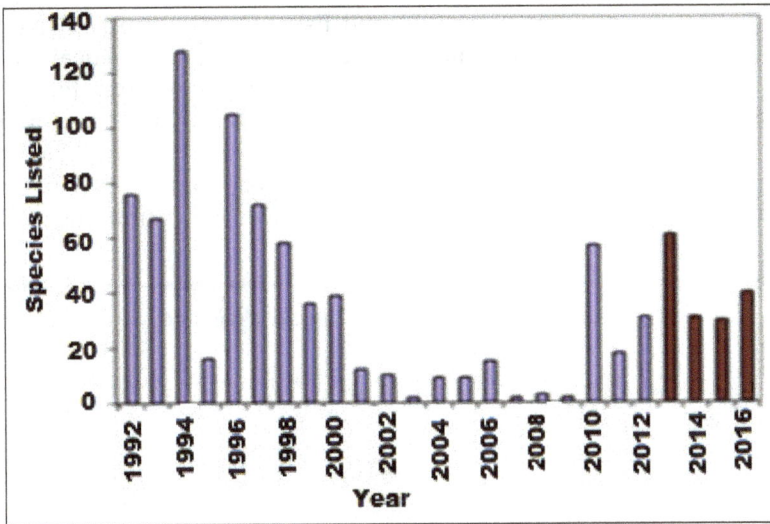

Figure 2: Listing of Threatened Species per year 1992-2016 (IUCN, 2015)

Reasons for Threat

Heavy and uncontrolled biotic interference has affected the natural forest environment thereby adversely influencing the wildlife. Many species of tropic flora and fauna have either reached extinction or nearing extinction. For better survival of wild animals and plants, it is necessary that they must be allowed to live in their own habitat. The causes of threats to species may be either natural or artificial (Pangman, 1995).

The natural causes responsible for endangering species are habitat loss, landslides, droughts, floods, storms, earthquakes, diseases [For example, a white tiger named "Shivani" suffering from Jaundice died at Kamla Nehru Zoo, Indore on 8th February 2016 (Nair, 2015)], avalanches predators, harsh winter or severe summer, jungle fire, pest infestation, radiations etc. Other threats include exotic and aggressive weeds, air, water pollution, lack of pollinators. According to UN report wild bees, butterflies and other insects that pollinate plants are shrinking towards extinction because of the change in the farming pattern leading to loss of diversity, habitat loss to cities, disease and global warming.

The artificial causes include grazing, commercial exploitation, industrialization, forestry, urbanization, scientific and educational research, construction of roads and dams increasing townships, internal tourism, mining and pressure of introduced species, hunting [For example, three persons were arrested for hunting migratory birds in Kolleru Wildlife Sanctuary Mandavalli, Andra Pradesh (Srinivas, 2016)], anthropogenic activities [For example, Carcasses of Endangered Olive Ridley Sea Turtle were spotted along the shoreline of Gahirmatha Marine Sanctuary, Kendrapara, Odisha which was suspected that the turtles are perishing due to trawl fishing (PTI, 2016)]; Illegal use of drug Diclofenac injection in animals meant for human use drastically brought down the vulture population in India (Singh, 2014). Run off from fertilizer rich field causes nutrient enrichment of water, oil spills in sea, intensive agriculture. The current rate extinction is 1,000-10,000 times the rate of normal background extinction. If the current rate of species extinctions goes on unabated, 50% species are liable to die out by the end of the 21st century (Lambacher, 2013).

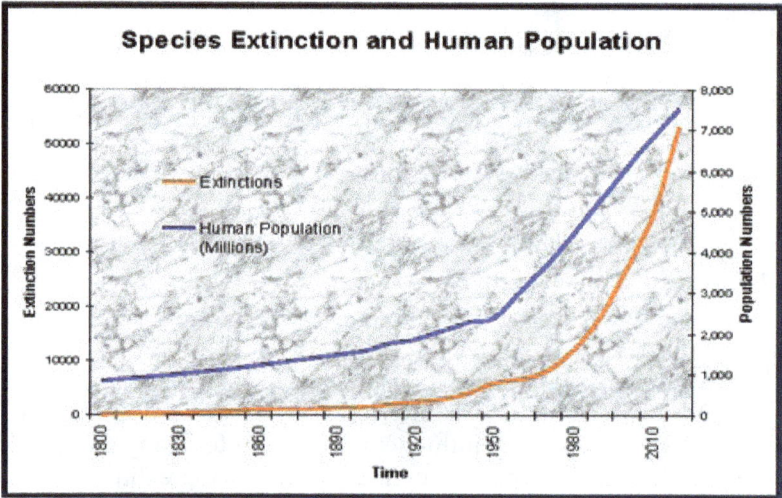

Figure 3: The rise in human population and in the number of extinct species between 1800 and 2010 (Glikson, 2012).

Figure 4: Migratory Kolleru Bird, Ladakh (Srinivas, 2016)

Figure 5: Endangered and Vulnerable Black-necked Cranes (Peerzada, 2016)

Figure 6: White Tiger "Shivani" dies at Indore Zoo due to disease contracted Jaundice (PTI, 2016)

Figure 7: Rhinos to be relocated within the Dudhwa National Park, Uttar Pradesh (PTI, 2016)

Figure 8: Decomposed carcasses of Olive Ridley turtles in beaches off Gahirmatha Coast, Odisha (PTI, 2016)

Figure 9: Stuffed leopards in the Museum of Natural History in Paris (Reuters, 2016)

Figure 10: NASA's Satellite "Whale Watch Tool" to help decrease whale mortality due to collisions with shipping and fishing gear (IANS, 2015)

Figure 11: Use of 'death' drug diclofenac reducing vulture population (Singh, 2014)

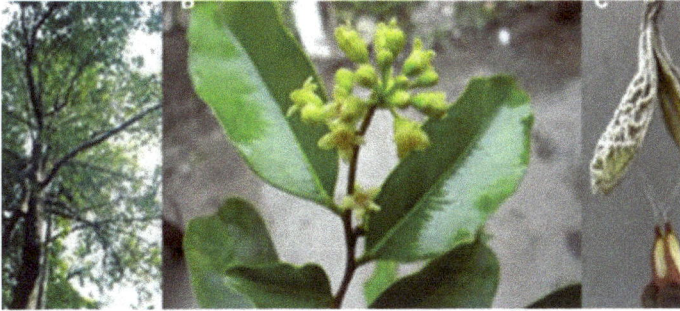

Figure 13: *Aquilaria malaccensis.* (**A**) Mature tree, (**B**) flowering twig, (**C**) mature fruit. Natural populations of *A. malaccensis* have been depleted due to over-harvesting of mature trees for its precious 'agarwood' (Choudhury and Khan, 2010).

Figure 14: *Convolvulus massonii* highly threatened by habitat loss (Valavanidis and Vlachogianni, 2011)

Figure 15: *Arabis kennedyae* a critically endangered plant species endemic to *Cyprus* (Andreou *et al.*, 2015)

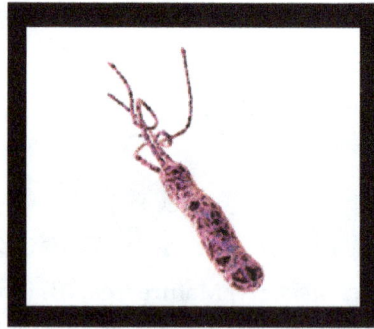

Figure 16: This helix-shaped bacteria called *Helicobacter pylori* is disappearing from our modern stomachs. Once thought of as a "bad bug", doctors are learning that this bacteria influences our immune system, our weight, and even our height (Jacobson, 2014).

Figure 17: Specific habitat and substrate requirements limit occurrence and distribution of *Hypocreopsis amplectens* in Australia and New Zealand (Johnston *et al.*, 2010).

Red Data Book

It is a book written on endangered animals, birds and flowering plants prepared by the survival service commission of International Union for conservation of Nature and Natural Resources (IUCN). Red data book according to Korneck and Sukopp (1988) is a standard list of endangered species. The best available knowledge at the time of observation about the incidence, distribution, frequency and temporal and spatial trends in these parameters influences the assessment of the status of individual species in the reference area. Continuity in acceptance, understanding and application of the classification criteria is instrumental in the monitoring with a large temporal prospective (Hilton-Taylor, 2000). Of the total 66,581 species globally assessed, the IUCN has classified 4,898 as Critically Endangered, 834 as Extinct, 69 as Extinct in the Wild. In the lower risk categories, there were 7,323 species in Endangered, 11,029 in Vulnerable and 5,204 in Near Threatened categories (IUCN, 2015)

Conservation Strategies:

Conservation aims to protect the natural world and sustain biodiversity by carefully preserving and managing existing habitats and restoring areas which have been damaged or degraded.

1. All the threatened species should be protected. All the possible varieties, old or new, of food forage and timber plants, livestock, aquaculture and microbes are conserved. Organizations like Wildlife SOS 24-hour Helpline Rescue has a dedicated team that works round the clock to protect the wildlife in distress. For example, Wildlife SOS working for animal protection seized 2,100 ducks from a livestock transport truck (Qureshi, 2016).

2. site).

    a. **In-situ Conservation:** Wild relatives of all the economically important organisms are conserved in protected areas such as in a Zoo, National Parks, Sanctuaries [For example, the number of Black-necked Cranes has gone up from 38 in 1997 to 95 in 2016 at Changthang Cold dessert sanctuary Ladakh (Peerzada, 2016)], Biosphere Reserves, or preserving endangered plants through the use of seed banks.

3. There are two types of conservation strategies: *in situ* (on site) and *ex situ* (off **Ex-situ Conservation**-Many endangered species are bred in captivity to preserve their numbers and in some cases it is possible to reintroduce them to the wild. Some species, like the Golden arrow poison frog, have even been deliberately removed from the wild to protect them from the spread of disease and ensure that a small population is preserved. Plant species are often cultivated in nurseries and preserved via the use of seed banks.

4. **Anti-poaching Measures**-In remote areas guards are sometimes employed to protect endangered species, such as the mountain gorilla, from poachers. This can be a way of involving local communities in

the protection of their wildlife whilst also providing some employment opportunities. For example, first female anti-poaching unit from South Africa, "Black Mamba Unit" has helped arrest six poachers and removed more than 1,000 snares that were laid to trap wild animals, according to the United Nations (Gunther, 2015).

5. **Wildlife Corridors**-Where habitats have been fragmented by divisions such as roads, urban areas or farmland, populations become isolated and are unable to move throughout their natural range to find sufficient resources and mates. Wildlife corridors help to reconnect habitat fragments and maintain genetic diversity. For example, three Rhinos from Dudhwa Rhino Rehabilitation Area (RRA) were shifted to the newly developed additional RRA at Bhadi Tal in Belrayan Range, Dudhwa National Park, Uttar Pradesh (PTI, 2016).

6. **Laws and Policies**-Some endangered species are protected by law or trade in them is restricted. CITES (The Convention on International Trade in Endangered Species) is an international agreement between governments to ensure that trade in wild animal and plant specimens does not threaten their survival.

7. Critical habitats for feeding/ breeding/ resting/ nursing species should be identified and safeguarded.

8. Trafficking of protected species should be prevented. For example, French customs seized seven stuffed leopards in 2015 and delivered them to the Museum of Natural History in Paris on 16th February 2016 (Reuters, 2016).

9. Surveillance by Satellite: Satellite data from NASA to be used in a new online tool to protect endangered blue whales. "The Whale Watch Tool" set to be released later by the National Oceanic and Atmospheric Administration (NOAA), which will help decrease whale mortality due to collisions with shipping and fishing gear (IANS, 2015).

10. Unique ecosystems should be preserved on priority bases.

11. Introduction of alien species should be controlling.

12. **Reducing pollution:** Reducing carbon emissions

13. **Public awareness:** Volunteering with wildlife charities and conducting workshops, conferences, symposiums etc. In areas where humans and animals are competing for space or resources, particularly in poorer developing countries, it is important that conservation work takes into account the needs of local people and works alongside them in protecting their native species.

14. Boycotting endangered animal products.

15. Tissue culture techniques are used to preserve highly endangered plants and animals. A living collection of endangered economic and rare species are maintained.

16. **Programmes:** Man and Biosphere Programme, is an International biological programme of UNESCO (United Nations Educational, Scientific and Cultural Organisation) which was started in 1971. MAB has studied human environment, impact of human interference and biotic and abiotic environments and conservation strategies for the present as well as future. For example, Project Tiger (1973) etc. Vana Surakshya Samitis and Ama Jungle Yojna are few plans taken up by government for forest conservation. Forest Ecosystem Climate Proofing Project has been launched to minimise and mitigate the effects of climate change along with enhancing the biodiversity. To protect forest, Supreme Court has made judgment to evict Kudremukh Iron Ore Company Limited (KIOCL) in Kudremukh National Park, Western Ghats of Karnataka against illegal mining (Sridhar, 2015).

The Indian government has enacted numerous laws regarding wildlife protection:

- The Prevention of Cruelty to Animals Act, 1960
- The Prevention of Cruelty to Draught and Pack Animals Rules, 1965
- The Experiments on animals (Control and Supervision) Rules, 1968
- The Wild Life Protection Act, 1972
- The Performing Animals Rules, 1973
- The Transport of Animals Rules, 1978
- The Performing Animals (Registration) Rules, 2001
- The Prevention of Cruelty to Animals (Slaughter House) Rules, 2001

Other Laws and Policies

Endangered Species Act (ESA)

The Endangered Species Act was established in 1973 to conserve the Nation's natural heritage for the enjoyment and benefit of current and future generations by conserving species that are in danger of extinction.

Marine Mammal Protection Act (MMPA)

The Marine Mammal Protection Act was established in 1972 to protect marine mammals by prohibiting take of marine mammals in U.S. waters and by U.S. citizens on the high seas, and the importation of marine mammals and marine mammal products into the U.S.

National Environmental Policy Act (NEPA)

The National Environmental Policy Act was enacted in 1969. NEPA requires Federal agencies to integrate environmental values into their decision-making processes by considering the environmental impacts of their major proposed actions.

The Migratory Bird Treaty Act

This Act, originally passed in 1918, provides protection for migratory birds. Under the Act, it is unlawful to take, import, export, possess, buy, sell, purchase, or barter any migratory bird. Feathers or other parts, nests, eggs, and products made from migratory birds are also covered by the Act. Take is defined as pursuing, hunting, shooting, poisoning, wounding, killing, capturing, trapping, or collecting.

The Eagle Protection Act

Bald Eagle protection began in 1940 with the passage of the Eagle Protection Act. Later amended to include the Golden Eagle, the Act makes it unlawful to import, export, take, sell, purchase, or barter any Bald Eagle or Golden Eagle, their parts, products, nests, or eggs. "Take" includes pursuing, shooting, poisoning, wounding, killing, capturing, trapping, collecting, molesting, or disturbing the eagles.

The Wild Bird Conservation Act

In 1992, the United States passed the Wild Bird Conservation Act. By October 1993, the law prohibited the import of all CITES-listed-birds (almost 1,000 species) except for those included in an approved list either by country of origin or wild-caught birds or by specific captive breeding facilities. For wild-caught approved birds, a management plan that provides for conservation of the species and its habitat is required. In addition, it established a moratorium on trade of any non-Cites species. Exemptions include game birds and bird species indigenous to the 50 United States and the District of Columbia.

The Act establishes an Exotic Bird Conservation Fund, to be funded by penalties, fines, donations, and any additional appropriations. The Fund is to be used to assist exotic bird conservation projects in their native countries. Particular attention is given to species subject to an import moratorium or quota in order to asset those countries in developing and implementing conservation management programs, law enforcement programs, or both.

Conclusion

Support for endangered species is very crucial as we are in a period of decline and there are going to be increasing consequences. Many species are shrinking toward extinction and the world needs to do something because we are responsible for this predicament. For this subtle problem, we have to take some hefty decisions to combat this havoc. Although various petitions and treaties have been created and signed to help save the lives of endangered life, it is up to each individual to do their part to aid in saving species in danger. There is a need to educate people about wildlife conservation and habitat protection and develop esteem for endangered species and stop illicit act against them. If these species are not protected now, it will be very difficult to manage them. It is the responsibility of everyone to protect these endangered creatures so that they may one day thrive again. Species recovery or management plans should be used to identify critical habitats of species which should then be given priority in the conclusion of management agreements. If we do not stop the loss of endangered species it will be a nemesis for our future generation.

References

1. Andreou, M., Kadis, C, Delipetroub, P. and Georghiou, K. (2015). Conservation biology of Chionodoxa lochiae and Scilla morrisii (Asparagaceae): Two priority bulbous plant species of the European Union in Cyprus. *Global Ecology and Conservation*. 3: 511-525.

2. Carr, G.D. (2005). Why save endangered species. U.S. *Fish & Wildlife Services*. 1-20.

3. Choudhury, B. and Khan, M. L. (2010). Conservation and Management of Endangered Plant Species: A Case Study from Northeast India. *Bioremediation, Biodiversity and Bioavailability*. 4(1): 47-53.

4. Department of Environment, Land, Water & Planning, Flora and Fauna Guarantee Act 1988, Threatened List, May 2015 (http://www.depi.vic.gov.au/__data/assets/pdf_file/0004/302827/201506-FFG-threatened-list.pdf).

5. Glikson, A. (2012). Mass extinction of species and climate change. *Earth and palaeoclimate science*. 1-8.

6. Gunther, J. (2015). On Patrol with the Black Mambas. *The New York Times*.

7. Hilton-Taylor, C. (2000). IUCN Red List of Threatened Species. The IUCN Species *Survival Commission*. 1-57.

8. IANS. (2015). Online tool to protect blue whales. *The Hindu*.

9. IUCN Red List of Threatened Species, 2015 (http://www.iucnredlist.org/)

10. Jacobson, R. (2014). Can we save our body's ecosystem from extinction? *PBS NEWSHOUR*.

11. Johnston, P.R., May, T.W, Park, D. and Horak, E. (2010). *Hypocreopsis amplectens* sp. nov., a rare fungus from New Zealand and Australia. *New Zealand Journal of Botany*. 45(4): 715-719.

12. Klemm, C. and Shine, C. (1993). Biological Diversity Conservation and the law: Legal Mechanisms for conserving species and ecosystems. *IUCN Environmental Law Centre, IUCN Biodiversity Programme*. 1-285.

13. Korneck, D. and Sukopp, H. (1988). Rote Liste der in der Bundesrepublik Deutschland ausgestorbenen, verschollenen und gefahrdeten Farn-und Blutenpflanzen und ihre Auswertung fur den Arten-und Biotopschutz [Red Data Book of th extinct and endangered vascular plant species of the Federal Republic of Germany including its evaluation]. *Schriftenreihe für Vegetationskunde*. 19: 1-210 (in German).

14. Lambacher, J. (2013). The Politics of the Extinction Predicament-Democracy, Futurity and Responsibility. Thesis (PhD). University of Washington.

15. Nair, P. (2015). Last surviving tiger cub of the two dies in Indore Zoo. *The Times of India*.

16. Pangman, J.K. (1995). Guide to Environmental Issues. *Diane Publishers*. 1-82.

17. Peerzada, A. (2016). Kashmir's "lovebirds" thriving. *The Hindu*.

18. PTI. (2016). Decomposed carcasses of Olive Ridley Sea Turtle found. *The Hindu*.

19. PTI. (2016). Three Dudhwa rhinos to be relocated within the park. *The Hindu*.

20. Qureshi, S. (2016). Agra Police rescues 2100 ducks being smuggled to Chandigarh. *India Today*.

21. Reuters. (2016). On the prowl. *The Hindu*.

22. Singh, R. (2014). Use of 'death' drug diclofenac reducing vulture population. *The Times of India*.

23. Sridhar, V.K. (2015). Supreme Court: Mining, Forest Encroachments and Rehabilitation from Kudremukh National Park. *Social Change and Development*. 1: 62-76.

24. Srinivas, R. (2016). Three held for hunting migratory birds in Kolleru. *The Hindu*.

25. Udgata, H.B. (2011). Wildlife Conservation with Peoples' Participation (A discussion on the occasion of 57th Wildlife Week, 2011). *Orissa Review*. 54-58.

26. Valavanidis, A. And Vlachogianni, T. (2011). Ecosystems and Biodiversity Hotspots in the Mediterranean Basin Threats and Conservation Efforts. *Science advances on Environment, Toxicology & Ecotoxicology issues*. 1-24.

27. Weinbauer, M.G. and Rassoulzadegan, F. (2007). Extinction of microbes: evidence and potential consequences. *Endangered Species Research*. 3: 205-215.

28. Whitman, W., Coleman, D. and Wiebe, W. (1998). Prokaryotes: the unseen majority. *Proc Natl Acad Sci USA*. 95: 6578-6583.

15

Impact of Environmental Changes on Biodiversity

Shikha Jaggi and Abhay Singh Yadav*

Department of Zoology, Kurukshetra University,
Kurukshetra-136119, Haryana, India
E-mail: abyzkuk@gmail.com

Abstract

Large scale imbalance to the environment owing to human activities is threatening our biological diversity to the point of extinction. Environment and biodiversity are inseparable. Biodiversity is under constant threat due to habitat degradation, including pollution, habitat destruction and fragmentation, over-exploitation of species, introduction of invasive species, diseases and global climate change. The impact of all these environmental alterations on biodiversity has been discussed in this chapter. The efforts to safe-guard threatened biodiversity at the global level are failing because many species usually face more than one threat. These environmental changes may act synergistically that a species may not be able to cope up with either by moving to a new suitable location or by adapting genetically to the change. Global climate patterns are already becoming erratic because of release of large amount of green house gases into the environment by human activities. Projected global temperature increase may render many species unable to adjust their ranges and tolerance limits and these may become extinct. Another impact will be due to melting of polar ice, low-lying coastal communities may get submerged and polluted. Sustainable development, reducing overconsumption of resources and slowing human population growth are key factors to deal with the biodiversity depletion.

Keywords: Biodiversity, Environmental change, Habitat, Pollution, Species

Introduction

Biodiversity represents the number, variety and variability of living organisms. It comprises inter-species diversity, intra-species diversity along with diversity among ecosystems. Biodiversity includes all ecosystems i.e. managed ecosystems (plantations, farms, aquaculture sites, urban parks, croplands etc.) and unmanaged ecosystems (wild lands, nature preserves or national parks). Biodiversity is an extremely important aspect of ecosystem. Biodiversity forms the foundation of the vast array of ecosystem services that critically contribute to human well being (Swingland, 2001). Biodiversity helps in maintaining the balance of the ecosystem through recycling and storage of nutrients, combating pollution, stabilizing climate, protecting water resources, forming and protecting soil and maintaining ecobalance (Rathore and Jasrai, 2013). Biodiversity promotes recreation, tourism, cultural value, education and research. It provide us with array of biological resources such as food items, medicines, ornamental plants, wood products, breeding stock which contribute to the economy (Bovarnick *et al.*, 2010).

According to the evolutionary theory, greater the diversity that exists within taxa, the more likely it is to survive environmental change. But humans have been the main root of recent swift ecological and evolutionary changes. Human activities have led to the destruction of ecosystems. Although ecosystem would be evolving regardless of human influence but human intervention leads to species extinction at an accelerated and dangerous rate. Human activities like urbanization, deforestation, damming of aquatic bodies, introduction of alien species etc. are causing rapid environmental changes like pollution, global warming, habitat fragmentation and loss etc. Environmental changes are the important drivers for biodiversity loss. In the past, environmental changes had an extensive impact on biodiversity and will continue to affect the biodiversity patterns in future. As the environmental changes will become more severe, higher will be the harmful impacts on species distributions, population sizes, migration events, timing of reproduction and disease outbreaks (Sharma and Mishra, 2011). Biodiversity is crucial for the well being of life on earth and hence there is an urgent need to protect natural environment and biodiversity. Economic policies must be created to conserve the biodiversity and to protect the habitat and species.

Discussion

Pollution

Pollution is a human created vision of the state of our environment. Due to industrialization, all forms of pollution pose a serious threat to biodiversity. Pollution impacts all the trophic levels from primary producers to top predators thereby, interfering with the structure and functioning of ecosystem (Todd *et al.*, 2010). There is wide variety of pollutants present in the environment. These include pesticides, oils, heavy metals, plastics etc. The pollutants can have ecological impacts at the (1) individual level i.e. in terms of natality, mortality, growth rate etc. (2) community level i.e. in terms of species diversity, food webs, species interactions etc. (3) ecosystem level i.e. in terms of primary productivity, secondary productivity, nutrient cycling etc. (4) landscape level i.e. in terms of transfer of nutrients, soil, spatial heterogeneity of organisms etc. (Edwards, 2002).

Pesticides

Pesticides are a major factor affecting the biodiversity globally. They can be directly lethal to organisms or cause alteration in their habitat or accumulate in food chain (Isenring, 2010). Chemical pollution including fertilizers, pesticides and heavy metal in water and terrestrial ecosystem is the primary pressure among 187 globally threatened bird species (Isenring, 2010). It has been reported in Europe that countries with more intensive agriculture have greater population decline among farmland birds (Donald *et al.*, 2002). Geiger *et al.* (2010) reported that use of insecticides and fungicides has negative effect on biodiversity in European farmlands and at the same time reduce the biological control potential. Beketov *et al.* (2013) reported that use of pesticides reduce the regional biodiversity of stream invertebrates in Europe (Germany and France) and Australia (Southern Victoria).

Seeds treated with fungicides and organophosphate insecticides including disulfoton, fenthion and parathion posses a potential risk for bird species foraging in fields (Prosser and Hart, 2005). An organophosphate, monocrotophos have killed 6,000 Swainson's hawks in a small area of Argentine pampas and over 1,00,000 bird deaths have been reported worldwide (Goldstein *et al.*, 1999). Herbicides and avermectin residues affect the bird biodiversity indirectly by lowering food abundance (Vickery *et al.*, 2001). There are many species of birds like grey patridge, corn bunting, yellowhammer, red-backed shrike, skylark, tree sparrow and yellow wagtail which are at risk due to indirect effects in UK (Central Science Laboratory, Game Conservancy Trust, RSPB and Department of Zoology, 2005). Sub-lethal doses of the pesticides can also affect nervous system of organisms causing alteration in behavior. It has been documented that after spraying azinphos-methyl, an organophosphate parent bird made fewer feeding trips in an orchard (Bishop *et al.*, 2000). Bioaccumulation of toxic pesticides is of a major concern to top predators such as mammals and raptors etc. In France, bioaccumulation of bromadiolone in prey tissues caused secondary poisoning in foxes (Berny *et al.*, 1997). Herbicides remove a particular plant food source and changes microclimate thereby affecting mammals like common shrew, wood mouse and Badger (Hole *et al.*, 2005).

Population of beneficial insects such as bees, beetles etc. might decline due to the use of broad-spectrum insecticides like carbamates, organophosphates and pyrethroids (Isenring, 2010). In UK, bee colonies are reported to be poisoned due to the use of carbamate (bendiocarb) and pyrethroids (Isenring, 2010). Low doses of imidacloprid negatively affects the foraging behavior of bees (Yang *et al.*, 2008) and learning capacity of bees is reduced due to its long term exposure (Decourtye *et al.*, 2003). Use of the pesticides at higher levels contaminates the surface water which affects fishes and aquatic invertebrates. Yadav *et al.* (2013) reported that butachlor interferes with the cellular activities in fishes at genetic level and induces chromosomal aberrations. In another study, it was revealed that a sublethal dose of chloropyrifos induces growth reatardation and depletion in protein contents of fresh water fish, *Cirrhinus mrigala* (Cheema *et al.*, 2014).

In US, red legged California frog and its habitat is endangered by hexazinone, a triazine herbicide (Isenring, 2010). Forson and Storfer (2006) reported thatuse of sodium nitrate and atrazine impacts the immune system of tiger salamander (*Ambystoma tigrinum*) by significantly decreasing the peripheral leukocyte levels. Organic systems in central Europe having reduced pesticide input by 97% had enhanced soil fertility and biodiversity (Maeder *et al.*, 2002).

Groundwater pollution due to pesticides is a serious concern. During one year survey in India, organochlorine pesticides were found to contaminate 58% of drinking water samples drawn from different handpumps and wells around Bhopal region (Kole and Bagchi, 1995). Data on reproductive performance were collected from 1016 couples in which males were exposed to pesticides in cotton fields in India. Fertility of males decreased significantly. The frequency of still births, neonatal deaths and congenital defects were significantly higher in offsprings of exposed males as compared to offsprings of control group (Rupa *et al.*, 1991). A study on workers (n=365) involved in manufacture of hexachlorocyclohexane (HCH) revealed neurological symptoms. Remarkably elevated concentration of HCH residues were observed in serum of all exposed workers (Nigam *et al.*, 1993). The best estimate for potential risk of chemicals is measurement of chemical content in total diet. In a study by Kannan *et al.* (1992), the average intake of HCH and DDT was found to be 115 and 48 mg per person, respectively in Indians, which were higher than those reported in other developed countries.

Oil spills

Release of millions of gallons of liquid petroleum hydrocarbon from tankers, coastal refineries, offshore platforms, drilling rigs and wells into the environment i.e. marine areas due to human activity causes oil pollution. After an oil spill has occurred, oil contaminants persist in the marine environment for years affecting the community structure (Kingston, 2002). There are various detrimental effects of petroleum products on aquatic organisms. Oil spill affects the biota in short term as well as long term. Mortality is a short term effect on flora and fauna of aquatic ecosystem. Long term effects are in form of genetic abnormalities and alteration in reproductive behavior (Baliarsingh *et al.*, 2014). Oil spillage can affect the ecosystem adversely through the following ways:-(1) formation of thick, smooth and slippery oil coating on the water surface, which might obstruct the photosynthesis of phytoplanktons affecting the productivity of aquatic system. Oil could also enter the lungs or cover blow holes causing health issues in dolphins. (2) oil can mix with the suspended matter and sink to the bottom which destroys benthal organisms and interfere with their spawning areas. (3) oil spillage can potentially damage the food chain as the soluble material may be ingested by the aquatic organisms and lead to bioaccumulation of toxicants (Bury, 1972).

Mortality and elimination of *Pisaster* spp. (sea-stars) and *Strongylocentrotus* spp. (sea-urchins) has been observed due to diesel oil pollution and tube feet of the sea-urchins may be inactivated by as small as 0.1% of oil emulsion (Smith, 1970). Bury (1972) reported that oil spillage of approximately 2,000 gallons of diesel fuel into Hayfork Creek, California resulted in death of thousands of aquatic insects, fishes, amphibians and reptiles. Luiselli and Akani (2003) studied both direct and indirect impacts of oil spillage on the diversity and functioning of turtle communities in the Niger Delta, Nigeria. Approximately 50% decline in species diversity and a strong reduction in turtle specimens which could withstand catastrophic pollution was observed after oil pollution. Various species from Nalabana Bird Sanctuary migrate to Chilika lagoon. Oil pollution in lagoon might adversely affect by coating their plumage, mating and misaligning their feathers affecting the insulation ability and reducing buoyancy (Baliarsingh *et al.*, 2014). Along with that organic additives in diesel and cutting oils may also cause nervous abnormalities in birds (Smith, 1970). The oil toxicity effects the population level of aquatic birds by reducing the egg viability (Islam and Tanaka, 2004). Oil contamination also causes significant genetic damage in aquatic animals as reported in *Mytilus galloprovincialis* from Galician coast due to Prestige oil spill in 2002 (Laffon *et al.*, 2006). Severe eye irritation with subsequent blindness has been reported among seals from Antarctica and Cornwall due to oil pollution (Smith, 1970).

Plastics

Plastic pollution is a global tragedy for biodiversity. Plastics are the synthetic polymers that are entering the environment at an accelerated rate from the last two decades of the 20th century (Moore, 2008). Plastic debris can be broadly classified into two groups: (1) Macro (>5mm) and (2) Micro (<5mm). Though macro-debris might be traced for its origin easily but it is really difficult to find the origin of micro-debris (Moore, 2008). Plastic debris is reported to affect approximately 267 marine species globally of which 86% are sea turtles, 44% are sea birds and 43% are mammals (Laist, 1997). There is a potential risk that plastic debris sinks to the bottom and act as barrier inhibiting gas exchange between the overlying waters and the pore waters of sediments (Islam and Tanaka, 2004) and resulting in hypoxic or anoxic condition at sea floor thus affecting the life of benthic organisms (Goldberg, 1994). Intertidal sediments of the world's largest ship-breaking yard at Alang-Sosiya, India have accumulated approximately 81mg small plastic fragments per kg of sediment (Reddy *et al.*, 2006). Aggregation of plastic debris by sessile organisms in the marine ecosystem might act as vector for the transport of alien species. Drifting plastic debris is another mode for introduction of alien species (Derraik, 2002). Gregory (1978) reported that *Membranipora tuberculata*, a bryozoans attached to plastic debris crossed the Tasman Sea from Australia to New Zealand (Gregory, 1978).

Since plastics are known to adsorb hydrophobic pollutants, its ingestion is highly lethal. 44% of all seabirds' species such as albatross, fulmars, sea-turtles etc. mistake the floating plastic for food and ingest it (Moore, 2008). According to a study by Franeker *et al.* (2011), 95% of fulmars have plastic in their stomach. Among 54 loggerhead sea turtles captured from Adriatic Sea, 35.2% turtles were found to ingest marine debris of which soft plastic ingestion was maximum among turtles (68.4%) followed by ropes, Styrofoam and monofilament lines (42.1%, 15.8% and 5.3%, respectively) (Lazar and Gracan, 2011). Different species of fish were found to have guts filled with plastic debris. The pattern of selective ingestion of white plastic spherules was among them indicating the selective feeding habit of fishes (Carpenter *et al.*, 1972). Ingestion of plastic particles decreases the storage capacity of stomach thus reducing the meal size in chickens (Ryan, 1988). Other harmful effects of ingestion of plastics include altered migration and reproductive effort (Connors and Smith, 1982), blocked gastric enzyme secreation, lowered steroid hormone levels, delayed ovulation (Azzarello and Van-Vleet, 1987), internal injury of gastro-intestinal tract (Zitko and Hanlon, 1991). Ingestion of plastic and polythene materials in cattles can cause indigestion, impaction, tympany, polybezoars and immune-suppression (Singh, 2005). Soil micro-and mesofauna could ingest the microplastic present in soil and accumulate in detrital food web. As microplastic can adsorb the harmful soil contaminants, it could locally concentrate them in soil (Rillig, 2012).

Plastic entanglement with discarded fishing or other ring shaped materials is a very serious threat to marine animals. The young fur seals are fascinated by floating plastic debris. They drive and role about in it (Mattlin and Cawthorn, 1986). Many fur seal pups have been observed to grow with plastic debris collars which tighten with time and damages the arteries (Weisskopf, 1988) leading to death of seal. Once the seal dies, the plastic collar is free to be picked by another victim. The viscious cycle continues affecting the seal population (Mattlin and Cawthorn, 1986). The decline in populations of Antartic fur seals (*Arctocephalus gazella*) (Croxall *et al.*, 1990), northern sea lion (*Eumetopias jubatus*), endangered Hawaiian monk seal (*Monachus schauinslandiI* (Henderson, 1990; 2001), Australian fur seals (*Arctocephalus pusillus doriferus*) (Hanni and Pyle, 2000) and northern fur seal (Fowler, 1987) have been reported due to entanglement in plastics.

Heavy metals

Heavy metal pollution has likely played an important role in global biodiversity decline. The pollutants such as heavy metals disturb the delicate balance between members of community by eliminating the most sensitive populations. Extremely low diversity of higher plant species and saprophytic fungi was reported in bare industrial area possessing Zn, Pb, Cu and Cd (Vangrosvold *et al.*, 1996). Hernandez and Pastor (2008) investigated the effect of heavy metal on plant biodiversity in grasslands of Central Spain. Soil heavy metals and sodium concentration had negative effects on plant biodiversity. Greatest effect on biodiversity was shown by Zn, followed by Cd and Cu. Floral diversity is greatly impacted by water-soluble concentration of Cu and Ni. Decrease in concentration of Cu and Ni in upper horizons of mine soil resulted in increased floral diversity (Bagatto and Shorthouse, 1999). Soil Pb, Zn and Cd concentration revealed a negative correlation with various plant biodiversity markers (Vidic *et al.*, 2006). Agoramoorthy *et al.* (2008) reported that average concentration of metal contaminants like Cu, Fe, Mg, Mn, Zn, Hg, Pb and SN from industrial sources affected the halophytic plant biodiversity of Tamil Nadu, India.

Heavy metal pollution drastically reduces the functional diversity of microbial communities in soil (Kandeler *et al.*, 1996). Yao *et al.* (2006) observed reduced population of microbial communities in Cu-polluted red soils. Soil Zn and Fe levels also correlated negatively with number of nematode genera (Gyedu-Ababio *et al.*, 1999) while Cu levels intensely suppressed the bacterial growth and number of bacteriovorous nematodes (Bouwman *et al.*, 2001). Nahmani and Lavelli (2002) reported that reduced density and diversity of earthworms and other macrofaunal communities is linked with total soil Zn concentration. Heavy metal pollution has also contributed to global amphibian decline. Ficken and Byrne (2013) reported that distribution of three frog species i.e. *Crinia signifera, Limnodynastes tasmaniensis* and *Limnodynastes ewingii* correlated negatively with total levels of metal contaminants in wetlands located along the Merri Creek Corridor in Victoria, South-Eastern Australia. They also found that species richness negatively correlated with sediment concentration of copper, nickel, lead, zinc, cadmium and mercury.

Heavy metals from industrial and agricultural sources has contaminated freshwater systems and resulted in decline of amphibian and freshwater vertebrate population. Sublethal doses of metal contaminants have variety of harmful effects like reduced growth rates, delayed metamorphosis and impaired behavioural responsiveness (Greig *et al.*, 2010). Industries have contributed a large amount of heavy metals like Cu, Cr, Pb, Zn to Jakarta Bay and resulted in reduced biodiversity of marine benthic organisms (Takarina and Adiwibowo, 2011). Heavy metal bioaccumulate in organisms in form of cations which possess capacity to bind with short carbon atoms (Aslam and Tanaka, 2004). Heavy metals bioaccumulate in various organs of fresh water fishes including gills, liver, kidney, flesh etc. that are present in water systems contaminated with heavy metals like Cr, Ni, Cd and Pb (Vinodhini and Narayanan, 2008). Heavy metal pollution from industries had a significant effect on the biotic community of Kenting National Park, Taiwan. Earthworm, snail, crab and bat bioaccumulated cadmium; invertebrates, amphibians and reptiles had high levels of mercury while high concentration of cadmium, mercury and tin was reported among plant species (Hsu *et al.*, 2006). Incorporation of heavy metal changes the cell division of phytoplankton and hence very large number of cells was reported in phytoplanktons (Davies, 1978).

Global warming

It has been accepted that the world is heating up as the earth's surface average temperature increased by 0.6 °C in the 20th century. According to Intergovernmental Panel on Climate Change, temperature could rise 1.4 ° to 5.8 ° above 1990 average by 2100 (Sharma, 2011). Global warming is one of the most dangerous threats to the planet's biodiversity (Malcolm *et al.*, 2006).

Since oceans and seas constitute a large portion of our planet, they are mostly affected by the process of changes caused by global warming. Global warming not only results in increased temperature of large water masses such as oceans, seas, lakes and ponds but it also causes hydrological events that changes physical and chemical properties of water. Species in the oceans and in the freshwater are at higher risk from global warming as they are highly sensitive to warming temperatures. Coral bleaching i.e. whitening of the coral is one of the consequences of global warming. Abnormal increase in sea water temperatures results in expelling single celled symbiotic alga called zooxanthellae. Since the zooxanthellae contributes to nutrient production, their expulsion influences coral growth and make coral prone to diseases (Wilkinson, 1999). As fishes are exotherms, their body temperature is influenced by the global temperature. It has been reported that different aspects of fish physiology like growth, reproduction and activity are directly related to changes in body temperature (Ficke *et al.*, 2007). Hence, rising global temperature can alter the fish physiology by altering thermal tolerance, growth, metabolism, food consumption, reproductive success and maintenance of homeostasis with varying external environment (Fry, 1971). Temperature fluctuations during early fish development might disrupt vital developmental process during morphogenesis and may induce various types of deformities (Eissa *et al.*, 2009). Increased carbon dioxide levels have been reported to cause sensory and behavioural impairment among various marine species (Nilsson *et al.*, 2012).

Increased water temperature due to global warming have a remarkable effect on the timing and strength of stratification in lotic system (Gaedke *et al.*, 1998). Due to rising climatic temperature, Lake Geneva in Switzerland has not experienced a complete mixing since 1986 (Gerdaux, 1998). Higher temperatures and insolation could increase water loss from lentic systems resulting in net reduction in water level (Allan *et al.*, 2005). Due to lower lake levels, temperate fishes often fail to reproduce successfully as they depend on littoral vegetation for reproductive success (Moyle and Cech, 2004).

Another possible consequence of global warming is the skewed sex ratio among various animals like reptiles, birds and fishes (Eissa and Zaki, 2011). As in the case of reptiles including tuatara (Huey and Janzen, 2008) and turtles (Janson, 1994), increased global temperatures would lead to feminizations of populations. Increase in global temperature is favouring the growth of pathogenic organisms leading to biodiversity erosion. Large scale warming has resulted in disappearance of 67% of the 110 species of *Atelopus*, in the mountains of Costa Rica, due to the outbreaks of a pathogenic chytrid fungus, Batrachochytrium dendrobatidis (Pounds *et al.*, 2006). Global warming also affects the crop yields. With every 1° C increase in night temperature, the rice yield decreased by 10% in the growing season (Peng *et al.*, 2004).

Deforestation

Deforestation has been increasing significantly over the past decades to the point where it has now reached an alarming rate. It is associated with loss of species due to habitat destruction and destabilization of food chains resulting in the collapse of the ecosystem. Between 1950 and 2000, 9.1% of the total species in Madagascar are in danger of extinction due to deforestation (Allnutt *et al.*, 2008). Amazonia has experienced massive deforestation which has led to loss of plant and animal biodiversity (Vieira, 2008). There has been massive deforestation in Indian Himalayas. By 2100, there is high risk that approximately quarter of endemic species, including 366 endemic vascular plant taxa and 35 endemic vertebrate taxa could be wiped out (Pandit *et al.*, 2007). Higher level of species loss was observed after 99.8% deforestation in Singapore. 26% of the vascular plant flora, 28% of resident avifauna and 44% of freshwater species were lost due to deforestation (Turner *et al.*, 1994). About 99.6% loss of Puerto Rico's primary forest is related to 12% loss of land bird species (Brash, 1987). A study by Fondo and Martens (1998) revealed that mangrove deforestation decreases the macrofaunal diversity in Gazi Bay, Kenya. Due to overdependence of rural livelihood on their traditional resources, mangrove forest of the Niger Delta is experiencing steady deforestation and subsequent loss of biodiversity (Mmom and Arokoyu, 2009).

Habitat fragmentation and loss

Habitat fragmentation and loss are recognized as one of the greatest existing threats to biodiversity. Habitat fragmentation can cause immediate loss of biodiversity along with time-delayed extinctions (Krauss *et al.*, 2010). Habitat loss reduces the trophic chain length, number of specialist species. Habitat loss also alters the breeding success, dispersal success, predation rate and foraging success rate (Fahrig, 2003). Iwata *et al.* (2003) reported that habitat alteration had an overall detrimental effect on biodiversity of stream communities in a tropical rain forest in Bornea. Habitat fragmentation caused by humans destroys species, disrupts community interactions and ultimately interferes with evolutionary processes (Levin, 1999). In 80 years, fragmented area of montane forest at San Antonia in the Columbian Andes had lost around a third of the total species (Kattan *et al.*, 1994). Due to loss of coral habitat, *Gobiodon* species C is facing global extinction in Kimbey Bay, Papua New Guinea (Munday, 2004).

Insects are highly prone to adverse effects caused by forest fragmentation. Forest fragmentation induces changes in species richness and abundance among insects (Didham *et al.*, 1996). Rathcke and Jules (1993) observed that plant-pollinator relationship is critically affected by fragmentation. When new patches of habitat were introduced within a continuous habitat, beta diversity will probably increase as new species will be observed in new patches (Rudnicky and Hunter, 1993). Jennersten (1998) observed that two fragments had lower visiting insect species than a continuous stretch of mainland habitat.

Higher trophic levels are more susceptible to habitat loss and fragmentation than lower trophic levels (Krauss *et al.*, 2010). Cohen *et al.* (1993) reported the negative impact of excess sediment pollution on the biodiversity of Lake Tanganyika, Africa. They observed that ostracodes and fish population was less diverse in highly disturbed sites than in less disturbed ones. Approximately 85% of globally threatened bird species are at risk due to habitat loss and degradation (BirdLife International, 2000). According to Gaston *et al.* (2003), land use changes have resulted in loss of 1/5th to 1/4th of pre-agricultural birds. A highly isolated 87-ha forest patch at Rio Palenque in Ecuador showed a loss of 25 species of birds in just 5 years (Leck, 1979). Bascompte *et al.* (2002) predicted the negative effect of habitat loss on the growth rate of population. Degradation of wintering sites at lower latitudes had affected the migratory breeding populations of relatively pristine high latitude regions (Faaborg, 2003). Aguilar *et al.* (2008) reported that habitat fragmentation leads to genetic erosion by shifting the mating patterns towards increased selfing. The age of the fragment also determines the variability. The higher the number of generations elapsed after habitat fragmentation, the higher the negative effect on the magnitude of heterozygosity. Sole *et al*. (2004) analyzed that after attainment of certain value of habitat loss, biodiversity collapses suddenly.

Invasive alien species

Alien species are species whose introduction and spread outside their natural habitat occurs. An alien species must arrive, survive and thrive to become an invasive species. Alien species invasions are growing pressure on the natural world and act as direct drivers of biodiversity loss across the globe as movement of species between continents, results in homogenization of the previously dissimilar communities and the results in loss of diversity at larger scales. International shipping, aquaculture and fisheries, aquarium industry, sea food trade, recreational boating, diving practices and floating debris offers transport opportunities to the alien species (Molnar *et al*., 2008). In a Hawaiian study, approximately 23% species were non-indigenous or cryptogenic in Pearl Harbour (Coles *et al*., 1999). California scientists have reported 212 alien marine, estuarine and freshwater in San Francisco Bay and Delta (Cohen and Carlton, 1998), while 159 alien marine species have been found in New Zealand (Cranfield *et al*., 1998). Out of total 252 introduced and cryptogenic marine and estuarine species in Australia, more than 150 alien species have been identified in Port Phillip Bay alone (Hewitt *et al*., 1999). *Parthenium hysterophorus* (commonly known as congress grass) is a noxious and rapidly spreading weed in Asia, America, Africa and Australia. Due to its allelopathic nature, crop production is drastically reduced. Its aggressive dominance in wasteland, road sides, railway sides, water courses, overgrazed pastures and cultivated fields threatens the biodiversity (Kaur *et al*., 2014).

The invasive alien species have resulted in transformation of natural habitats around the world. The exotic species might dominate the natural species in terms of number of species, number of individuals and biomass and high and accelerating rate of invasion. These species might displace native species; alter community structure, food webs structure and changes fundamental processes like nutrient cycling and sedimentation rates. Invasive species decreases the richness and diversity due to competition, predation, and parasitism and might cause extinction of native species. Invasion of Asian class *Potamocorbula amurensis* in San Francisco Bay and the Eurasian Zebra mussel *Dreissena polymorpha* in Great lakes have transformed the community structure and function. They have become dominant in terms of number and replaced other benthic organisms and cleared planktonic communities from overlying waters (Ruiz *et al*., 1997). Some invasive species can even have serious health concerns. Invasive chytrid fungus, *Batrachochytrium dendrobatidis* causes chytidiomycosis in amphibians and contributes to global amphibian declines (Fites *et al*., 2013). Invasive alien species affect the native species through predation, as in the case of feral cats which have invaded particularly islands (Nogales *et al*., 2004). Initially feral cat introductions were made to control rodent and rabbit population but it had devastating impact on island fauna i.e. mammals, reptiles and birds (Nogales *et al*., 2004). Hybridization within invasive species further adds to the invasive success by affecting genetic diversity. Japanese knotweed is a virulent invasive hybrid which spreads faster than its parents, outcompeting other plants and altering ecosystem with effects on other species. It causes damage to building foundations, excludes the native species, and dies in winter leaving river banks vulnerable to erosion (Parepa *et al.*, 2014). Unfortunately, there is lack of information on invasive alien species at a global scale. There is a need for compilation of information on alien species that possesses threat to ecosystems, habitats or species so that prevention and mitigation activities could be initiated.

Dams

A dam is a physical barrier built across a river or stream to impound water and contribute towards water resource management. Though dams were built with positive intensions in mind i.e. water storage, irrigation, generation of hydroelectricity, prevention of floods, navigation, irrigation etc. (McCall, 2008) but their negative impacts on the elements of biodiversity have overbalanced the positive goal (Lin, 2011).

Damming a river has a variety of effects on the freshwater ecosystem. Dams threaten the water quality and water dependent biodiversity. A group of researchers from National University of Singapore predicted that dam-related activities will submerge and dismantle 1,70,000 hectares of forests. They also projected that damming would cause deformation and extinction of 22 flowering plant and 7 vertebrate species in Indian Himalayas (Grumbine and Pandit, 2013).

Undammed rivers bring natural floods which increase the nutrient and moisture level of flooded areas. The increased level of nutrients due to flooding serves as food for stream residents. Hence, various aquatic animals plan their reproductive cycles according to annual flood seasons (Dynesius and Nilsson, 1994). Moreover, flood also provides shallow backwater areas that serve as protection sites for young ones. Therefore, surrounding areas of dam free river banks are rich in biodiversity. Dammed rivers reduce the flood rates and modify the flooding character of river thereby affecting the ecology, agriculture and biodiversity (Bednarck, 2001).

Dams reduce the flow of water in rivers and moreover cementing walls impede the migration of fishes. Damming block their migration to spawning areas and contribute to decline or even extinction of species such as Chinook salmon and steelherd trout in USA and Atlantic salmon in Europe by increased nitrogen gas at the bottom of dam (Nilson *et al.*, 2005). In another study, it was reported that migration of 3 mahseer fish species which are long distance migrants has been obstructed by Tehri dam (Mathur *et al.*, 2012). Isolation of species due to a damming barrier results in reduced genetic diversity and therefore put species at greater risk for disease (Lin, 2011). Various other environmental problems i.e. increased salinity and increased algal blooms are caused by reduced water flow in rivers may also affect the ecosystem. River damming leads to reduced water tables and homogenized river flow thereby reducing the variability of ecosystem.

Older dams release water that is stored at the bottom of dam, which is typically colder and adversely affects the species adapted to warmer temperatures. Such an effect is sometime referred to as 'cold water pollution' (Lin, 2011). Hence, dams change the overall temperature regimes and transport of sediments favoring the generalist species and loss of specialist species leading to biotic homogenization (Lin, 2011). As dam reservoirs are stratified with respect to the oxygen level, the bottom layer lacks oxygen leading to anaerobic degradation of biomass and generation of methane, a greenhouse gas. Damming also results in deposition of heap of sediments, limiting the storage capacity and at the same time it might lead to bursting of dam and expiration of river. Entrapment of nutrients due to dam construction also results in excessive growth of aquatic weeds and low dissolved oxygen, ultimately having detrimental effect on fishes. Dams are one of the greatest global threats to biodiversity.

Conclusion

Alteration of global environment has triggered an extinction event in the history of life. Due to human activities, the local and global diversity of the planet is greatly threatened. Various factors causing biodiversity loss are pollution, global warming, deforestation, habitat fragmentation and loss, species introductions and damming of aquatic bodies. Pollution has been a major factor affecting the floral and faunal biodiversity. It has already caused major changes at all trophic levels. It has altered the structure and function of aquatic, terrestrial and avian communities over large areas. Bioaccumulation of toxic pollutants alters the foraging behavior, reproductive ability, learning capacity and physiology of different species. It causes severe nervous disorders among different organisms. Another important environmental factor is deforestation which destabilizes food chains and leads to loss of species by habitat destruction. Habitat fragmentation and loss has large consistent negative effect on biodiversity leading to genetic erosion. It decreases the proportion of appropriate habitat in landscape and isolation effect influences the population size of species. It affects the breeding success, dispersal success, predation rate and foraging success rate. It has elevated adverse effect on higher trophic levels than lower ones. Invading alien species is another important aspect affecting biodiversity. Invaded species act as vectors of diseases, alters ecosystem processes, reduces crop production, transforms natural habitats, reduces biological diversity and promotes extinction. Redistribution of species on earth at such a swift pace is challenging the ecosystem, threatening human health and straining economies. It is both economically as well as ecologically damaging and the costs will continue to worsen. Nowadays, construction of dams over water bodies in expected to increase worldwide. Developing countries are building dams at an alarming rate as they produce clean hydroelectric power and assist to improve the irrigation and sanitation facilities. Damming disturbs the flow rate of water bodies, blocks fish migration, and alters water temperature and quality. Dams might also cause depletion of biodiversity.

The problems due to environmental changes are likely to exacerbate and pose significant risk to biodiversity in coming years. Immediate actions in form of conservation strategies are necessary in order to conserve the threatened species for future generations. Public awareness and efforts to conserve biodiversity should be encouraged. Concerned and informed citizens can participate by recognizing the issues and preventing them. Best possible utilization of the available resources should initiated from local to international and global scale for broader interest of mankind. The concept of thinking globally and acting locally can mitigate the threats caused by environmental changes. Management practices that maintain integrity of ecosystems, halt extinctions and enhance recovery of the threatened species should be promoted.

References

1. Aguilar, R., Quesada, M., Ashworth, L., Diego, Y.H. and Lobo, J. (2008). Genetic consequences of habitat fragmentation in plant populations: susceptible signals in plant traits and methodological approaches. *Molecular Ecology.* 17(24): 5177-5188.

2. Allan, J.D., Palmer, M. and Poff, N.L. (2005). Climate change and freshwater ecosystems. In: Lovejoy, T. E., Hannah, L. (Eds.), *Climate change and biodiversity* (pp. 274-295), Yale University Press, New Haven, CT.

3. Allnutt, T.F., Ferrier, S., Manion, G., Powell, G.V., Ricketts, T.H., Fisher, B.L. and Lees, D.C. (2008). A method for quantifying biodiversity loss and its application to a 50-year record of deforestation across Madagascar. *Conservation Letters.* 1(4): 173-181.

4. Azzarello, M.Y. and Van-Vleet, E.S. (1987). Marine birds and plastic pollution. *Marine Ecology Progress Series.* 37: 295-303.

5. Bagatto, G., and Shorthouse, J.D. (1999). Biotic and abiotic characteristics of ecosystems on metaliferous mine tailings near Sudbury, Ontario. *Canadian Journal of Botany.* 77: 410-425.

6. Baliarsingh, S.K., Sahoo, S., Acharya, A., Dalabehera, H.B., Sahu, K.C. and Lottiker, A.A. (2014). Oil pollution in Cilika lagoon: an anthropogenic threat to biodiversity. *Current Science.* 106(4): 516-517.

7. Bascompte, J., Possingham, H. and Roughgarden, J. (2002). Patchy populations in stochastic environments: critical number of patches for persistence. *American Naturalist.* 159(2): 128-137.

8. Bednarck, A.T. (2001). Undamming rivers: a review of the ecological impacts of dam removal. *Environment Management.* 27(6): 803-814.

9. Beketov, M.A., Kefford, B.J., Schäfer, R.B. and Liess, M. (2013). Pesticides reduce regional biodiversity of stream invertebrates. Proceedings of the National Academy of Sciences. 110(27): 11039-11043.

10. Berny, P.J., Buronfosse, T., Buronfosse, F., Lamarque, F. and Lorgue, G. (1997). Field evidence of secondary poisoning of foxes (Vulpes vulpes) and buzzards (Buteo buteo) by bromadiolone, a 4-year survey. *Chemosphere.* 35(8): 1817-1829.

11. BirdLife International. (2000). Threatened birds of the world. Lynx Edicions and BirdLife International. Barcelona and Cambridge, UK.

12. Bishop, C.A., Ng, P., Mineau, P., Quinn, J.S. and Struger, J. (2000). Effects of pesticide spraying on chick growth, behavior, and parental care in tree swallows (*Tachycineta bicolor*) nesting in an apple orchard in Ontario, Canada. *Environmental Toxicology and Chemistry.* 19(9): 2286-2297.

13. Bouwman, L.A., Bloem, J., Römkens, P.F.A.M., Boon, G.T. and Vangronsveld, J. (2001). BeneWcial eVects of the growth of metal tolerant grass on biological and chemical parameters in copper and zinc contaminated sandy soils. Minerva Biotecnologica. 13: 190-226.

14. Bovarnick, A., Alpizar, F. and Schnell, C. (2010). The importance of biodiversity and ecosystem in economic growth and equity in Latin America and the Caribbean: an economic valuation of ecosystem. *United Nations Development Programme.*

15. Brash, A.R. (1987). The history of avian extinction and forest conversion on Puerto Rico. *Biological Conservation.* 39(2): 97-111.

16. Bury, R.B. (1972). The effects of diesel fuel on a stream fauna. *Calif Fish and Game.* 58(4): 201-205.

17. Carlton, J.T. (2001). Introduced species in US coastal waters: environmental impacts and management priorties, Arlington, Virginia: Pew Oceans Commission.

18. Central Science Laboratory, Game Conservancy Trust, RSPB and Department of Zoology. (2005).

19. Cheema, N., Bhatnagar, A. and Yadav, A.S. (2014). Changes in growth performance and biochemical status of fresh water fish *Cirrhinus mrigala* exposed to sublethal doses of chlorpyrifos. *International Journal of Agricultural and Food Science Technology.* 5(6): 619-630.

20. Cohen, A.N. and Carlton, J.T. (1998). Accelerating invasive rate in a highly invaded estuary. *Science.* 279: 55-58.

21. Cohen, A.S., Bills, R., Cocquyt, C.Z. and Caljon, A.G. (1993). The impact of sediment pollution on biodiversity in Lake Tanganyika. *Conservation Biology.* 7(3): 667-677.

22. Coles, S.L., DeFelice, R.C. Eldredge, L.G. and Carlton J.T. (1999). Historical and recent introductions of non-indigenous marine species into Pearl Harbor, Oahu, Hawaiian Islands. *Marine Biology.* 135: 147-158.

23. Connors, P.G. and Smith, K.G. (1982). Oceanic plastic particle pollution: suspected effect on fat deposition in red phalaropes. *Marine Pollution Bulletin.* 13: 18-20.

24. Cranfield, H.J., Gordon, D.P., Wiilian, R.C., Marshall, B.A., Battershill, C.N., Francis, M.P., Nelson, W.A., Glasby, C.J. and Read, G.B. (1998). Adventive marine species in New Zealand. National Institute of Water & Atmosphere. Technical Reports 34, Wellington.

25. Croxall, J.P., Rodwell, S. and Boyd, I.L. (1990). Entanglement in manmade debris of Antarctic fur seals at Bird Island, South Georgia. *Marine Mammal Science.* 6: 221–233.

26. Decourtye, A., Lacassie, E. and Pham-Delègue, M.H. (2003). Learning performances of honeybees (Apis mellifera L) are differentially affected by imidacloprid according to the season. *Pest Management Science.* 59(3): 269-278.

27. Derraik, J.G.B. (2002). The pollution of the marine environment by plastic debris: a review. *Marine Pollution Bulletin*. 44: 842-852.

28. Didham, R.K., Ghazoul, J., Stork, N.E. and Davis, A.J. (1996). Insects in fragmented forests: a functional approach. *Trends in Ecology and Evolution*. 11(6): 255-260.

29. Donald, P.F., Pisano, G., Rayment, M.D. and Pain, D.J. (2002). The Common Agricultural Policy, EU enlargement and the conservation of Europe's farmland birds. *Agriculture, Ecosystems & Environment*. 89(3): 167-182.

30. Dynesius, M. and Nilsson, C. (1994). Fragmentation and flow regulation of river systems in northern third of the world. *Science*. 266(5186): 753-762.

31. Edwards, C.A. (2002). Assessing the effects of environmental pollutants on soil organisms, communities, processes and ecosystem. *European Journal of Soil Biology*. 38(3-4): 225-231.

32. Eissa, A.E. and Zaki, M.M. (2011). The impact of global climatic changes on the aquatic environment. *Procedia Environmental Sciences*. 4: 251-259.

33. Eissa, A.E., Moustafa, M., El-Husseiny, I.N., Saeid, S., Saleh, O. and Borhan, T. (2009). Identification of some skeletal deformities in some freshwater teleost raised Egyptian aquaculture. *Chemosphere*. 77: 419-425.

34. Faaborg, J. (2003). Saving migrant birds: developing strategies for the future. Austin TX. University of Texas Press.

35. Fahrig, L. (2003). Effects of habitat fragmentation on biodiversity. *Annual Review of Ecology, Evolution, and Systematics*. 34: 487-515.

36. Ficke, A.D., Myrick, C.A. and Hansen, L.J. (2007). Potential impacts of climate change on freshwater fisheries. *Reviews in Fish Biology and Fisheries*. 17: 581-613.

37. Ficken, K.L.G. and Byrne, P.G. (2013). Heavy metal pollution negatively correlates with anuran species richness and distribution in south-eastern Australia. *Austral Ecology: a journal of ecology in the Southern Hemisphere*. 38(5): 523-533.

38. Fites, J.S., Ramsey, J.P., Holden, W.M., Collier, S.P., Sutherland, D.M., Reinert, L.K., Gayek, A.S., Dermody, T.S., Aune, T.M. and Oswald-Richter, L.A. (2013). The invasive chytrid fungus of amphibians paralyzes lymphocyte responses. *Science*. 342(6156): 366-369.

39. Fondo, E.N. and Martens, E.E. (1998). Effects of mangrove deforestation on macrofaunal densities, Gazi Bay, Kenya. *Mangroves and salt marshes*. 2(2): 75-83.

40. Forson, D.D. and Storfer, A. (2006). Atrazine increases ranavirus susceptibility in the tiger salamander, *Ambystoma tigrinum*. *Ecological Applications*. 16(6): 2325-2332.

41. Fowler, C.W. (1987). Marine debris and northern fur seals: a case study. *Marine Pollution Bulletin*. 18: 326-335.

42. Franeker, J.A.V., Blaize, C., Danielsen, J., Fairclough, K., Gollan, J., Guse, N., Hansen, P.L., Heubeck, M., Jensen, J.K., Guillou, G.L., Olsen, B., Olsen,

K.O., Pedersen, J., Stienen, E.W.M. and Turner, D.M. (2011). Monitoring plastic ingestion by the northern fulmar Fulmarus glacialis in the North Sea. *Environmental Pollution*. 159(10): 2609-2615.

43. Fry, E.E.J. (1971). The effect of environmental factors on the physiology of fish, In: Hoar, W.S., Randall, D.J. (Eds.), *Fish physiology: environmental relations and behavior*, Academic press, New York. 1-98.

44. Gaedke, U., Ollinger, D., Kirner, P. and Bauerle, E.L. (1998). The influence of weather conditions on the seasonal plankton development in a large and deep lake. In: George, D. G. (Ed.) *Management of lakes and reservoirs during global climate change*, Kluwer Academic Publishers, Dodrecht, Netherlands. 71-84.

45. Gaston, K.J., Blackburn, T.M. and Goldewijk, K.K. (2003). Habitat conversion and global avian biodiversity loss. *Proceedings of the Royal Society of London B*. 270: 1293-1300.

46. Geiger, F., Jan Bengtsson, Frank Berendse, Wolfgang W. Weisser, Mark Emmersond, Manuel B. Morales, Piotr Ceryngier, Jaan Liira, Teja Tscharntke, Camilla Winqvist, Sönke Eggers, Riccardo Bommarco, Tomas Pärt, Vincent Bretagnolle, Manuel Plantegenest, Lars W. Clement, Christopher Dennis, Catherine Palmer, Juan J. Oñate, Irene Guerrero, Violetta Hawro, Tsipe Aavik, Carsten Thies, Andreas Flohre, Sebastian Hänke, Christina Fischer, Paul W. Goedhart, Pablo Inchausti (2010) Persistent negative effects of pesticides on biodiversity and biological control potential on European farmland. Basic and Applied Ecology. 11(2): 97-105.

47. Gerdaux, D. (1998). Fluctuations in lake fisheries and global warming. In: George, D. G. (Ed.) *Management of lakes and reservoirs during global climate change*, Kluwer Academic Publishers, Dodrecht, Netherlands. 263-272.

48. Goldberg, E.D. (1994). Diamonds and plastics are forever? *Marine Pollution Bulletin*. 28: 466.

49. Goldstein, M.I., Lacher, T.E., Woodbridge, B., Bechard, M.J., Canavelli, S.B., Zaccagnini, M. E., Cobb, G.P., Scollon, E.J., Tribolet, R. and Hopper, M.J. (1999). Monocrotophos-induced mass mortality of Swainson's Hawks in Argentina, 1995–96. *Ecotoxicology*. 8(3): 201-214.

50. Gregory, M.R. (1978). Accumulation and distribution of virgin plastic granules on New Zealand beaches. *New Zealand Journal of Marine and Freshwater Research*. 12: 399-414.

51. Greig, H.S., Niyogi, D.K., Hogsden, K.L., Jellyman, P.G. and Harding, J.S. (2010). Heavy metals: confounding factors in the response of New Zealand freshwater fish assemblages to natural and anthropogenic acidity. *Science of the Total Environment*. 408: 3240-50.

52. Grumbine, R.E. and Pandit, M.K. (2013). Threats from India's Himalaya dams. *Science*. 339(6115): 36-37.

53. Hanni, K.D. and Pyle, P. (2000). Entanglement of pinnipeds in synthetic materials at South-east Farallon Island, California, 1976–1998. *Marine Pollution Bulletin*. 40: 1076–1081.

54. Henderson, J.R. (1990). A review of Hawaiian monk seal entanglement in marine debris. In: Shomura, R.S., Yoshida, H.O. (Ed.), *Proceedings of the Workshop on the Fate and Impact of Marine Debris*. 27–29 November 1984, Honolulu. US Department of Commerce. 326-336.

55. Henderson, J.R. (2001). A pre-and post-MARPOL Annex V summary of Hawaiian monk seal entanglements and marine debris accumulation in the Northwestern Hawaiian Islands, 1982–1988. *Marine Pollution Bulletin*. 42: 584-589.

56. Hewitt, C., Campbell, M., Thresher, R. and Martin, R. (1999). Marine biological invasions of Port Philip Bay, Victoria. CRIMP Technical Report 20, CSIRO Marine Research, Hobart, Tasmania.

57. Hole, D.G., Perkins, A.J., Wilson, J.D., Alexander, I.H., Grice, P.V. and Evans, A.D. (2005). Does organic farming benefit biodiversity? *Biological conservation*. 122(1): 113-130.

58. Huey, R.B. and Janzen, F.J. (2008). Climate warming and environmental sex determination in tuatara: the last of the sphenodontians? *Proceedings of the Royal Society B*. 275(1648): 2181-2183.

59. Isenring, R. (2010). Pesticides reduce biodiversity. *Pesticides News*. 88: 1-7.

60. Islam, M.S. and Tanaka, M. (2004). Impacts of pollution on coastal and marine ecosystems including coastal and marine fisheries and approach for management: a review and synthesis. *Marine Pollution Bulletin*. 48(7-8): 624-649.

61. Iwata, T., Nakano, S. and Inoue, M. (2003). Impacts of past riparian deforestation on stream communities in a tropical rain forest in Bornea. *Ecological Applications*. 13(2): 461-473.

62. Janson, F.J. (1994). Climate change and temperature-dependent sex determination in reptiles. *Proceedings of the National Academy of Sciences of the United States of America*. 91(16): 7487-7490.

63. Jennersten, O. (1998). Pollination in *Dianthus deltoides* (Caryophyllaceae): effects of habitat fragmentation on visitation and seed set. *Conservation Biology*. 2(4): 359-366.

64. Kandeler, E., Kampichler, C. and Horak, O. (1996). Influence of heavy metals on the functional diversity of soil microbial communities. *Biology and Fertility of Soils*. 23: 299-306.

65. Kannan, K., Tanabe, S., Ramesh, A., Subramanian, A. and Tatsukawa, R. (1992). Persistent organochlorine residues in foodstuffs from India and their implications on human dietary exposure. *Journal of Agricultural and Food Chemistry*. 40(3): 518-524.

66. Kattan, G.H., Alvarez-López, H. and Giraldo, M. (1994). Forest fragmentation and bird extinctions: San Antonio eighty years later. *Conservation Biology*. 8(1): 138-146.

67. Kaur, M., Aggarwal, N.K., Kumar, V. and Dhiman, R. (2014). Effects and management of *Parthenium hysterophorus*: a weed of global significance. *International Scholarly Research Notices*. 2014: 12.

68. Kingston, P.F. (2002). Long-term environmental impact of oil spills. *Spill Science & Technology Bulletin.* 7(1): 53-61.

69. Kole, R.K. and Bagchi, M.M. (1995). Pesticide residues in the aquatic environment and their possible ecological hazards. *Journal of the Inland Fisheries Society of India.* 27(2): 79-89.

70. Krauss, J., Bommarco, R., Guardiola, M., Heikkinen, R.K., Helm, A., Kuussaari, M., Lindborg, R., Ockinger, E., Pärtel, M., Pino, J., Pöyry, J., Raatikainen, K.M., Sang, A., Stefanescu, C., Teder, T., Zobel, M. and Steffan-Dewenter, I. (2010). Habitat fragmentation causes immediate and time-delayed biodiversity loss at different trophic levels. *Ecology Letters.* 13(5): 597-605.

71. Laffon, B., Rábade, T., Pásaro, E. and Méndez, J. (2006). Monitoring of the impact of Prestige oil spill on *Mytilus galloprovincialis* from Galician coast. *Environment International.* 32(3): 342-348.

72. Laist, D.W. (1997). Impacts of marine debris: entanglement of marine life in marine debris includinga comprehensive list of species with entanglement and ingestion records. In: Coe, J.M., Rogers, D.B. (Eds.), *Marine Debris— Sources, Impacts and Solutions. Springer-Verlag, New York.* 99–139.

73. Lazar, B. and Gracan, R. (2011). Ingestion of marine debris by loggerhead sea turtles, Caretta caretta, in the Adriatic Sea. *Marine Pollution Bulletin.* 62(1): 43-47.

74. Leck, C.F. (1979). Avian extinctions in an isolated tropical wet-forest preserve, Ecuador. *The Auk.* 96: 343-352.

75. Levin, S.A. (1999). Fragile dominion: complexity and the commons. Reading MA: Perseus Books.

76. Lin, Q. (2011). Influence of dams on river ecosystem and its counter measures. *Journal of Water Resources and Protection.* 3: 60-66.

77. Luiselli, L. and Akani, G.C. (2003). An indirect assessment of the effects of oil pollution on the diversity and functioning of turtle communities in the Niger Delta, Nigeria. *Animal biodiversity and conservation.* 26(1): 57-65.

78. Mäder, P., Fliessbach, A., Dubois, D., Gunst, L., Fried, P. and Niggli, U. (2002). Soil fertility and biodiversity in organic farming. *Science.* 296(5573): 1694-1697.

79. Malcolm, J.R., Liu, C., Neilson, R.P., Hansen, L. and Hannah, L. (2006). Global warming and extinction of endemic species from biodiversity hotspots. *Conservation Biology.* 20(2): 538-548.

80. Mathur, V.B., Choudhury, B.C., Melkani, V.K. and Uniyal, V.K. (2012). Cumulative Impacts of Hydropower Dams on Alaknanda& Bhagirathi Rivers on Aquatic and Terrestrial Ecosystems. *Wildlife Institute of India.* 2012.

81. Mattlin, R.H. and Cawthorn, M.W. (1986). Marine debris—an international problem. *New Zealand Environment.* 51: 3–6.

82. McCall, J.M. (2008). Primary production and marine fisheries associated with the Nile outflow. *Earth & Environment.* 3: 179.

83. Mmom, P.C. and Arokoyu, S.B. (2009). Mangrove forest depletion, biodiversity loss and traditional resources management practices in the Niger Delta, Nigeria. *Research Journal of Applied Sciences, Engineering and Technology*. 2(1): 28-34.

84. Molnar, J.L., Rebecca, L.G., Revenga, C. and Spalding, M.D. (2008). Assessing the global threat of invasive species to marine biodiversity. *Frontiers in Ecology and the Environment*. 2008: 6.

85. Moore, C.J. (2008). Synthetic polymers in the marine environment: a rapidly increasing, a long-term threat. *Environmental Research*. 108(2): 131-139.

86. Moyle, P.B. and Cech, J.J. (2004). Fishes: an introduction to ichthyology (5th Edn.). Prentice Hall, Englewood Cliffs, NJ.

87. Munday, P.L. (2004). Habitat loss, resource specialization, and extinction on coral reefs. *Global Change Biology*. 10(10): 1642-1647.

88. Nahmani, J. and Lavelle, P. (2002). ffVects of heavy metal pollution on soil macrofauna in a grassland of northern France. *European Journal of Soil Biology*. 38: 297–300.

89. Nigam, S.K., Karnik, A.B., Chattopadhyay, P., Lakkad, B.C., Venkaiah, K. and Kashyap, S.K. (1993). Clinical and biochemical investigations to evolve early diagnosis in workers involved in the manufacture of hexachlorocyclohexane. *International Archives of Occupational and Environmental Health*. 65(1): S193-S196.

90. Nilson, C., Reidy, C.A., Dynesius, M. and Revenga, C. (2005). Fragmentation and flow regulation of the world's large river systems. *Science*. 308(5720): 405-408.

91. Nilsson, G.E., Dixson, D.L., Domenici, P., McCormick, M.I., Sorensen, C., Watson, S.A. and Munday, P.L. (2012). Near future carbon-dioxide levels alter fish behavior by interfering with neurotransmitter function. *Nature Climate Change*. 2: 201-204.

92. Nogales, M., Martin, A., Tershy, B.R., Donlan, C.J., Veitch, D., Puerta, N., Wood, B. and Alonso, J. (2004). A review of feral cat eradication on islands. *Conservation Biology*. 18(2): 310-319.

93. Pandit, M.K., Sodhi, N.S., Koh, L.P., Bhaskar, A. and Brook, B.W. (2007). Unreported yet massive deforestation driving loss of endemic biodiversity in Indian Himalaya. *Biodiversity and Conservation*. 16(1): 153-163.

94. Parepa, M., Fischer, M., Krebs, C. and Bossdorf, O. (2014). Hybridization increases invasive knotweed success. *Evolutionary Applications*. 7(3): 413-420.

95. Peng, S., Huang, J., Sheehy, J.E., Laza, R.C., Visperas, R.M., Zhong, X., Centeno, G.S. and Cassman, K.G. (2004). Rice yields decline with higher night temperatures from global warming. *Proceedings of the National Academy of Sciences of the United States of America*. 101(27): 9971-9975.

96. Pounds, J.A., Bustamante, M.R., Coloma, L.A., Consuegra, J.A., Fogden, M.P.L., Foster, P.N., Marca, E.L., Masters, K.L., Viteri, A.M., Puschendorf, R., Ron, S.R., Azofeifa, G.A.S., Still, C. J. and Young, B.E. (2006). Widespread amphibian extinctions from epidemic diseases driven by global warming. 439: 161-167.

97. Prosser, P. and Hart, A.D.M. (2005). Assessing potential exposure of birds to pesticide-treated seeds. *Ecotoxicology*. 14(7): 679-691.

98. Rathcke, B.J. (1993). Habitat fragmentation and plant–pollinator. *Current Science*. 65: 273-277.

99. Rathore, A. and Jasrai, Y.T. (2013). Biodiversity: importance of climate change impacts. *International Journal of Scientific and Resesarch Publications*. 3(3): 1-5.

100. Reddy, M.S., Basha, S., Adimurthy, S. and Ramachandraiah, G. (2006). Description of the small plastics fragments in marine sediments along the Alang-Sosiya ship-breaking yard, India. *Estuarine, Coastal and Shelf Science*. 68(3-4): 656-660.

101. Rillig, M.C. (2012). Microplastic in terrestrial ecosystem and the soil? Environmental Science & Technology. 46: 6453-6454.

102. Rudnicky, T.C. and Hunter, M.L. (1993). Reversing the fragmentation perspective: effects of clearcut size on bird species richness in Maine. *Ecological Applications*. 3: 357-366.

103. Ruiz, G.M., Carlton, J.T., Grosholz, E.D. and Hines, A.H. (1997). Global invasions of marine and estuarine habitats bu non-indigenous species: mechanisms, extent and consequences. *American Zoologist*. 37: 621-632.

104. Rupa, D.S., Reddy, P.P. and Reddi, O.S. (1991). Reproductive performance in population exposed to pesticides in cotton fields in India. *Environmental research*. 55(2): 123-128.

105. Sharma, D.K. and Mishra, J.K. (2011). Impact of environmental changes on biodiversity. *Indian Journal of Scientific Research*. 2(4): 137-139.

106. Sharma, U. (2011). Effect of global warming on biodiversity in India. National Conference on Forest Biodiversity: Earth's Living Treasure 22nd May. 76-84.

107. Singh, B. (2005). Harmful effect of plastic in animals. *The Indian Cow*. 4: 10-17.

108. Smith, N. (1970). The problem of oil pollution of the sea. *Advances in Marine Biology*. 8: 215-306.

109. Sole, R.V., Alonso, D. and Saldana, J. (2004). Habitat fragmentation and biodiversity collapse in neutral communities. *Ecological Complexity*. 1: 65-75.

110. Swingland, I.R. (2001). Biodiversity, Definition of. *Encyclopedia of Biodiversity*. 1: 377-391.

111. Todd, P.A., Ong, X. and Chou, L.M. (2010). Impacts of pollution on marine life in Southeast Asia. *Biodiversity and Conservation*. 19(4): 1063-1082.

112. Turner, I.M., Tan, H.T.W., Wee, Y.C., Ibrahim, A.B., Chew, P.T. and Corlett, R.T. (1994). A study of plant species extinction in Singapore: lessons for the conservation of tropical biodiversity. *Conservation Biology*. 8: 705-712.

113. Vangronsveld, J., Colpaert, J.V., and Van Tichelen, K.K. (1996). Reclamation of a bare industrial area contaminated by nonferrous metals: physicochemical and biological evaluation of the durability of soil treatment and revegetation. *Environmental Pollution*. 94: 131–140.

114. Vickery, J.A., Tallowin, J.R., Feber, R.E., Asteraki, E.J., Atkinson, P.W., Fuller, R.J. and Brown, V.K. (2001). The management of lowland neutral grasslands in Britain: effects of agricultural practices on birds and their food resources. *Journal of Applied Ecology*. 38(3): 647-664.

115. Vidic, T., Jogan, N., Drobne, D. and Vilhar, B. (2006). Natural revegetation in the vicinity of the former lead smelter in Zerjav, Slovenia. *Environmental Science and Technology*. 40: 4119-4125.

116. Vieira, I.C.G., Toledo, P.D., Silva, J.D. and Higuchi, H. (2008). Deforestation and threats to the biodiversity of Amazonia. *Brazilian Journal of Biology*. 68(4): 949-956.

117. Weisskopf, M. (1988). Plastic reaps a grim harvest in the oceans of the world (plastic trash kills and maims marine life). *Smithsonian*. 18: 58.

118. Wilkinson, C.R. (1999). Global and local threats to coral reef functioning and existence: review and predictions. *Marine and Freshwater Research*. 50(8): 867-878.

119. Yadav, A.S., Bhatnagar, A. and Kaur, M. (2013). Aberrations in the chromosome *of Cirrhinus mrigala* (Hamilton) upon exposure to butachlor. *Iranian Journal of Toxicology*. 7(21): 858-865.

120. Yang, E C., Chuang, Y.C., Chen, Y.L. and Chang, L.H. (2008). Abnormal foraging behavior induced by sublethal dosage of imidacloprid in the honey bee (Hymenoptera: Apidae). *Journal of economic entomology*. 101(6): 1743-1748.

121. Yao, H.Y., Liu, Y.Y., Xue, D. and Huang, C.Y. (2006). Effect of copper on phospholipid fatty acid composition of microbial communities in two red soils. *Journal of Environmental Science*. 18: 503-509.

122. Zitko, V. and Hanlon, M. (1991). Another source of pollution by plastics: skin cleaners with plastic scrubbers. *Marine Pollution Bulletin*. 22: 41-42.

16

Possible Threat of Radiations of Cell Phones and their Towers on Existing Biodiversity

Dev Narayan Gautam, S.Marmat, Taj N. Qureshi and H. S. Rathore

School of Studies in Zoology and Biotechnology,
Vikram University, Ujjain-456010, Madhya Pradesh, India
E-mail: hrvuz2000@rediffmail.com

Abstract

Cell phone i.e mobile phone and towers for their operation are present in metro cities and in villages. Cell phone has become as essential instrument for everybody hence it got respectable place in our life. We cannot think of living without it. All living forms from simple organisms and simple plants up to human beings and trees i.e. whole biosphere is exposed to radiations emitted from cell phones and their towers. Scientists have studied and reported adverse health effects of such radiations. An attempt is made to present available reports in this paper. Now heap of evidence is warning us that cell phones and tower radiation can damage at various levels. Genotoxic effect is most threatening. After millions of years our DNA has evolved and other species have also came into existence. Today biodiversity is seen but it is likely to be decline slowly. Pollution and overpopulation are already affecting existing biodiversity. Under such pressure cell phone and tower radiations further enhance chances of loss of biodiversity. Time has come to think, discuss and formulate policies to prevent excess and undesirable exposures towards biosphere.

Key Words: Cell Phone, Tower, Radiations, Genotoxicity, Biodiversity)

Introduction

Scientific innovations have both two aspects i.e beneficial and harmful. Cell phone is no exception. Cell phone has become essential item in our lives. We like it, want it and have developed love and faith in this instrument without knowing possible harms due to its use. Now we cannot live without a cell phone even for an hour. It became popular fast and is providing much facilities day by day. Cell phone and their towers emit radiations. Network is worldwide, hence problem is also global. Scientists have tried to find biological effects of radiations. Intrestingly list of ill effects is very lengthy. Radiations are emitted in the form of electromagnetic waves which are harmful for animal, human, plants and microbes. Radiations are emitted from towers in the range of 935MHz to 960 MHz. Rate of human body absorption is 4W/ Kg/BW.

History of Cell Phone

In India mobile cellular phone was initially launched on 31 july 1995 and the first service provider was Modi Groups.

Biological Effects

These are categorized under different sub heading as under and also presented in detail in the tabular form.

Carcinogenic Effects

Increased cases of cancer and mortality caused by high frequency of electromagnetic fields have been reported.

Genotoxic Effects

DNA and chromosomal damage in human females has been found in the presence of electromagnetic fields. Chromosomal aberration has also been reported in *Allium cepa*. While Winter wheat (*Triticum aestivum* L.) showed genetic diversity too in the presence of high frequency of electromagnetic fields.

Cardiac Effects

In the presence of electromagnetic field, distributed heart beat was observed in frog *Rana pipiens*.

Reproductive Effects

Reduce reproductive potential in frogs is on record. Parental disabilities and infertility in mice and birds have been reported because of the presence of mobile tower. Reduced semen quality and poor male reproduction is also on record.

Neurobiological and Neurobehavioral Effects

Electromagnetic radiation can exert an aversive behavioral response in rats, bats and birds such as sparrows. Therefore microwave and radiofrequency pollution constitutes a potential cause for the decline of animal populations and deterioration of health of plants growing near phone masts. Electromagnetic field reduced potentials of brain functions also. It has been supported by observed changes in biological rhythms of central nervous system. Mammalian brain showed nerve cell damage, memory loss, headache, dizziness and tremors in the presence of electromagnetic waves.

Morphological Effects

In the high frequency of electromagnetic fields biological clock of the *Penicillium claviforem* has been found to be changed. Cancerous Tumor II growth and abnormal photoperiodism in trees has been observed. In the leguminous plants, nodule formation has been found disturbed because of the interference of the electromagnetic waves.

Hematological Effects

Harmful health effects have been reported among populations residing near mobile towers.

Miscellaneous Effects

Electromagnetic waves exert ill effects on biomass, enzyme activity, and hormonal balance. It has adverse effects on physiology of humans, plants and animals. Electromagnetic fields cause higher polarization of water too. Current state of knowledge about biological effects of radiations of cell phones and their towers as scientific reports are presented in tabular form (Table 1).

Conclusion

Above mentioned reports provide an overview of the biological effects of radiations emitted by cell phones and their towers. No doubt cell phones are useful instrument but it should be used consciously. General public should be educated about its harmful effects. Pregnant females and children should not be allowed to use it freely. Also, towers should be stationed in remote areas not in crowded human and animal population. It should not be used by ill people.

In a recent review it is suggested that trees, algae, vegetation, insects and reptiles are also susceptible towards radio frequency electromagnetic field from cell phone towers and wireless devices. During the course of evolution of millions of years DNA of each species has evolved hence there is biodiversity which is declining day by day. We human beings represent apex of animal evolution. Time has come to make prudent use of all radiation emitting devices including cell phones. It is difficult to forecast that cell phone will turn out to be a friend or foe in future.

Table 1: Biological effects of cell phone and their towers.

S. N.	YEAR	MODEL	NATURE OF STUDY	FINDINGS / SUGGESTIONS	CONCLU-SION	REFE-RENCE
1.	1986	Frog (Rana pip-iens)	Modification of heart function with low intensity Electromagnetic energy.	The data indicated that rate of change of beat is influenced by exposure to EM energy at incident average power densities of 3 micro watt/cM2.	+VE	Fery and Eichert
2.	1996	Govt. Report	Guideline for evaluating the environmental effects of radiofrequency radiations	Govt.report Washington D.C.,federal communications commission	+VE	Federal communications commission
3.	1996	Human	Cancer incidence, mortality and proximity to TV towers	An association was found between increase childhood leukemia incidence and mortality	+VE	Hocking et al.
4.	1997	Mice	RF radiation-induced changes in the prenatal development of mice	A progressive decrease in the number of the newborns, however, crown-rump length, the body weight, and the number of the lumbar,sacral,and coccygeal vertebrae, was not affected.	+VE	Magras and Xenos
5.	1997	Mice Human (in-vitro)	A report on non-ionizing radiation In vitro effects of low level,	Digital mobile phone radiation boosted cancer rate in mice	+VE	Micro-waves news
6.	1999		low frequency electromagnetic fields on DNA damage in human leukocyte by comet assay	Male were pooled for each flux density, with one exception, there was a significant increase in the DNA damage from the control value. When compare with similar study on females the DNA damage level was significantly higher in the females as compared to the male for each flux density.	+VE	Ahuja et. al.
.	2000	Human	Probable health effects associated with base station in communities: The need of health surveys.	Exposure caused significant and dose response increased in brain cancer, leukemia and other cancers, cardiac, neurological and reproductive health effects.	+VE	Cherry
8.	2003	Human	Report on radiation, mobile phones, base stations and your health.	Informative report on radiation, mobile phones, base stations and health	+VE	Hoong
9.	2003	Rat	Nerve cell damage in mammalian brain after exposure to micro waves from GSM mobile phones	Neuronal damage in the cortex, hippocampus and basal ganglia in the brains of exposed rats.	+VE	Salford et al.

S. N.	YEAR	MODEL	NATURE OF STUDY	FINDINGS / SUGGESTIONS	CONCLU-SION	REFE-RENCE
10.	2003	Human (in-vitro)	Exposure of human peripheral blood lymphocytes to electro-magnetic fields associated with cellular phones leads to chromo-somal instability	Genotoxic effect of the electromagnetic radiation is elicited via non-thermal pathway	+VE	*Muschevich et al.*
11.	2003	Fungi (Penicilli-um claviforme)	Effect of magnetic field on the biological clock in *penicillium claviforme*	The magnetic field distinctly affected the morphology of the newly formed coremia, which were identical in light and in darkness. In the applied conditions the magnetic field replaced the inductive effect of light, activating the expression of the biological clock in darkness.	+VE	Binczycka et al.
12.	2004	Human	The influence of being physically near to a cell phone transmission mast on the incidence of cancer	Proportion of newly developing cancer cases was significantly higher among those patients who had lived during the past ten years at a distance of up to 400 meters from the cellular transmitter site, which has been in operation since 1993, compared to those patients living further away, and that the patients fell ill on average 8 years earlier.	+VE	*Eger et al.*
13.	2004	Human	HF-radiation levels of GSM cel-lular phone towers in residential areas	High frequency radiation had thermal effect and heating of the body tissue.	+VE	Haumann et al.
14.	2005	White stork (Ciconia ciconia) Bird	Possible effects of electromagnet-ic fields from phone masts on a population of white stork (*Cico-nia ciconia*).	Microwaves disturbed with the reproduction of white storks.	+VE	Balmori
15.	2006	Human (In-vi-tro)	Effects of electromagnetic radi-ation from a cellular phone on human sperm motility: an *In vitro* study	Data suggested that EMR emitted by cellular phone influenced human sperm motility. In addition to these acute adverse effects of EMR on sperm motility, long-term EMR exposure might lead to behavioural or structural changes of the male germ cells. These effects may be observed later in life, and they are to be investigated more seriously.	+VE	*Erogul et al.*

S. N.	YEAR	MODEL	NATURE OF STUDY	FINDINGS / SUGGESTIONS	CONCLU-SION	REFE-RENCE
16.	2006	*Human*	Neurobehavioral effects among inhabitants around mobile phone base stations	Inhabitants living nearby cell phone base stations are at risk for developing neuropsychiatric problems and some changes in the performance of neurobehavioral functions either by facilitation or inhibition.	+VE	Rassoul *et al.*
17.	2006	*Frog (Rana temporaria)*	The incidence of electromagnetic pollution on the amphibian decline	Electromagnetic pollution is a caused deformation and decline of some amphibian populations.	+VE	Balmori
18.	2006	*Human*	Health effects of mobile phone transmitter masts and the planning application by orange mast in St. Michael's church, Aberystwyth	Radiation from mobile base station may cause cancer and makes tumour's grow more aggressively after killing off cells of all kinds including cancer cells	+VE	Busby and Coghill
19.	2006	*Human*	A review of the mechanism of interaction between the extremely low frequency electromagnetic field and human biology.	Existing report are reviewed	+VE	Sadafi *et al.*
20.	2007	*Maize (Zea mays)*	The effects of high frequency electromagnetic waves on the vegetal organisms	Long term exposure declined pigment content	+VE	Ursache *et al.*
21.	2007	*Bird (Passer domsticus)*	A Possible effects of Electromagnetic radiations from mobile phone base station on the number of breeding house sparrows(*Passer domsticus*)	Long term exposure to higher levels of radiation negatively affects the abundance or behavior of house sparrows in the wild.	+VE	Everaert and Bauwens
22.	2007	Human	Biological effects from electromagnetic field exposure and public exposure standards	It included childhood leukemia, brain tumors, geno-toxic effects, neurological effects and neurodegenerative diseases, immune system deregulation, allergic and inflammatory responses, breast cancer, miscarriage and some cardiovascular effects.	+VE	Hardell and Sage

S.N.	YEAR	MODEL	NATURE OF STUDY	FINDINGS / SUGGESTIONS	CONCLUSION	REFERENCE
23.	2008	Plant (Pinus sylve-stris)	Effects of extremely high frequency electromagnetic fields on the microbiological community in rhizosphere of plant	Exposure resulted in the proliferation of agronomically useful microorganisms including nitrogen-fixing ones.	+VE	Ratushnyak et al.
24.	2008	Human	Theoretical estimation of power density levels around mobile phone base station	Incident power density of the radiation flux is possible by knowing the antenna's technical data from the constructor and by defining the position of the exposed person.	+VE	Al-Bazzaz
25.	2008	E.coli	DNA and chromosome damage-a crucial non thermal biological effect of microwave radiation.	The impact of RF-EMF exposure on DNA/chromosome damage plays a crucial role in the health impact of this type of radiation, it is important to realize the described genotoxic effects.	+VE	Shiroff
26.	2008	Human (In vivo)	Increased frequency of micro nucleated exfoliated cells among humans exposed in vivo to mobile telephone radiations	Slight increase in mean frequency found to be positively correlated	+VE	Yadav and Sharma
27.	2009	Human	Cell phone and brain tumours: A review including the long term epidemiologic data	Using a cell phone for near about 10 years approximately doubles the risk of a brain tumor on the same side of the head as that preferred for cell phone use.	+VE	Khurana et al.
28.	2009	Human	Mobile phone base stations-effects on wellbeing and health	Long term exposure near base stations affects wellbeing. Symptoms most often associated with exposure were headaches, concentration difficulties, restlessness tremor and sleeping problems.	+VE	Kundi and Hutter
29.	2009	Human	How susceptible are genes to mobile phone radiation?	Long term exposure of radiation from cell phones are responsible for DNA damage, Chromosomal alteration, brain cancer and neurobehavioural changes.	+VE	Adlkofer et al.
30.	2009	Human	Biological effects of cell tower radiation on human body	Continuous exposure of microwaves cause serious health problems.	+VE	Kumar and Kumar

S. N.	YEAR	MODEL	NATURE OF STUDY	FINDINGS / SUGGESTIONS	CONCLU-SION	REFE-RENCE
31.	2009	Birds (Passer domesticus and Peacock)	Electromagnetic pollution from phone masts: effects on wildlife	Electromagnetic radiation may hurt wildlife. Micro-waves and radiofrequency pollution constitutes a potential cause for the decline of animal population and deterioration of health of plants living near phone masts.	+VE	Balmori
32.	2009	Maize (Zea mays)	Effects of radio frequency radia-tion on root tip cells of zea mays	The mitotic index and chromosomal aberration fre-quency showed linear increasing for radiofrequency radiation treatment of increased exposure time.	+VE	Racuciu
33.	2009	Human	Genotoxic effects of radio fre-quency electromagnetic field	Genotoxic action may be mediated by micro thermal effects in cellular structures, formation of free radi-cals, or an interaction with DNA-repair mechanisms.	+VE	Ruediger
34.	2009	Enzyme	Effects of 50HZ Electromagnetic Fields on acid phosphatase ac-tivity.	The results showed that extremely low frequency, EMF have significant influence on enzyme activity.	+VE	Prashanth et al.
35.	2010	Wheat (Triticum aestivum L.)	Genetic diversity of winter wheat (Triticum aestivum L.) growing near a high voltage transmission line	The genetic diversity of individuals of wheat plants grown under transmission lines increased, although seed germination and the amount of total soluble protein did not change.	+VE	El-Bakatoushi
36.	2010	Plant (Lens clu-naris medik.)	Cytotoxic effects of an electro-magnetic field on the meriste-matic root cells of lantils (Lens clunaris Medik)	EMF exposure increased chromosomal abnormalities and intracellular cA2+ deposition, accompanied by deformation of cytoplasm and plasma membrane, however, reduced mitosis.	+VE	Eren et al.
37.	2010	Plant(Allium cepa)	Genotoxic effects of electromag-netic field from high voltage power line on some plants	Electromagnetic fields from high voltage power lines increased the mitotic index and chromosomal aber-rations.	+VE	Aksoy et al.
38.	2010	Frog (Rana tem-poraria)	Mobile phone mast effects on common frog (Rana temporaria) Tadpoles: The city turned in to laboratory.	Radiation emitted by phone mast in a real situation affected the development and caused an increase in mortality of exposed tadpoles.	+VE	Balmori

S. N.	YEAR	MODEL	NATURE OF STUDY	FINDINGS / SUGGESTIONS	CONCLU-SION	REFE-RENCE
39.	2010	*Human*	Report on cell phone tower radiation	Non thermal effects are 3 to 4 times more harmful than the thermal effects	+VE	Kumar
40.	2010	*Human*	Indian Govt. report	Inter ministerial committee on EMF radiation	+VE	Report of the inter-ministerial committee
41.	2010	*Human*	Epidemiological evidence for a health risk from mobile phone base station	Adverse health impact.	+VE	Khurana *et al.*
42.	2010	Human	Biological effects from exposure to Electromagnetic radiation emitted by cell tower base station and other antenna arrays.	Headaches, skin rashes, sleep disturbances,depression,decreased libido and other neurophysiological effects have found.	+VE	Levitt and Lai
43.	2011	*Human*	Effects of electromagnetic radiation from mobile phones towers on human body	Affected fat and muscles	+VE	Kumar and Pathak
44.	2011	*Human*	Cell phone and their impact on male fertility: Fact or fiction	Affected sperm biology	+VE	Hamada *et al.*
45.	2011	Urban plant (Oak and Beech)	Why our urban trees are dying	A more obvious effect of radio waves on the biological clock is on the shedding of leaves and fruit in some trees. If the electromagnet act as growth hormones, this could stimulate the production of undifferentiated callus in the phloem to form "cancer like" growths.	+VE	Goldsworthy
46.	2011	*Maize (Zea mays)*	Effects of pre showing electromagnetic treatment on seed germination and seedling growth in maize (Zea mays L.)	The presence of chlorophyll pigments and carotenoids was assessed and were alterationsfound in the concentration of chlorophyll a, chlorophyll b and carotene in corn seedlings grown from electromagnetically treated seeds.	+VE	Elizabeth *et al.*

S. N.	YEAR	MODEL	NATURE OF STUDY	FINDINGS / SUGGESTIONS	CONCLU-SION	REFE-RENCE
47.	2011	*Human*	Effects of combined magnetic fields on human sperm parameters	Motility was significantly affected by the exposure of spermatozoa to magnetic field, but sperm structural parameter remained intact.	+VE	Falahati *et al.*
48.	2012	*Review with Many models*	Impact of radio-frequency electromagnetic field (RF-EMf) from cell phone tower and wireless devices on bio system and eco-system.	RF-EMF exposure can change neurotransmitter function, blood brain barrier, morphology, electrophysiology, cellular metabolism, calcium efflux, and gene and protein expression in certain type of cell at lower intensities. Impacts on frog, honey bees, house sparrows, and bats are also recorded.	+VE	Sivani and Sudarsanam
49.	2012	*Human*	Effect of mobile tower radiation in and case studies from different countries pertaining the issue.	People living near cell tower receive strong signal strength but at the expense of health.	+VE	Kaushal *et al.*
50	2012	*Human (In-vitro)*	Mobile phone electromagnetic waves and its effect on human ejaculated semen: an *in vitro* study	Sample exposed to EMR showed a significant decrease in sperm motility and viability, increase in reactive oxygen species (ROS) and DNA fragmentation index (DFI) compared to unexposed group mobile phone.	+VE	Veerachari And Vassan
51.	2013	*Human*	Study of cell tower radiation levels in residential areas.	Continuous exposure to microwave radiation from cell phone towers cause serious health problems over the years.	+VE	Saeid
52.	2013	*Human*	Effects of mobile phone and mobile phone tower radiations on human health.	The ongoing use of mobile phone has raised the healtheffects. Electromagnetic radiations emmited by mobile phone and mobile phone towers are harmfull by the people who are living near by the transmissions tower.	+VE	Bhat *et al.*
53.	2013	*Wistar Rat*	Cell phone radiation exposure on brain and associated biological systems.	The regular and long term use of microwave devices (mobile phone and microwave oven) at domestic level can have negative impact upon biological system especially on brain. Increased ROS play an important role by enhancing the effect of microwave radiations which may cause neurodegenerative diseases.	+VE	Kesari *et al.*

S. N.	YEAR	MODEL	NATURE OF STUDY	FINDINGS / SUGGESTIONS	CONCLU-SION	REFE-RENCE
54.	2013	*Mice*	Effects of electromagnetic field on Red Blood Cells of adult male *Swiss albino* mice	The result showed altered blood, cell morphology, and scanning electron micrograph of RBCs. RBC Count and haemoglobin concentration was reduced up to day 42 of exposure.	+VE	Singh *et al.*
55.	2013	*Human*	Cancer induction molecular pathways and HF-EMF irradiation	The microenvironment that exist during chronic inflammation can contribute to cancer progression. The data support the proposition that long term HF EMF exposures associated with improper use of cell phone can potentially cause cancer.	+VE	Ledoigt and Belpomme
56.	2013	*Ants*	Ants can be used as bio indicators to reveal biological effects of electromagnetic waves from some wireless apparatus.	Numerical and statistical results allowed detecting any effect of a radiating source on these organisms.	+VE	Cammaerts And Johansson
57.	2013	*Human (In-vitro)*	The ability of safe space cell phone patch to neutralize the Harmful biological effects of electromagnetic fields generated from cell phones.	Energy from the safe space can prevent biological effect on DNA by enhancing the natural re-winding process.	+VE	Rein
58.	2013	*Human*	Heart-response and the vitaplex-: the chronotropic myocardial response as an indication of the health promotic properties of dimensional design's subtle energy field storage technology.	The vitaplex is created with an imprinted holographic circuit which generates subtle energetic signals designed to balance the body and mind.	+VE	Nokken
59.	2013	*Testing Report for safe use*	Bio resonance testing report	Safe space cell phone patch test provided guidance notes.	+VE	Nokken
60.	2013	*DNA (In-vitro)*	Effect of dimensional design technology to prevent power line radiation damage	Preventive measures to restore properties of DNA.	+VE	Rein

S. N.	YEAR	MODEL	NATURE OF STUDY	FINDINGS / SUGGESTIONS	CONCLU-SION	REFE-RENCE
61.	2014	Plant (Vicia faba)	Micronucleus induction by 915MHZ radio frequency radiation in Vicia faba root tips	Increased micronuclei frequency up to ten fold, suggest chromosomal damage.	+VE	Gustavino et al.
62.	2014	Rat	Effects of cell phone radiation on the levels of T3, T4 and TSH and histological changes in thyroid gland in rats treated with Allium sativum extract.	Microwaves can caused weight lost, however, presence of allicin and vitamins a and b in garlic could compensate this effect partially.	+VE	Hajioun et al.
63.	2014	Human	Study of radiation exposure due to mobile towers and mobile phones	Cardiac effect varied from person to person, some people are more susceptible to mobile phone radiation.	+VE	Aghav
64.	2014	Human	Interaction of mobile phone radiation with biological system in veterinary and medicine	EM fields influenced several biological functions of cells and tissues, modulated intracellular reactive oxygen species level and the cell cycle progression.	+VE	Fathi and Farahzadi
65.	2014	Birds (Rock dove) and Human	Effects of exposure to EM radiations on living beings and environments.	Adverse effects on human head, eyes, ears, body and on animals. birds like rock dove, pigeons, magpie were found disappeared near cell towers.	+VE	Jain and Bagai
66.	2014	Leguminus plants	Effects of mobile phone radiation on nodule formation in the leguminous plants	Radiation emitted from mobile phone interfered with both morphological and the biochemical processes and affected the growth and nodule formation in the plants.	+VE	Sharma and Parihar
67.	2014	Plants (Oxalis corniculate)	Electromagnetic radiation influencing stomatal patterning in Oxalis corniculata L.	Oxalis plants manifested significant impacts of electromagnetic radiation from cell towers.	+VE	Geeta and Singh
68.	2014	Mice	Mobile tower exposure affects memory and motor co-ordination test, showed significant changes in the treated mice in mice	Results of short term memory and long term memory as compared with without treated mice.	+VE	Maurya and Upadhyay
69.	2014	Human	Impact of cell tower radiation on human health.	Results related with health issue due to impact of EM wave's radiation.	+VE	Karkare and Tewatri

S. N.	YEAR	MODEL	NATURE OF STUDY	FINDINGS / SUGGESTIONS	CONCLU-SION	REFE-RENCE
70.	2014	*Human*	Study of cell tower radiation and it's health hazards on human body	Inhabitants living within 50M were having more health complaints than those living outside 50M. It was also found that females had more complaints than males.	+VE	Pachuau and Pachuau
71.	2015	*Allium cepa L.*	Effects of magnetic field on germination, seedling growth and cytogenetic of onion (*Allium cepa L.*)	Significant change in mitotic activity and chromosomal aberration.	+VE	Hozayn et al.
72.	2015	*Water*	Magnetic and electric effects on water	Higher molecular polarisation	+VE	Chaplin

References

1. Adlkofer F, Belyaev IY, Richter K and Shiroff VM. (2009). How susceptible are genes to mobile phone radiation? Brochure series by the *competence initiative for the protection of humanity Environment and Democracy.* 3:1-61.

2. Aghav SD. (2014). Study of radiation exposure due to mobile towers and mobile phones.*Indian Streams Research Journal.*3(12):1-6.

3. Ahuja YR, Vijayashree B,Saran R, Jayashri EL, Manoranjani JK and Bhargava SC. (1999). *In-vitro* effects of low level, low frequency electromagnetic fields on DNA damage in human leucocytes by comet assay. *Indian Journal of Biochemistry and Biophysics.* 36:318-322.

4. Aksoy H,Unal F and Ozcan S. (2010). Genotoxic Effects of Electromagnetic Fields from High Voltage Power Lines on Some Plants. *International Journal of Environtal Research.* 4(4):595-606.

5. Al-Bazzaz SHS. (2008). Theoretical estimation of power density level around mobile telephone base stations. *Journal of Science and Technology.* 13(2):3-16.

6. Balmori A. (2005). Possible effects of electromagnetic fields from phone masts on a population of White Stork(*Ciconia ciconia*). *Electromagnetic Biology and Medicine.* 24:109-119.

7. Balmori A. (2006). The incidence of electromagnetic pollution on the amphibian decline: Is this an important piece of the puzzle? *Toxicological and Environmental Chemistry.* 88(2): 287-299.

8. Balmori A. (2009). Electromagnetic pollution from phone masts. Effects on wildlife. *Pathophysiology.* 16:191-199.

9. Balmori A. (2010). Mobile phone mast effects on common frog (*Rana temporaria*) tadpoles: The city turned in to a laboratory. *Electromagnetic Biology and Medicine.* 29:31-35.

10. Binczycka BP, Fiema J and Nowak M. (2003). Effect of the magnetic field on the biological clock in *Penicillium claviforme. Acta Biologica Cracoviensia.* 45(2): 111-116.

11. Bhat MA, Kumar V and Gupta GK. (2013). Effects of mobile phone and mobile phone tower radiations on human health. *Indian Journal of Recent Scientific Research.* 4(9):1422-1426.

12. Busby C and Coghill R. (2006). Health effects of mobile phone transmitter masts and the planning application by orange PLC for a mast in St. Michael's church Aberystwyth, Green Audit Aberystwyth. admin@ greenaudit.org.

13. Cammaerts MC and Johansson O. (2013). Ants can be used as bio indicators to reveal biological effects of electromagnetic waves from some wireless apparatus. *Electromagnetic Biology and Medicine.* 1–7.

14. Chaplin M. (2015). Magnetic and electric effects on water. www.isbu.ac.uk

15. Cherry N. (2000). Probable health effects associated with base station in communities: The need of health surveys 109-114 Lincoin University Christchurch www.land-sbg.gv.at /cell tower.

16. Eger H, Hagen KU, Lucas B, Vogel P and Voit H. (2004). The influence of being physically near to a cell phone transmission mast on the incidence of cancer. *Umwelt Medizin Gesellschaft.*17 (4): 1-7.

17. El-Bakatoushi R. (2010). Genetic diversity of winter Wheat(*Triticum Aestivum* L.) growing near a high voltage transmission line. *Romanian Journal of Biology-Plant Biology.* 55(2):71–87.

18. Elizabeth IA, Claudia HA, Arturo DP and Alfredo CO. (2011). Effect of pre-sowing electromagnetic treatment on seed germination and seedling growth in maize (*Zea mays* L.). *Agronomia Colombiana.* 29(2):213-220.

19. Eren P, Vardar F, Birbir Y, İnan D and Unal M. (2010). Cytotoxic effects of an electromagnetic field on the meristematic root cells of Lentils (*Lens clunaris* Medik.). *Fresenius Environmental Bulletein.* 19(3): 481-488.

20. Erogul O, Oztas E,Yildirim I,Kir T, Aydur E, Komesli G, Irkilata HC, Irmak MK and Peker AF. (2006). Effects of Electromagnetic Radiation from a Cellular Phone on Human Sperm Motility: An *In-vitro* Study. *Archives of Medical Research.* 37:840-843.

21. Everaert J and Bauwens D. (2007). A possible effects of electromagnetic radiation from mobile phone base station on the number of breeding house Sparrows (*Passer domesticus*). *Electromagnetic Biology and Medicin.* 26: 63-72.

22. Falahati SA, Anvari M and Khalili MA. (2011). Effects of combined magnetic fields on human sperm parameters. *Iranian Journal of Radiation Research.* 9 (3):195-200.

23. Fathi E and Farahzadi R. (2014). Interaction of Mobile telephone radiation with biological systems in veterinary and medicine. *Journal of Biomedical Engineering & Technology.* 2(1):1-4.

24. Federal communication commission. (1996). Guideline for evaluating the environmental effects of radiofrequency radiation, Washington D.C.

25. Fery AH and Eichert ES. (1986). Modification of heart function with low intensity electromagnetic energy. *Journal of Bioelectricity.* 5(2): 201-210.

26. Geeta and Singh PK. (2014). Electromagnetic radiation influencing stomatal patterning in *Oxalis Corniculata* L. *International Journal of Engineering& Advanced Technology.* 3(5):147-150.

27. Goldsworthy A. (2011). Why our Urban trees are dying. *MCSA News.* 6-11.

28. Gustavino B, Carboni G, Petrillo R, Rizzoni M and Santovetti E. (2014). Micronucleus induction by 915 MHz radiofrequency radiation in *Vicia faba* root tips,University of Rome Tor Vergata, Via *della Ricerca Scientifica*, 00133 Rome (Italy).

29. Hajioun B, Jowhari H and Mokhtari M. (2014). Effects of cell phone radiation on the levels of T3, T4 and TSH, and histological changes in thyroid gland in rats treated with *Allium sativum* extract. *African journal of Biotechnology.* 13(1):163-169.

30. Hamada AJ,Singh A and Agarwal A. (2011). Cell phone and their Impact on Male fertility: Fact or Fiction. *Reproductive Science Journal.* 5:125-137.

31. Hardell L and Sage C. (2007) Biological effects from electromagnetic field exposure and public exposure standards. *Biomedicine and Pharmacotherapy.* 20:1-6.

32. Haumann T, Munzenberg U, Maes W and Sierck P. (2004). Hf-radiation levels of GSM cellular phone towers in residential areas. www.hbelc.org

33. Hocking B, Gordon IR, Grain HL and Hatfield GE. (1996). Cancer incidence and mortality and proximity to TV towers. *Medicine Journal of Australia.* 165:601-605.

34. Hoong NK. (2003). Report on radiation, mobile phones, base stations and your health. Published by *Malaysian communication and multimedia commission.*

35. Hozayn M, El-Mahdy AAA and Abdel Rahman HMH. (2015). Effect of magnetic field on germination, seedling growth and cytogenetic of onion (*Allium cepa* L.). *African Journal of Agriculture Research.* 10(8):849-857.

36. Jain A and Bagai D. (2014). Effects of exposure to electromagnetic radiations on living beings and environment. Proceedings of 4th SARC International Conference, 30th March-2014, Nagpur, India.

37. Karkare S and Tewatri K.(2014). Impact of cell tower radiation on human health VESIT, International Technological Conference-Jan. 3-4,2014.

38. Kaushal M, Singh T and Kumar A. (2012). Effects of mobile tower radiations and case studies from different countries pertaining the issue. *International Journal of Applied Engineering Research.* 7(11):1-4.

39. Kesari KK, Siddiqui MH, Meena R, Verma HN And Kumar S. (2013). Cell phone radiation exposure on brain and associated biological systems. *Indian Journal of Experimental Biology.*51:187-200.

40. Khurana VG, Hardell, Everaert J, Bortkiewicz A, Carlberg M and Ahonen M. (2010). Epidemiological evidence for a health risk from mobile phone base stations. *International Journal of Occupational and Environmental Health.* 16:263-267.

41. Khurana VG, Teo, Kundi M, Hardell L and Carlberg M. (2009). Cell phones and brain tumors:a review including the long-term epidemiologic data. *Surgical Neurology.* 72:205–215.

42. Kumar G. (2010). Report on Cell phone tower radiation, submitted to Secretary, DOT, Delhi India.

43. Kumar N and Kumar G. (2009). Biological effects of cell tower radiation on human body. *ISMOT/09/C/318.*

44. Kumar S and Pathak PP. (2011). Effects of electromagnetic radiation from mobile phones towers on human body. *Indian Journal of Radio and Space Physics.* 40:340-342.

45. Kundi M and Hutter H. P. (2009). Mobile phone base station-Effects on wellbeing and health. *Pathophysiology.* 16:123-135.

46. Ledoigt G and Belpomme D. (2013). Cancer induction molecular pathways and HF-EMF irradiation. *Advanced in Biological Chemistry.* 3:177-186.

47. Levitt BB And Lai H. (2010). Biological effects from the exposure to electromagnetic radiation emitted by cell tower base stations and other antenna arrays. *Environmental Review.* 18:369-395.

48. Magras IN And Xenos TD. (1997). Rf radiation-induced changes in the prenatal development of mice. *Bioelectromagnetics.* 18:455-461.

49. Maschevich M, Folkman D, Kesar A, Barbul A, Korenstein R, Jerby E and Avivi L. (2003). Exposure to human peripheral blood lymphocytes to electromagnetic fields associated with cellular phones leads to chromosomal instability. *Bioelectromagnetics.* 24:82-90.

50. Maurya SK and Upadhyay SK. (2014). Mobile tower exposure affects on memory and motor Co-ordination on Mice. *G-Journal of Environmental Science and Technology.* 1(5):103-107.

51. *Micro wave news* 1997, A report on non ionizing radiation, 17 (3):1-20.

52. Nokken MB. (2013). Bio resonance testing report. www. safespacetechnology.

53. Nokken MB. (2013). Heart-response and the Vitaplex: The chronotropic myocardial response as an indication of the health promoting properties of dimensional design's subtle energy field storage technology. www. Safespacetechnology.

54. Pachuau L and Pachuau Z. (2014). Study of cell tower radiation and its health hazards on human body. *Journal of Applied Science.* 6(1): 01-06.

55. Prashanth KS, Chouhan TRS and Nadiger Snehalatha.(2009). Effect of 50Hz electromagnetic fields on acid phosphatase activity. *African Journal of Biochemistry Research.* 3(3):60-65.

56. Racuciu M. (2009). Effects of radiofrequency radiation on root tip cells of *Zea mays. Roumanian Biotechnological Letters.* 14(3):4365-4369.

57. Rassoul GA, EL-Fateh OA, Salem MA, Micheal A,Farahat F, El-Batanouny M and Salem E. (2006).Neurobehavioral effects among inhabitants around mobile phone base stations. *Neurotoxicology.* 636:1-7.

58. Ratushnyak A A, Andreeva MG, Morozova OV, Morozov GA, and Trushin MV. (2008). Effect of extremely high frequency electromagnetic fields on the microbiological community in rhizosphere of plants. *International Agrophysics.* 22: 71-74.

59. Rein G. (2013). The ability of the Safe space © cell phone patch to neutralize the harmful biological effects of electromagnetic fields generated from cell phones, Laboratory testing report, Quantum Biology Research Labs PO Box 157 Northport, NY 11768.

60. Rein G. (2013). Effect of dimensional design technology to prevent power line radiation damage, Laboratory testing report, Quantum Biology Research Lab Ridgway, CO 81432.

61. Report of the Inter-ministerial committee on EMF radiation. (2010). Govt. of India. Ministry of Communication and Information Technology Department of Telecommunications.

62. Ruediger HW. (2009). Genotoxic effects of radiofrequency electromagnetic fields. *Pathophysiology.* 602:1-14.

63. Sadafi HA, Mehboodi Z and Sardari D. (2006). A Review of the Mechanisms of interaction between the extremely low frequency electromagnetic fields and human biology progress in electromagnetic research symposium, Cambridge, USA,.99-103.

64. Saeid SH. (2013). Study of the cell tower radiation levels in residential areas. Proceeding of the 2013 International Conference on electronics and communication system.

65. Salford LG, Brun AE, Eberhardt JL, Malmgren L and Persson BRR. (2003). Nerve cell damage in mammalian brain after exposure to microwaves from GSM mobile phones. *Environmental health perspectives.* 111(7): 881-883.

66. Sharma S and Parihar L. (2014). Effect of mobile phone radiation on nodule formation in the Leguminous Plants. *Current World Environment.* 9(1):145-155.

67. Shiroff V M. (2008). DNA and Chromosome Damage: A Crucial Non-Thermal Biological Effect of Microwave Radiation. In: A Brochure series by the competence initiative for the protection of Humanity, Environment and Democracy, 3:29-43.

68. Sivani S And Sudarsanam D. (2012). Impact of radiofrequency electromagnetic field (RF-EMF) from cell phone tower and wireless devices on bio system and ecosystem-a review. *Biology and Medicine.* 4(4):202-216.

69. Singh H, Kumar C and Bagai U. (2013). Effect of electromagnetic field on red blood cells of adult male *Swiss albino* mice. *International Journal of Theoretical & Applied Sciences.* 5(1):175-182.

70. Ursache M, Mindru G, Creanga DE, Tufescu FM and Goiceanu C. (2007). The Effects of high frequency electromagnetic waves on the Vegetal organisms. *Romanian Journal of Physics.* 54(1–2):133–145.

71. Veerachari SB and Vasan SS. (2012). Mobile phone electromagnetic waves and it's effect on human ejaculated semen: An *in-vitro* study. *International Journal of Infertility and Fatal Medicine.* 3(1):15-21.

72. Yadav AS and Sharma MK. (2008). Increased frequency of micronucleated exfoliated cells among humans exposed *in vivo* to mobile telephone radiations. *Mutation Research.* 650 (2): 175-180.

17

Conservation and Management of Biological Diversity: Global Challenges and Opportunities

Anil Sharma, Rinku, S. K. Patel and G. S. Singh*

*Institute of Environment and Sustainable Development, Banaras Hindu University,
Varanasi-221005, Uttar Pradesh, India
E-mail: gopalsingh.bhu@gmail.com*

Abstract

Biodiversity provides various ecosystem services and enormous economic benefits, through natural ecosystems, and plays an important role in modifying ecosystem structure, function and stability. Biodiversity is unevenly distributed on the earth, varying on spatio-temporal and geo-climatic scales which availability expected from five to more than fifty million species. The predicted extent of species loss has drawn attention worldwide, increasing efforts raised rapidly to assess and conserve biodiversity at multifarious scales. Fundamental processes of speciation, endemism, coexistence, extinction, and different susceptibility of taxa and habitats are not sufficiently understood. Accuracy of assessments of the total number of native species and current rates of extinction remains uncertain, and the impact of species losses on ecosystem function and stability is still a contradictory matter among ecologists. The study of biodiversity is an interdisciplinary science with a growing concepts, hypotheses, methodologies, and internalization of human sociological aspects. Biodiversity has been protected by the indigenous communities that lived in their adjacent area, either as sacred sites or as a consequence of sustainable management practices using empirical practices that flourish through generations. The ecological basis of these indigenous approaches developed through a clear understanding by the traditional communities confirms economic, social and environmental sustainability on one side and sustainable conservation and management practices on the other side. The growing challenges centred on policy linked with biodiversity conservation and management practices should be strengthen.

Keywords: Biodiversity, changing climate, conservation and management, policy issues

Introduction

Biodiversity is the root of survival and sustenance, economic well-being, and comprises all life forms, ecosystems and ecological processes, acknowledging the hierarchy at genetic, taxon and ecosystem levels (McNeely *et al.,* 1990). The current assessments (May, 1988) of the total number of species on earth vary from 5 to 50 million, with a more conventional figure of 13.6 million species (Hawksworth *et al.,* 1995). Only 1.76 million species have yet been defined and granted scientific names. Thus, our knowledge of diversity is unusually inadequate. Biodiversity has drawn attention to the globe because of the rising awareness of its importance and the estimated mass depletion. This article discusses on the benefits and role, accumulation, distribution and loss, and assessment and conservation of biodiversity. It will be clear that there are more assessments than observed data, and more hypotheses than solid concepts (Singh, 2002). The methodologies for the assessment and conservation of biodiversity also remain insufficient.

Benefits and role

Apart from the moral values and cultural aspects, biodiversity provides various provisioning services in the form of timber, food, fibre, industrial enzymes, food flavours, fragrances, cosmetics, emulsifiers, dyes, plant growth regulators and pesticides (Costanza *et al.,* 1997; Mannion, 1995). Biodiversity is of immense value to human health, although only 1100 of the world's 365,000 known species of plants have been studied for their medicinal properties (Dobson, 1995).

Ethno medicinal value of biodiversity to human health

India consists of 15 Agro-climatic zones and 17,000-18,000 flowering plants species of which 6,000-7,000 are expected to have ethnomedicinal values in traditional healthcare systems of medicine like Ayurveda, Siddha, Unani and Homeopathy (NMPB, 2014), of which 95% of the 400 plant species used in formulating medicine by various industries are extracted from wild inhabitants (Uniyal et al., 2000). In India, Himalaya has the rich plant diversity, even Himalaya is the home of various flora comprising over 8000 angiosperms, 44 gymnosperms, 600 pteridophytes, 1737 bryophytes, 1159 lichens, and other species has been a base of medicine for millions of people in the India and elsewhere in the world (Singh *et al.,* 1996). Himalaya provides over 1748 (32.2% of India) floral species of well-known ethnomedicinal importance. The unique diversity of medicinal plants in the region is manifested by the presence of a number of native (31%), endemic (15.5%) and threatened elements: 14% of total Red Data plant species of the IHR and 3.5%of total medicinal plants in different threat categories of Conservation Assessment and Management Plan (CAMP) (Samant *et al.,* 1998).

Based on the assessment of Dobson *et al.,* 1995 the following important points were noted: (i) every 125 plant species studied at the Herb Research Foundation, Boulder, produced a major drug in the US of at least $ 200 million per year; (ii) of the 118 (out of the top 150) medical drugs in the US, 74% are based on plants, 18%

on fungi, 5% on bacteria and 3% on vertebrates; (iii) out of top 10 medical drugs in the US, 9 are based on natural plant products. In 1990, sales of medical drugs with active constituents of plant origin amounted to about $ 1550 million; (iv) 80% of the world's population depends on ethno-medicinal plant; (v) ingredients from Gingko leaves are used by 80% Europeans to prevent senile dementia, and (vi) 3-4 potentially valuable drugs are lost due to loss of one tree species every year, at a total cost of $ 600 million.

Biodiversity conservation

Conservation of species from disappearance is need of hour and named them or recognised their probable uses and role, advocates that it is wise to take a preventive approach based on the existing practices, and make serious challenges to conserve them (Ramakrishnan *et al.*, 1998; Singh, 2002). Protected area networks are the most popular and well acknowledged approach for biodiversity conservation in the face of environmental change (Chape, 2005).

Convention on Biological Diversity (CBD) 2010, encouraged that at least 10% of each of the global ecological regions should be properly managed. Currently, protected areas cover 12.7% of global terrestrial system and 1.6% of the marine system (Bertzky *et al.*, 2012). The main objective of the global protected-area network is to protecting biodiversity and ecosystem goods and services (*Rodrigues et al.*, 2004). CBD recommended that 17% of the world's terrestrial surface and 10% of the marines be nominated as protected areas by 2020 (CBD, 2010), and consequences suggest that protected areas could reaching 15%–29% of global land by 2030 (McDonald and Boucher, 2011). The concept and use of conservation corridors arose to protect linked areas and raise the movement of species among them (Bennett, 1990).

The management of corridors can give distribution opportunities for diversity of species between protected areas (Simberloff *et al.*, 1992; Gilbert-Norton *et al.*, 2010). The aims of corridors to sustaining the biological strength of fragmented forest landscapes (Hilty *et al.*, 2006) and increasing the conservation role of protected areas in the face of changing climate (Hannah, 2008), especially in mountain regions (Loarie *et al.*, 2009). A social–ecological system approach considering landscape, play an important role for conservation and management of protected areas because it highlights the complexity of interactions in every landscape. In this view, adding ecological consent with social analyses will offer a better conservation approach that integrates the social practices that influence conservation choices (Ban *et al.*, 2013).

Increasing technology and information transfer between nations has been the most important mechanisms to report environmental problems. However, direct measurement of the effect of knowledge sharing and benefits to biodiversity have slow down. Although, in global development, practical assistance tries to foster technological expansion among countries (Sawada *et al.*, 2010), whereas other researches connects precise conservation approaches that comprise technology transfer with fruitful conservation results, at least project level (e.g., Howe & Milner-Gulland, 2012). Technology transfer has become a crucial element of global environmental policy (e.g., United Nations Framework Convention on Climate Change UNFCCC). The need for technology and knowledge transfer has been recognized by the CBD to advancing scientific, technical and organisational capacities to provide the fundamental understanding to plan and implement suitable actions" (CBD, 1992).

India has a rich tradition of biodiversity conservation (Ramakrishnan, 1996). Traditional human relationships like beliefs, faith, taboos, customs and priorities have significant role in conservation of habitats and biodiversity (Jain *et al.,* 2000).There are 668 protected areas including 102 national parks, 515 wildlife sanctuaries, 47 conservation reserves and 4 community reserves covering a total of 1,61,221.57 km^2 of geographical area which is approximately 4.9% of the country. In addition there are 18 biosphere reserves, 5 natural world heritage sites and 25 Ramsar wetland sites in India (MoEF, 2014). Conservation of natural resources under community based conservation system is a basic tool to reduce the loss of biodiversity. Several programmes have been applied, for the conservation and management of natural resources in the Indian Himalaya under the protected area network (Rawat *et al.,* 2010).

Some examples of traditional customs of biodiversity conservation is still in practice, which comprise the idea of sacred groves, sacred species and sacred landscape (Ramakrishnan *et al.,* 2002). The indications propose that sacred grove concept of biodiversity conservation had adopted by various indigenous communities worldwide (Singh, 2000; Kala, 2011). Sacred sites/groves are natural places protected by indigenous communities for cultural or religious purpose and are found on all over the globe (Bhagwat & Rutte, 2006). Local communities maintains both access and use of resources from their sacred sites and check destruction of sacred sites by people (Colding & Folke, 2001). Harvest of non-timber forest products (e.g., firewood, fodder, medicinal plants, and fungi) in a sustainable manner is sometimes allowed, depending on the cultural or religious significance of a sacred site (Shen *et al.,* 2012). The local people managed the sacred groves with a great passion and holiness. The traditional institutional mechanisms have been assisting the native people in maintaining the sacred groves.

Biodiversity conservation in mountain ecosystem

The ultimate topographic characteristics of mountainous regions, such as slope, aspect and altitude, provide distinguished spatial shapes for mountain ecosystems and processes (Radcliffe, 1982). Prominent floral regions are based mainly on altitudinal and climatic variations, while the difference in aspect enhances habitat heterogeneity and brings micro-environmental variation in to the vegetation pattern (Clapham, 1973).The Himalayas, the mountains of Central Asia, south-west China, the Caucasus, East Africa and the Andes are recognized as globally important biodiversity hotspots (Figure-1).

Mountain biodiversity is, however, under threat and a number of endangered plant species are on the verge of extinction because plant species respond in a very complex way to environmental changes (Gordon *et al.,* 2002; Holtmeier and Broll, 2005; Miller *et al.,* 2006; Thuiller, 2007). A magnitude of the narrow ecological amplitudes shown by many mountain and alpine species, but it also reveals increasing grazing pressure or gathering for food or other uses (Hobbs and Huenneke, 1992). As a result, mountain regions are predicted to be places for prompt species extinction, particularly under the risk of global warming (Kullman, 2010). Mountain ecosystems need proper management against climatic and anthropogenic impacts for their future sustainability (Kessler, 2000; Halloy and Mark, 2003; Holzinger *et al.*, 2008; Erschbamer *et al.*, 2011).

In the Himalayas, for example, there is extensive traditional use of species, often resulting in overexploitation and overextraction, combined with a lack of scientific recording which makes the preparation of conservation planning a challenging job. The Himalaya differ from other mountain systems, for example the European Alps, people still possess traditional healthcare system and ethnobotanical knowledge. Ethnobotanical knowledge has been largely lost in the Alps by contrast, and there is also a lower population density at high altitudes. The main land-use problem in the Alps at the current time is land abandonment, rather than degradation through extensive use (Gehrig-Fasel *et al.*, 2009; Niedrist *et al.*, 2009).

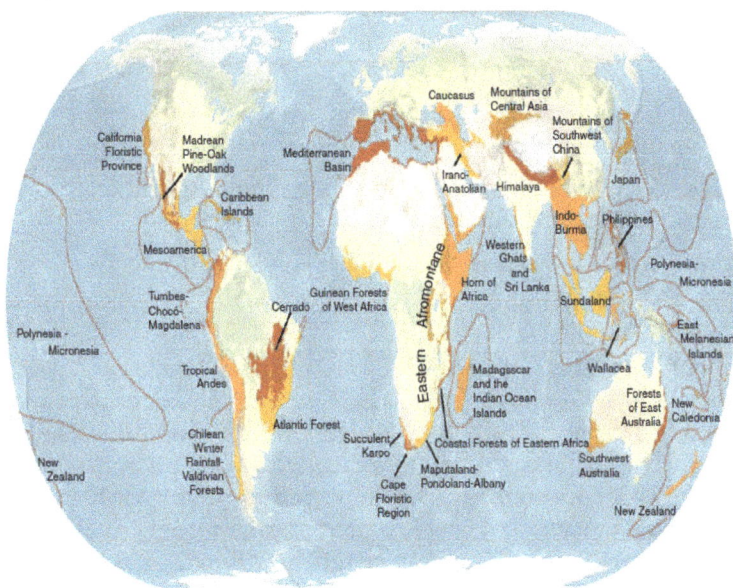

Figure-1: Biodiversity hot spots around the globe (Myers *et al.*, 2000). Source: http://www.conservation.org/where/priority_areas/hotspots/Pages/hotspots_main.aspx

Achieving the goal of sustainability into use and management of plant resources in mountain areas is a challenging job, especially in remote mountain ranges such as the Himalayas, Hindu Kush and Karakoram where there are both geographical and geopolitical restrictions. These mountain ranges are also situated in geopolitically undeveloped and democratically young countries such as India, Nepal, Pakistan and Afghanistan where, in the majority of cases, policy-makers and politicians pay little attention to the scientific evidence on plant diversity and threats to its survival when taking decisions related to natural resource management. In addition, parts of these geopolitical boundaries have faced various governmental or tribal conflicts and unrest, e.g. the Hindu Kush Mountains in Afghanistan. Such conflict decreases the opportunity for documentation of present biodiversity and the implementation of conservation and management (Khan *et al.*, 2013).

Major conservation issues

Notwithstanding its significance, India faces various risks to its biological and other diversity as a result of habitat alteration and fragmentation, unsustainable extraction and use of resources, impacts from tourism, and others (Chettri, 2000). As a result of the diverse cultural, governmental, and administrative settings conservation issues and priorities differ. These include:

(a) **Overexploitation of resources** – Illegal trading of wildlife (e.g. musk bile) and plant species (e.g. Cordyceps, Daphne); uncontrolled collection of fuelwood and medicinal plants; and timber harvesting by organizations as well as individuals;

(b) **Land-use practices** – Unregulated grazing by livestock; habitat fragmentation by invasion, and over-use of resources; conversion of forest area into agricultural fields.

(c) **Livelihood options** – Unregulated/unplanned tourism activities; transhumance; and people-wildlife conflicts and their management issues;

(d) **Policies** – Weak implementation of existing policies and laws; undefined and weak policies and regulations for international trade.

The above transboundary issues present further challenges:

(i) a complete information database is required to develop intervention policies and to adequately address transboundary issues including payment for ecosystem services and upstream-downstream linkages; Various legal and policy aspects affect resource use and conservation tools (including community rights on the use of resources) differently in the countries within the landscape; (ii) other livelihood options are limited and have limited opportunities for scaling up; and (iii) physical and financial limitations prevent interacting and regular communication of information and appropriate practices among countries within the landscape.

Mountain Biodiversity Conservation in Changing Climate

Climate change poses key challenges to biodiversity conservation. As atmospheric CO_2 increases over the next century, it is predicted to become the first or second greatest driver of global biodiversity loss (Sala *et al.*, 2000; Thomas *et al.*, 2004).

Diverse mountain biotas

It is estimated that, globally, alpine flora constitute about 4% of total flowering plant diversity, while the vegetated area is at most 3% of the ice-free land surface (Korner, 1995). Montane species diversity is estimated to be 2-3 times higher than expected from the montane area alone. On a global scale, whole mountain regions clearly exceed the species diversity of the plains (Barthlott *et al.*, 1996), largely a result of the compression of climatic zones across elevation gradients over very short geographical distances. Within the margins of a tropical mountain we may find humid forest in the foothills, montane cloud forest, alpine heathland, and a distinct flora resembling life forms. In addition to the merging of contrasting climatic belts, the causes of the species richness in the mountains are (i) latitudinal diversity of the land surface, i.e., a varied topography yielding a variety of habitats with contrasting life conditions, and (ii) the spatial fragmentation of mountains leading to the separate evolution of autochthonous biota (Korner, 2004).

Imperative of mountain biodiversity conservation

There are a number of explanations for the protection of biota. The most fundamental explanation is the ethical one, which indicates the right to exist for any species in the best conditions possible. This was the base of most conservation priorities, relatively it is the cultural heritage intention, which accounts for the fact that human societies have created unique biota, domesticated plants and animals, and added a 'cultural' element to biological richness. On the other hand, science is able to add further explanations, which neither diminish nor interchange the ethical and cultural intentions. Among them are the motives of ecosystem functioning and the economic motive (Korner, 2004). The first motive relates to the importance of biological richness for ecosystem processes, such as productivity, nutrient, carbon and water cycle, soil protection and erosion control, and biotic interactions such as plant-animal (such as pollination) and any other organismic interdependency (host-parasite, prey-predator, and facilitation). Some of these factors also have economic value, such as protection from erosion, clean water supplies, and so forth, but there are also more specific economic values acquired to rich biota when it comes to specific target organisms (certain crop, medicinal, and timber plants).

Climate change severely affects people's well-being directly through extreme weather events and threats, and indirectly through its effects on biodiversity and ecosystem services (Xu *et al.*, 2009).The more varied a system is, the more likely there will be species or genotypes that can cope with such extreme events: events that may be abiotic (freezing, storms, or fire) or biotic (pest outbreaks, pathogens, or invasive species). In other cases, such important species are termed as keystone species. Their absence or presence is of key importance to ecosystem functioning and agronomic success (Korner, 2008).

Mountain biodiversity in a warmer world

For many years ecologists described mountain biota as limited by low temperatures. A lot of research was dedicated to explaining the constraints of temperature to life. Now, the very same experts are concerned that a reduction of these temperature constraints by global warming will be unfavourable to high-elevation biota; conclusions are confusing to general public. There has been sufficient time for evolutionary processes (selection) and migration effects to occur (Korner, 2003). It never makes sense to apply an agronomic yield or growth-oriented constraint concept to natural plant and animal accumulations, which are controlled by selection and reproduction rather than body mass or mass increase per unit of time. In this sense, any change in environmental conditions creates changes in community composition, with some species losing and others gaining ground.

Mountains are in a particularly good situation in comparison to any other environment. At least for slowly-migrating organisms, such as plants, mountains are in fact the best places on earth to cope with climatic warming. While mountains offer us early biological indications of change, the biodiversity consequences will not be extraordinary, except for in very low ranges (Korner, 2007). So competition is likely to increase. Many species varieties have migrates poleward and uphill in elevation in the last period (Parmesan and Yohe, 2003; Root *et al.,* 2003) and will certainly remain to do so. Due to the rich topography, there is a high degree of segregation into diverse micro-habitats, facilitating coexistence in a given landscape. Nevertheless, the land area shrinks, and small plants might lead to drastic declines in populations of large territorial animals. These alterations increase concerns about the efficiency of current biodiversity conservation approaches (Halpin, 1997; Hannah *et al.,* 2002; Peters and Darling, 1985; Scott *et al.,* 2002). Figure-2 summarises the various migratory options plants and animals have in a mountain landscape.

1. Lowland species, lacking close distance escapes from too warm situations, **2.**Foothill species migrating upslope, **3.**High-elevation species migrating towards summit regions, **4.**summit species with no upslope escape, but increasing competition from immigrants from lower elevations, **5.**short distance escapes in highland taxa using micro-habitat diversity in rugged terrain, changing community mosaics at a given elevation.

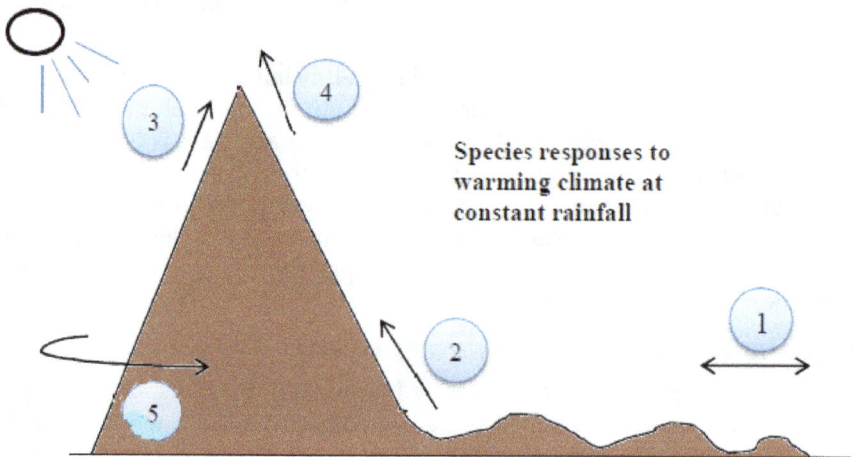

Figure-2: A schematic presentation of migration of species in response to warming climate (Source: Korner, 2008)

DIVERSITAS' Global Mountain Biodiversity Assessment (GMBA)

GMBA goals are set to enlighten and synthesise the understanding of biodiversity in mountainous region. This cross-cutting link of the global diversity research programme, DIVERSITAS, concentrates on the top montane range, the timberline ecotone, and all life above it. GMBA has two issued, one on the common state and causes of mountain biodiversity (Korner and Spehn, 2002) and other on land use changes (Spehn *et al.*, 2006). GMBA inspires research on the functional role of biodiversity in steep mountain topography with a particular importance on soil stability and hydrological implications, as symbolised in Figure-3. The current thrust area of GMBA is on developing tools for quantitative valuations of mountain biodiversity using electronic records (Spehn and Korner, 2009). The 'Kazbegi agenda' (Korner *et al.*, 2007), lead to an electronic mountain portal in collaboration with the Global Biodiversity Information System (GBIF).

Conservationists and managers of protected areas, as well as the global change research community, will profit from this new implement. By linking, the global topography database and climatological databases such as WorldClim (www. worldclim.org), with geo-referenced record data on organisms, it will be possible to explore the relative occupation of the possible environmental space (niche) today and approach projections for future ranges in a warmer world (Guisan *et al.*, 1998). Other applications are testing spatial phylogenetic trends among broader taxonomic groups or using the climatic understanding of species and functional types in the sense of 'trials by nature'.

Adaptation approaches

Biodiversity entities in natural ecosystems have been adapting naturally without much modification of those who benefit from their services. As the extent of climate change and other aspects of global change increases with time, the need for planned adaptation will become acute. Indigenous communities that directly or indirectly depend on natural resources have informal institutions and customary principles to ensure that external disturbance do not exceed natural resilience. Traditional approaches may have to be provided by formal adaptation processes to address the new risks to biodiversity. Better consideration of people's perception in relation to various climatic changes can support to scientific and policy consultations on adaptation to climate change. In this linking there is a necessity to understand experiences and responses of peoples to climatic change, because these reactions can reduce or intensify the impacts (Tripathi *et al.*, 2013; Bord *et al.*, 1998).

Sustainable practices in strategic adaptation are the base on the accessibility of sufficient information about the status of biodiversity; trends in ecological change, including climate change and its probable impacts on biodiversity; and the status of human population, capability, institutional ability, political assurance, and financial assets. Some of the adaptation options include the following (Sharma *et al.*, 2008): (i) institutional arrangement is responsible to addressing the climate-change issues and sensitive to the communal and economic preferences at local, national and regional levels, and (ii) research and development in agroforestry and community forestry to increase carbon sequestration, decrease soil erosion, improve water quality, and enhance livelihood options; (iii) operational framework should be designated for transboundary landscape approach to biodiversity conservation and protected area management (iv) establishment of stations to monitor the climate and long-term climatological time series and related infrastructure for networking and sharing data. (v) recognising and monitoring climate-sensitive organisms as indicators for early detection of climate-change indications and enabling intervention for practical adaptation, and (vi) sustainable management of rangelands and validation of climate-sensitive pastoralism not only to increase productivity but also to protect the ecosystem, decrease CO_2 emissions, and increase storage above and below ground.

Challenges and possible framework for biodiversity conservation

The major challenges to biodiversity conservation in India are inadequate policies and strategies; weak institutional, administrative, planning, and management capacities; inadequate management of data and information; unsustainable harvesting of resources; and to some extent poverty. A possible framework with ultimate aims, goals, and challenges is given in Table-1.

Table-1: Possible framework for biodiversity conservation

Level	Challenging tasks	Aim/Objective(s)	Strategy/Programmes
Global	CBD target	Observatory tools for biodiversity patterns in eco regions/nations	Develop accurate, time-bound, assessable goals following 2010 target guiding principle.
	Climate change	Proper understanding of biodiversity	Develop, test, and determine novel analytical tools and techniques through experimental studies.
Regional	International trade of animal products and illegal hunting	Long-term observation and collaboration among nations distributing transboundary landscapes	Develop inter connectivity between different ecosystems and corresponding laws through regional partnership.
National	Sustainable utilization of natural resources	Reduce unsustainable harvesting of natural resources.	Develop good governance skills and management.
	Equal benefit sharing and right to use genetic resource	Ensure the unbiased and rightful sharing of benefits from the use of genetic resources.	Report equal benefit sharing appropriately in law or as fundamental rights in the constitution.
	Sustainable eco-tourism	Conservation of natural and cultural diversity and development of socio-economic status	Develop social justice mechanisms to sustain indigenous communities.

Level	Challenging tasks	Aim/Objective(s)	Strategy/Programmes
Local	Sustainable Livelihood production	Increase expansion of socio-economic activities	Improve strong connections with processes working at regional and global scale.
	Grass land management	Management of traditional grass lands	Improve methods to reduce or avoid grazing burden.
	Human-wildlife conflict	Crop destruction and livestock depredation	Organize camps and training programmes.

Source: Chaudhary, 2008

Climate change and coping strategies for conservation in Himalaya

The Indian Himalayas form a huge mountain system with a geographical area of about 5, 91,000 sq.km covering 18% of the area of India. The protected area network in the Himalayan region consists of five biosphere reserves, 18 national parks, and 71 wildlife sanctuaries and covering 9.2% of the area of Indian Himalayas (Maikhuri and Rao, 2006). India is known as one of the 17th mega biodiverse regions of the world, and this position is mainly because of the Himalayas. The rich biodiversity of the region has a great extent to the indigenous and ethnic values of the society.

Global climate change has severe impacts on the biotic and abiotic environment and the socioeconomic settings and livelihood options of people in the Himalayas and neighbouring areas in the plains. Due to species composition and diversity, habitats, and the occurrence of rare and threatened species, as well as alien species at high altitude, will also be affected, thus threatening the conservation value of Himalayan environments (Maikhuri *et al.*, 2000; Nautiyal *et al.*, 2002; Maikhuri *et al.*, 2003). It is also affecting glacier, thereby freshwater supplies and other ecosystem services.

Impact of climate change on traditional hill agriculture

Traditional agricultural practices in hilly areas are an important land use practice. Due to the effect of climate change, agricultural crop production in the high-altitude region differs according to crop composition, soil conditions, and the cropping pattern. In high-altitude areas, the global alteration in the environment is leading to increase in temperature. However, it provides more prospects particularly for cash crops like tomatoes, cabbage, chillies, peas and medicinal plants (Maikhuri *et al.*, 2003). Notwithstanding, decrease in snowfall, rainfall, and melt water flows will produce a shortage of soil moisture that reduces any increase in crop production resulting from temperature increases. It is also noticed that slight changes in temperature have a key impact on the severity of diseases. *Amaranthus* crops are most susceptible to climate change as observed, when this crop was harshly attacked by a disease called *Hymenia rickervalis* at between 1,000 and 1,800 masl, whereas at between 2,200 and 2,800 masl it grew well.

Diseases such as rust and blight are common in cereals and potato crops, and legumes such as *Phaseolus spp.* become infected through soil- borne insects such as *Coleoptera species*. These insects harm the crops in early phases of seed growth, due to the unfavourable climatic conditions for the life cycle of the insects i.e., an increase in moisture or humidity in the lower regions (between 500-1,500 masl). High-altitude agriculture is definitely in conversion and temperature rise in the future may enhance agricultural productivity (Maikhuri *et al.*, 2008).

Climate change impact on forest and timberline vegetation

In the Central Himalaya, the predominant forest types are *Pinus wallichiana, Quercus* species, mixed (pine-oak), *Cedrus deodara*, mixed conifer, *Betula* and *Abies* and *Cupressus torulosa*, along with scrubland and alpine and low-altitude grasslands. In terms of biological diversity it is very important. The timberline, the most conspicuous and significant ecological margin where the sub-alpine forest ends, has been acknowledged as a region sensitive to environmental change and could be efficiently monitored for future impacts of climate change. In some places, timberline vegetation represents evergreen conifers entirely, while in some areas it is enclosed totally by deciduous broad-leaved trees (Purohit, 2003). The native species at the timberline are *Betula utilis, Abies pindrow*, and *Rhododendron companulatum*, have complex, distinct habitats of medicinal and aromatic plants and wild foods.

In this region there is a dominance of tree species such as *Abies pindrow, Betula utilis*, and *Acer caesium* because of their biological adaptation to extremely low temperatures. These species with narrow ecological niches may vanish if they fail to out compete with new species under warmer conditions. Mid-altitude species (1,600-2,000 masl), such as *Pinus roxburghii, Cedrus deodara, Cupressus torulosa, Quercus dialtata, Q. semicarpifolia, Q. leucotricophora*, and *Rhododendron arboretum* have a wider altitudinal range than alpine and sub-alpine species and hence vanishing of the former is less likely than of the later (Maikhuri *et al.*, 2003). Climate change affects both species negatively and could lead to a decline or a shift in their ranges towards higher altitudes. In addition, many important tree species in the timberline zone have already been listed in the rare and endangered categories, i.e., *Taxus baccata, Juniperus spp*, and *Betula utilis*. These species are overharvested, legally or illegally, to a great range and increased rates of devastation and the effect of changing climate have made the condition worse. *Betula utilis* has sociocultural and religious value and is also considered as a keystone species of the timberline ecosystem: it forms the upper limit of forest vegetation rising to an altitude of between 3,200-4,000 masl (Purohit 2003) *Betula utilis* needs a lot of light and grows very well where heavy snowfall occur (Maikhuri *et al.*, 2003).

Ecotourism and climate change

There is lot of openings of nature-based tourism or ecotourism, pilgrimages, and spiritual tourism activities present in Himalaya. It has been reported that, in areas where extensive tourist attraction occurs where ecosystems are fragile and have severe impacts. The nature and degree of such impacts depend on the intensity of tourism activity as well as the sensitivity of the affected ecosystems. Most of the studies showed that severe impacts of tourism on species and ecosystems arise from the structure and building construction activities involved, rather than from the recreational activities, as in the case of sacred tourism carried out at Badrinath,

Hemkund Saheb (Valley of Flowers), Gangotri, Bhojwasa, and Gaumukh.

Climate change could generate not only serious problems, but also chances for the tourism industry. Earlier, tourism rotated around trekking and pilgrimage in the Himalayan region. However, tourism in the form of adventure tourism, winter sports, and expeditions to glaciers, mountaineering, nature ramblers, and pilgrimages has expanded rapidly. This has had a negative impact on biodiversity of area, but a positive impact on the living standards of the people inhabiting as well as of those dependent on tourism. Tourism may provide better chances for income generation as other primary and secondary production sectors (i.e., agriculture, livestock, and non-timber forest product collection) decline. The culture and religion of traditional and local communities, however, are open to pressures that may have uncertain outcomes (Maikhuri and Rao, 2006).

Coping and mitigation strategies: The significance interventions

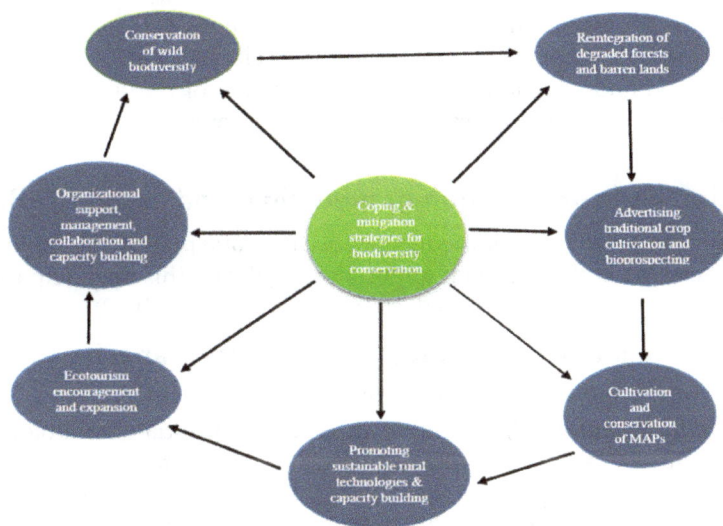

Figure-3: Coping and mitigation strategies for biodiversity conservation (Based on Maikhuri *et al.*, 2008)

Conservation of wild biodiversity: reinforcement of the protected area network

Redundancy associated with species richness is likely to increase the possibility of reimbursement of negative impacts caused by changing environmental conditions. Although we have a long past of strategic conservation (9.2% area of the Himalayas is legally protected), our knowledge of people, biodiversity, susceptibility, and their inter connections are very limited. Therefore, participatory research/management could turn through people's negative attitudes to positive attitudes towards protected areas as well as improving scientific knowledge related to potential uses of biodiversity for coping and mitigation (Maikhuri *et al.*, 2000).

Reintegration of degraded forests and barren lands

The failure of afforestation and reforestation efforts to develop degraded lands in the Himalayan mountains could be endorsed largely to unawareness of people's essential needs, and, hence, their lack of support. People's participation is now considered to be a requirement to the success of any land rehabilitation efforts in the Himalayas. The local people widely accepted the practice and structure developed for degraded land rehabilitation. Considering the diversity of ecosystems, indigenous knowledge, and socioeconomic conditions in the Himalaya, any rehabilitation strategy should be site-specific (Maikhuri *et al.*, 2003; Maikhuri *et al.*, 2007).

Advertising traditional crop cultivation and bioprospecting

In spite of the many assets of traditional crops, valuable genetic diversity, the rivet of stability for the ecosystem, is gradually being lost (Maikhuri *et al.*, 2000). Realistic multidisciplinary research efforts are needed to develop farming systems and select suitable crops in view of future climate change so that sufficient supplies of food and economic security, conservation of traditional crop wealth, sustainability of production systems, and environmental conservation are guaranteed.

Cultivation and conservation of medicinal and aromatic plants

The cultivation and use of medicinal and aromatic plants offer potentials for income generation in this region if undertaken well, and this may help reduce the existing pressure on resources (Singh, 1999; Maikhuri *et al.*, 2000).

Promoting sustainable rural technologies and capacity building

Technology transfer is an important instrument in the constant process of socioeconomic development, and limited access to appropriate technologies is one of the main causes of poverty, food insecurity, and natural resource exploitation in the Himalayas. Hence appropriate technologies suitable for high-altitude regions, such as protected cultivation, organic manures and bio-fertilizers, bioprospecting of wild edibles, off-farm, and other associated technologies are needed to provide sustainable options for income generation and thus reduce the existing pressures on forest and other bio resources (Maikhuri *et al.*, 2007).

Ecotourism encouragement and expansion

Ecotourism in Himalayan reserve such as the NDBR (Nanda Devi Biosphere Reserve) needs to be very careful consideration and, therefore, ecotourism plans need to be combined with other management plans, such as those for wildlife, fire, vegetation, and eco-development, to decrease the overall pressure on forests and biodiversity (Maikhuri and Rao, 2006).

Organizational support and capacity building to address climate change

There is insufficient capacity in many research and development institutions working on environmental and conservation issues in relation to climate change:

therefore, awareness raising and capacity building at individual and institutional levels are of ultimate importance. It is also important to enhance the capacities of local people who are likely to be susceptible to projected climate impacts (Figure-3).

Drylands Biodiversity Conservation

Biodiversity is the variety and variability in all life forms in ecosystems on earth: flora and fauna, microbes, their habitats and their genes. Biodiversity creates the mesh of which humans are involve: it controls the biological cycles of the earth and is liable for our survival and sustenance (Cardinale *et al.*, 2012). Drylands, including dry sub-humid, semi-arid, arid and hyper-arid lands, cover 41.3%2 of the earth's terrestrial area. Dry land biodiversity plays a crucial role for combating poverty, climate change and desertification. Dry lands are defined by water scarcity and categorized by periodical climatic extremes and erratic precipitation patterns. drylands areas, where the possible amount of water that is transferred from the land to the atmosphere is at least 1.5 times greater than the mean precipitation: a calculation known as the aridity index (UNCCD, UNDP and UNEP, 2009).

Dryland biodiversity is vital to sustainable development and to the employments of many deprived persons of the globe: the value of biodiversity to poverty eradication and economic welfare in the drylands may be high in many other ecosystems (Smith *et al.*, 2010). This is because drylands are considered by extreme climatic uncertainties in which biodiversity plays a key role in traditional rural risk management approaches.

Apart from non-dryland ecosystems, drylands have rich diversity. Dryland ecosystems comprise Mediterranean regions, like Mediterranean Basin or the Cape Floristic Region of South Africa, as well as cold deserts such as the Gobi in Mongolia and hot deserts like the Sahara, where both climate and latitude strongly influence biodiversity. Dryland biodiversity is also influenced by altitude, which ranges from low lying areas to high altitude drylands in countries. These diverse dryland ecosystems cover a greater diversity of flora and fauna that have grown to aggregate their distinct territories (Davies *et al.*, 2012). Extreme dry desert area of western Rajasthan (Singh and Saxena, 1998) cold area of north western Himalaya (Singh *et al.*, 1997) faced dual crisis of vegetation and livelihood. Water shortage has played an important role in inducing biological diversity in the drylands, but differences in topography, geology, soil type and quality have also been responsible elements in this phenomena. Other factors include the periodical patterns of precipitation, fires and herbivore pressure as well as the impact of human management over periods. Resulting, all these factors, drylands consist of a framework of habitats, which regulates the dispersal of living organisms (Bonkoungou, 2003). Conserving the sacred species and sacred landscape appeared to be one of age-old practices in desert area would be one of model for conservation (Singh and Saxena, 1998).

Status of dryland biodiversity

Data from the IUCN Red List (2000) show that, across all biomes, over 32% of species (out of 59,507 species) are threatened with disappearance. The causes of biodiversity loss are critical and inter-linked, include human population growth, modification of habitat for extension of farming and increase of urban areas. Invasion play an important role for the loss of native biodiversity and often reduce ecosystem services. Human-induced climate change and habitat destruction is changing species migration patterns, shifting the range that many species can inhabit and accelerating the range of exotic species (McNeely *et al.*, 2001).

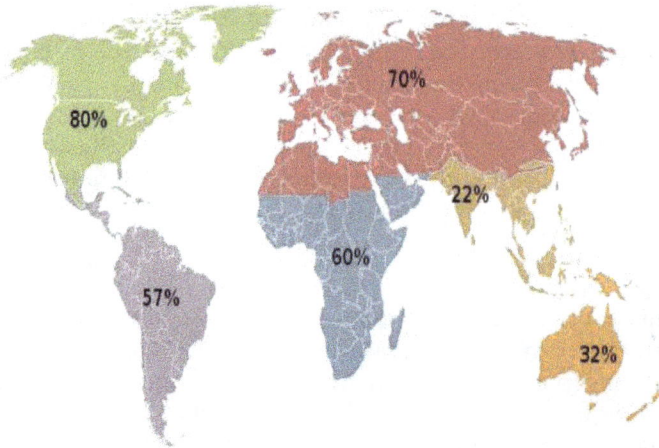

Figure-4: Percentage of mammals, birds and amphibians occurring in drylands for different realms   Source: (UNEP-WCMC 2011, realms based on Olson *et al.*, 2002).

Drylands are universally recognized for enigmatic wildlife species such as the Lion (*Panthera leo*), African elephant (*Loxodonta africana*), Bactrian camel (*Camelus ferus*) and bovines like the wild yak (*Bos mutus*) and the American bison (Bison bison). While these species may attract millions of tourists annually for wildlife 'safaris', particularly in East Africa. Drylands sustains crucial plant groups such as the cacti and succulents, trees such as acacia and baobab and many of the world's meadows. Overall, 10,000 mammals, birds and amphibian species can be found in drylands: 64% of all birds, 55% of mammals and 25% of amphibians. Comparatively, the richest terrestrial biome – tropical and sub-tropical moist broadleaf forests – comprises 70% of global terrestrial animal (Safriel *et al.*, 2005). Drylands provide habitats for 80% of North America's mammal, bird and amphibian species. In Europe, Asia and North Africa, around 70% and in sub-Saharan Africa and Latin America, around 60%. Immense number of species arising in dryland habitats, 4% of mammals and amphibians, and 3% of birds are endemic to drylands (Figure-4).

Drivers of dryland biodiversity loss

The Millennium Ecosystem Assessment provides information of several factors that influence on dryland biodiversity, comprising habitat modification, climate change, over-exploitation, grazing pressures, exotic species and unsustainable soil management practices (access use of fertilizers, pesticides). These factors are affected by policy failure, for example in relation to land privileges and natural resource governance, and lack of community asset, which impulse people into adopting unsustainable practices. Urbanization also leads to dryland biodiversity loss (MEA, 2005).

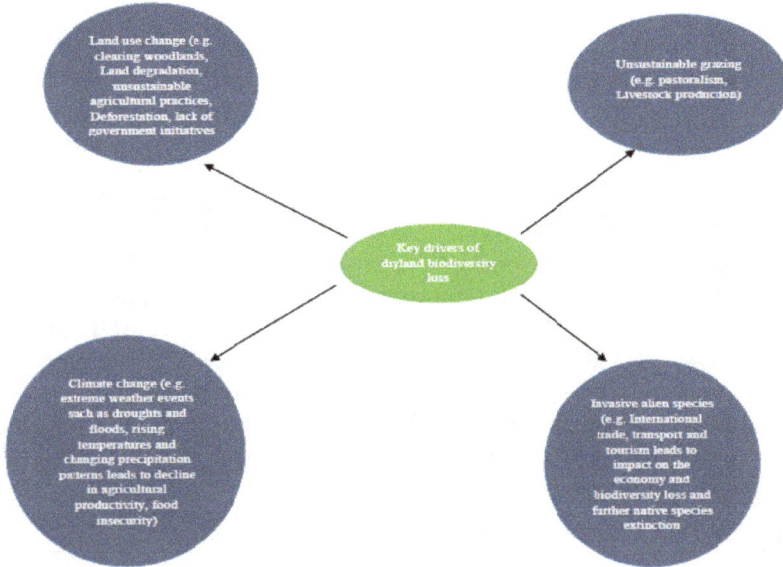

Figure-5: Key driving factors of dryland biodiversity loss

Land-use change

Land-use change enhances biodiversity loss directly, through tilling rangelands or clearing woodlands, and indirectly through land degradation. Unsustainable agricultural practices has led to extensive deforestation and modification of rangelands to agricultural lands, with severe environmental and economic concerns. Irrigated crop farming often uses limited water resources that are important value to a larger dryland ecosystem. Land-use change has a number of causal factors and can be initiated by government development programmes, fragile evaluation of current land uses and low concern for the cost of dryland ecosystem services.

Unsustainable grazing

Food production in many drylands has concerned on pastoralism: the widespread production of cattle, often arranging herd movement as a management plan. The basis of pastoral movement and the related common property tenure preparations have often been challenged and intensive efforts have been made to replace pastoralism with other forms of livestock production; this constrains stretches back for millennia (Scoones, 1995). The outcome of these changes has been both economically and environmentally detrimental and the strategy of livestock production has been a crucial role to land degradation.

Climate change

Climate change is documented to be an important factor of environmental change in the global drylands (IPCC, 2007). There will be major regional variances in the consequence of climate change, but generally it is estimated that climate change will lead to a decline in water availability and quality of 10% - 30% in the next 40 years, while extreme weather events such as droughts and floods will increase in number and/or intensity. Rising temperatures and changing precipitation patterns are predicted to lead to an expansion of drylands worldwide. Climate change is predicted to reduce agricultural productivity overall in the drylands and this will have severe impacts on food security. Climate change is also predicted to increase the rate of urban growth, with associated links to environmental impact (IPCC, 2007; Stern, 2007).

Invasive alien species

International trade, transport and tourism are leading to the introduction of invasive species, as species are relocated into new areas that are alien to them. Many species do not adapt to new environment and cannot persist there. However, some of these species do survive and even flourish in new environment, and they cause an invasion that has a worse impact on the economy and biodiversity of a particular region. These species are called as invasive and become a major cause of species extinctions, increasing the rate of extinction by about 1000% (Invasive Species Specialist Group, 2012) (Figure-5).

Managing and Conserving Dryland Biodiversity

Conservation of dryland biodiversity is not well studied, many identified factors of biodiversity loss are exists in the drylands. These factors include fast demographic shifts and urbanization, agricultural intensification, land use conversion, lack of proper policies and the introduction and extent of exotic species. Enhancing dryland development is expected to increase the rate of biodiversity loss. The arrangement of habitat loss and disintegration will decrease the chances for dryland biodiversity to adapt and survive, with the other impacts of climate change further accelerating the problem. Although approximately 9% of drylands getting proper protection, the protected areas are not characteristic of all the drylands (Davies *et al.*, 2012). Large areas of drylands are protected by the indigenous communities that lives in the area, either as sacred sites or as a consequence of sustainable management practices that grown through generations.

This traditional conservation is occasionally identified by government and is often neglected by government policies. Many traditional land management practices have confirmed to be more economically feasible than contemporary substitutes, although instantaneously providing conservation incentives. The ecological basis of these indigenous approaches developed through a clear understanding by the traditional communities of their close environment confirms both economic and environmental sustainability. The drylands offer more chances for attaining both conservation and development priorities concurrently. Drylands supports indigenous and community conserved areas, and traditional natural resource management approaches.

Knowledge gaps and research needs

Adaptations to changing climate for the biodiversity conservation and its sustainable use need to come from a strong understanding of the important territories and species and the nature of their resiliency to climatic stress. The following important points prioritized for research work and advancement of understanding of climate change and its possible impacts on biodiversity mitigation and adaptation were summarised by (Sharma *et al.*, 2008; Maikhuri *et al.*, 2008): (i) policies should be revised and reinforced to make them more delicate to the interaction in processes and the connections between the consequences of climate change to biodiversity; (ii) a comprehensive catalogue of species and ecosystems and appropriate records of traditional knowledge and practice on adaptation to climate change, including variability and extremes (iii) broad and in-depth assessment of the actions of alien, invasive species and serious landscape linkages to keystone species, PA network and effectiveness, adaptability of biodiversity entities, fire controlling regimes, and impacts on agricultural production are other important areas demanding serious consideration. (iv) extensive survey and record of the distribution range of flora and fauna, biodiversity within PAs, population trends of keystone/endemic or threatened species and other groups of animals should be carried out (v) capacity building is needed to carry out particular study in taxonomy, conservation biology, and impact assessments (vi) effort of different conservation initiatives must be coordinated and collaborative partnerships developed to cope with the present and future impacts of climate change (vii) an assessment should be carried out of ecosystem structures, functioning, and productivity and distribution of ecosystem goods and services (viii) a study should be carried out on the relationship between climate change and land use change to evaluate their impact on biodiversity, atmospheric CO_2 concentration, species composition, and carbon dynamics in different ecosystems (ix) economic assessment should be made of ecosystem services from conservation area (x) socioeconomic studies of land-use systems, food security, rights to use of resources, decision making processes, and governance that describe community resiliency to climate change should be carried out. (xi) documentation of traditional ecological knowledge as well as people's perceptions and experiences about the array and indicators of climate change and its impacts on forests, alpine meadows, agriculture, livestock, and humans through use of participatory attitudes (xii) effect of the climate change on periodical variability and consistency, and climate extremes affecting agricultural production, forests, and water resources (xiii) establishment of permanent sample plots in different forest types along elevational gradients for an effective and widespread monitoring programme to track the response to a changing climate at both community and species level. (xiv) capacity building for researchers and scientists involved in the field of climate change and modelling studies (xv) expansion of proper weather and meteorological stations in important and sensitive biomes and ecosystem types for regional projections of climatic factors to facilitate development of regional climate models (xvi) interfaces with policy issues, management, local communities, research and academic institutions on the broad aspects of options for adaptation, mitigation, and livelihoods.

In addition to above points, there is a need to understand holistic approach of landscape ecology in an integrated manner, so that a comprehensive plan could be framed to link sociological and economic factors. However, distinct model should be adopted for various geographical and climatic conditions. Different stratified models should be furnished for hilly regions and in plains, different geo-climatic conditions should be followed. Existing viable models practicing at different places should be promoted on long-term scale. Peoples participation and policy an obvious gaps should be minimized. There is need to summarise scattered knowledge existing at various levels so that a viable environmentally friendly need and action based priorities standardised.

Conclusion

Biodiversity is vital for human survival and economic benefits and for the ecosystem functioning and stability. Since biodiversity is unequally distributed on the earth, with wide global and regional patterns therefore balanced approach of scientific and community based practices would be a better option. The increasing awareness of importance and severe loss make it vital for rapid assessment and conservation of biodiversity, both at regional and global scale. Although, with increasing number of literature, there is a scarcity of field level data, which formulate into concepts and approaches for better conservational aspects of biodiversity. Effective strategies for people's participation in conserving biodiversity are appeared to be one of a substantial approach. India has a rich tradition of conservation, and with increasing efforts from the government, scientists and NGOs, should offer governance in developing proper methodologies and plans for biodiversity assessment and conservation. Technical know-how and do-how of the local communities settled closed to nature should be strengthen.

References

1. Balmford, A. and Gaston, K.J. (1999). Why biodiversity surveys are good value. *Nature.* 398: 204-205.

2. Ban, N.C., Mills, M., Tam, J., Hicks, C.C., Klain, S., Stoeckl, N. and Chan, K. M. (2013). A social–ecological approach to conservation planning: Embedding social considerations. *Frontiers in Ecology and the Environment.* 11: 194-202.

3. Barthlott, W., Lauer, W. and Placke, A. (1996) 'Global distribution of species diversity in vascular plants: Towards a world map of phytodiversity'. *Erdkunde.* 50: 317-327.

4. Bennett, A.F. (1990). Habitat corridors and the conservation of small mammals in a fragmented forest environment. *Landscape Ecology.* 4: 109-122.

5. Bertzky, B., Corrigan, C., Kemsey, J., Kenney, S., Ravilious, C., Besançon, C. and Burgess, N. (2012). Protected Planet Report 2012: Tracking Progress towards Global Targets for Protected Areas. International Union for Conservation of Nature and the United Nations Environment Programme's World Conservation Monitoring Centre.

6. Bhagwat, S.A and Rutte, C. (2006). Sacred groves: potential for biodiversity management. *Frontiers in Ecology and the Environment.* 10: 519-524.

7. Bonkoungou, E.G. (2003). Biodiversity in drylands: Challenges and opportunities for conservation and sustainable use. The Global Drylands Partnership, IUCN.

8. Bord, R.J., Fisher, A. and O'Connor, R.E. (1998). Public perceptions of global warming: United States and international perspectives. *Climate Research.* 11: 75-84.

9. Cardinale, B.J., Duffy. J.E., Gonzalez, A., Hooper, D.U., Perrings, C., Venail, P., Narwani, A., Mace, G.M., Tilman, D., Wardle, D.A., Kinzig, A.P., Daily, G.C., Loreau, M., Grace, J.B., Larigauderie, A., Srivastava, D.S. and Naeem, S. (2012). Biodiversity loss and its impact on humanity. *Nature.* 486: 59-67.

10. Chape, S., Harrison, J., Spalding, M. and Lysenko, I. (2005). Measuring the extent and effectiveness of protected areas as an indicator for meeting global biodiversity targets. *Philosophical Transactions of the Royal Society B.* 360: 443-455.

11. Chaudhary, R.P. (2008). Hindu Kush-Himalayas: Current Status, Challenges and Possible Framework for the Conservation of Biodiversity *.IMBC-Plenary Session III:* Climate Change and its Implications for Mountain. ICIMOD, Kathmandu, Nepal.

12. Chettri, N. (2000). Impact of Habitat Disturbances on Bird and Butterfly Communities along Yuksam-Dzongri Trekking Trail in Khanchendzonga Biosphere Reserve, PhD Thesis. Rajaramohanpur, Siliguri (India): North Bengal University.

13. Clapham, W.B. (1973). Natural ecosystems. New York: Macmillan/Collier-Macmillan. 536.

14. Colding, J. and Folke, C. (2001). Social taboos: "Invisible" systems of local resource management and biological conservation. *Ecological Applications.* 11: 584-600.

15. Convention on Biological Diversity (CBD). (1992). Convention on Biological Diversity. United Nations. 2010. COP 10 Decision X/2: X/2. Strategic Plan for Biodiversity 2011–2020. United Nations. (31 October 2013; www.cbd. int/decision/cop/default.shtml?id=12268).

16. Costanza, R. (1997). The value of the world's ecosystem services and natural capital. *Nature.* 387: 253-260.

17. Davies, J., Poulsen, L., Schulte-Herbrüggen, B., Mackinnon, K., Crawhall, N., Henwood, W.D., Dudley, N., Smith, J. and Gudka, M. (2012). Conserving Dryland Biodiversity. IUCN.

18. Dobson, A. (1995). Biodiversity and human health. *Trends in Ecology and Evolution.* 10: 390-391.

19. Erschbamer, B., Unterluggauer, P., Winkler, E. and Mallaun, M. (2011). Changes in plant species diversity revealed by long-term monitoring on mountain summits in the Dolomites (northern Italy). *Preslia.* 83: 387-401.

20. Gehrig Fasel, J., Guisan, A. and Zimmermann, N.E. (2009).Tree line shifts in the Swiss Alps: climate change or land abandonment? *Journal of Vegetation Science*. 18: 571-582.

21. Gilbert-Norton, L.R., Wilson, J., Stevens, R. and Beard, K.H. (2010). A meta-analytic review of corridor effectiveness. *Conservation Biology*. 24: 660-668.

22. Gordon, J.E., Dvora´k, I.J., Jonasson, C., Josefsson, M., Kocianova, M. and Thompson, D.B.A. (2002). Geo-ecology and management of sensitive montane landscapes. *Geografiska Annaler, Series A: Physical Geography*. 84: 193-203.

23. Guisan, A., Theurillat, J.P. and Kienast, F. (1998). 'Predicting the potential distribution of plant species in an Alpine environment'. *Journal of Vegetation Science*. 9: 65-74.

24. Halloy, S.R.P. and Mark, A.F. (2003). Climate-change effects on alpine plant biodiversity: a NewZealand perspective on quantifying the threat. *Arctic, Antarctic, and Alpine Research*. 35: 248-254.

25. Halpin, P.N. (1997). Global climate change and natural-area protection: management responses and research directions. *Ecological Applications*. 7: 828-843.

26. Hannah, L. (2008). Protected areas and climate change. *Annals of the New York Academy of Science*. 1134: 201-212.

27. Hannah, L., Midgley, G.F. and Millar, D. (2002). Climate change integrated conservation strategies. Global *Ecology and Biogeography*. 11: 485-495.

28. Hawksworth, D.L. and Kalin-Arroyo, M.T. (1995). In *Global Biodiversity Assessment* (ed. Heywood, V. H.), Cambridge University Press, Cambridge. 545-606.

29. Hilty, J.A., Lidicker, W.Z. and Merenlender, A.M. (2006). *Corridor ecology: the science and practice of linking landscapes for biodiversity conservation*. Island Press, Washington, D.C.

30. Hobbs, R.J. and Huenneke, L.F. (1992). Disturbance, diversity, and invasion: implications for conservation. *Conservation Biology*. 6: 324-337.

31. Holtmeier, F.K. and Broll, G. (2005). Sensitivity and response of northern hemisphere altitudinal and polar treelines to environmental change at landscape and local scales. *Global Ecology and Biogeography*. 14: 395-410.

32. Holzinger, B., Hulber, K., Camenisch, M. and Grabherr, G. (2008). Changes in plant species richness over the last century in the eastern Swiss Alps: Elevational gradient, bedrock effects and migration rates. *Plant Ecology*. 195: 179-196.

33. Howe, C. and Milner-Gulland, E.J. (2012). Evaluating indices of conservation success: a comparative analysis of outcome- and output-based indices. *Animal Conservation*. 15: 217-226.

34. Invasive Species Specialist Group. (2012). http://www.issg.org/about_is.htm

35. IPCC, Climate Change. (2007). Impacts, Adaptation and Vulnerability Working Group II Contribution to the Intergovernmental Panel on Climate Change Fourth Assessment Report. Summary for Policy Makers. Brussels, Belgium.

36. IUCN. (2000). IUCN Guidelines for the Prevention of Biodiversity Loss Caused by Alien Invasive Species. Species Survival Commission Invasive Species Specialist Group. IUCN, Gland. http://www.issg.org/pdf/guidelines_iucn.pdf

37. Jain, S.K. (2000). Human aspects of plant diversity. *Economic Botany.* 54: 459-470.

38. Kala, C.P. (2011). Traditional ecological knowledge, sacred groves and conservation of biodiversity in the Pachmarhi biosphere reserve of India. *Journal of Environmental Protection.* 2: 967-973.

39. Kessler, M. (2000). Elevational gradients in species richness and endemism of selected plant groups in the central Bolivian Andes. *Plant Ecology.* 149: 181-193.

40. Khan, S.M., Page, S.E., Ahmad, H. and Harper, D.M. (2013). Sustainable utilization and conservation of plant biodiversity in montane ecosystems: the western Himalayas as a case study. *Annals of Botany.* Page 1-23. doi:10.1093/aob/mct125.

41. Korner, C. (2008). Conservation of Mountain Biodiversity in the Context of Climate Change. *IMBC-Plenary Session 1*: Climate Change and its Implications for Mountain. ICIMOD, Kathmandu, Nepal.

42. Korner, C. (1995). 'Alpine plant diversity: A global survey and functional interpretations'. In Chapin, FS III; Korner, C. (eds) *Arctic and alpine biodiversity: Patterns, causes and ecosystem consequences,* pp45-62. Berlin: Springer.

43. Korner, C. (2003). 'Limitation and stress – always or never?' *Journal of Vegetation Science.* 14: 141-143.

44. Korner, C. (2004). 'Mountain biodiversity, its causes and function'. *Ambio Special Report.* 13: 11-17.

45. Korner, C. (2007). 'Climatic treelines: Conventions, global patterns, causes'. *Erdkunde.* 61: 315-324.

46. Korner, C. and Spehn, E.M. (eds.) (2002). *Mountain biodiversity. A global assessment.* New York: Parthenon.

47. Kullman, L. (2010). Alpine flora dynamics – a critical review of responses to climate change in the Swedish Scandes since the early 1950s. *Nordic Journal of Botany.* 28: 398-408.

48. Loarie, S., Duffy, P.B., Hamilton, H., Asner, G.P., Field, C.B. and Ackerly, D.D. (2009). The velocity of climate change. *Nature.* 462: 1052-1055.

49. Maikhuri, R.K., Rawat, L.S., Negi, V.S., Phondani, P., Bahuguna, A., Chamoli, K.P. and Farooquee, N. (2008). Impact of Climate Change and Coping Strategies in Nanda Devi Biosphere Reserve, Central Himalayas,

India. *IMBC-Technical Working Groups*: Group 1. Climate Change and its Implications for Mountain. ICIMOD, Kathmandu, Nepal.

50. Maikhuri, R.K., Nautiyal, S., Rao, K.S., Chandrasekhar, K., Govall, R. and Saxena, K.G. (2000). 'Analysis and resolution of protected area-people conflicts in Nanda Devi biosphere reserve, India'. *Environmental Conservation*. 27: 43-53.

51. Maikhuri, R.K. and Rao, K.S. (2006). Developing eco-tourism for Nanda Devi Biosphere Reserve: Strategies and action plan. New Delhi: G.B. Pant Institute of Himalayan Environment and Development and UNESCO.

52. Maikhuri, R.K., Rao, K.S., Patnaik, S., Saxena, K.G. and Ramakrishnan, P.S. (2003). 'Assessment of vulnerability of forest, meadows and mountain ecosystems due to climate change'. *ENVIS Bulletin*. 11(2): 1-9.

53. Maikhuri, R.K., Rawat, L.S., Negi, V. and Purohit, V.K. (2007). Eco-friendly appropriate technologies for sustainable development of rural ecosystems in Central Himalaya. New Delhi: G.B. Pant Institute of Himalayan Environment and Development.

54. Mannion, A.M. (1995). Biodiversity, biotechnology, and business. *Environmental conservation*. 22: 201-210.

55. May, R.M. (1988). How many species are there on earth? *Science*. 241: 1441-1449.

56. McDonald, R.I. and Boucher, T.M. (2011). Global development and the future of the protected area strategy. *Biological Conservation*. 144: 383-392.

57. McNeely, J.A., Miller, K.R., Reid, W.V., Mitter-meier, R.A. and Werner, T.R. (1990). Conserving the World's Biological Diversity. IUCN. Gland.

58. McNeely, J.A., Mooney, H.A., Neville, L.E., Schei, P. and Waage J.K. (eds.). (2001). A Global Strategy on Invasive Alien Species. IUCN Gland, Switzerland, and Cambridge, UK.

59. MEA (Millennium Ecosystem Assessment). (2005). Ecosystems and Human Well-being: Desertification Synthesis. World Resources Institute, Washington, DC.

60. Miller, R.M., Rodrıguez, J.P. and Aniskowicz-Fowler, T. (2006). Extinction risk and conservation priorities. *Science*. 313: 441.

61. MoEF. (2014). Annual Report 2014-15. Ministry of Environment, Forests and climate change. Govt. of India, New Delhi.

62. Myers, N., Mittermeler, R.A, Mittermeler, C.G, Da-Fonseca, G.A.B. and Kent, J. (2000). Biodiversity hotspots for conservation priorities. *Nature*. 403: 853-858.

63. Nautiyal, S., Rao, K.S., Maikhuri, R.K. and Saxena, K.G. (2002). 'Transhumant pastoralism and sustainable development: A case study in the buffer zone of the NDBR, India'. *Mountain Research and Development*. 23(3): 255-262.

64. Niedrist, G., Tasser, E., Luth, C., DallaVia, J. and Tappeiner, U. (2009). Plant diversity declines with recent land use changes in European Alps. *Plant Ecology*. 202: 195-210.

65. NMPB. National Medicinal Plants Board. (2014). http://www.nmpb.nic.in

66. Olson, D.M. and Dinerstein, E. (2002). The Global 200: Priority eco regions for global conservation. Annals of the Missouri Botanical Garden 89: 199-224.

67. Parmesan, C. and Yohe, G. (2003). A globally coherent fingerprint of climate change impacts across natural systems. *Nature.* 421: 37-42.

68. Peters, R.L. and Darling, J.D.S. (1985). The greenhouse-effect and nature reserves. *Bioscience.* 35: 707-717.

69. Purohit, A. (2003). Studies on structural and functional aspects of timberline vegetation in Nanda Devi Biosphere, Garhwal Himalaya. PhD thesis, H.N.B. Garhwal University, Srinagar, Garhwal.

70. Radcliffe, J.E. (1982). Effects of aspect and topography on pasture production in hill country. *New Zealand Journal of Agricultural Research.* 25: 485-496.

71. Ramakrishnan, P.S. (1996). Conserving the sacred: from species to landscapes. *Nature & resources.* 32: 11-19.

72. Ramakrishnan, P.S., Rai, R.K., Katwal, R.P.S. and Mehndiratta, S. (2002). *Traditional Ecological Knowledge for Managing Biosphere Reserves in South and Central Asia.* Oxford & IBH, New Delhi.

73. Ramakrishnan, P.S., Saxena, K.G. and Chandrasekhara, U. (1998). *Conserving the Sacred for Biodiversity Management.* Oxford & IBH Publication, New Delhi, UNESCO Vol.

74. Rawat V.S., Rawat, Y.S. (2010). Van Panchayats as an effective tool in conserving biodiversity at local level. *Journal of Environmental Protection.* 1: 278-283.

75. Rodrigues, A.S.L. (2004). Effectiveness of the global protected area network in representing species diversity. *Nature.* 428: 640-643.

76. Root, T.L., Price, J.T., Hall, K.R., Schneider, S.H., Rosenzweig, C. and Pounds, J.A. (2003). Fingerprints of global warming on wild animals and plants. *Nature.* 421: 57-60.

77. Safriel, U., Adeel, Z., Niemeijer, D., Puigdefabregas, J., White, R., Lal, R., Winslow, M., Ziedler, J., Prince, S., Archer, E. and King, C. (2005). Chapter 22: Dryland systems. In: Hassan, R., Scholes, R. and Ash, N. (eds.) Millennium Ecosystem Assessment. Vol. 1. Ecosystems and human well-being: Current state and trends. World Resources Institute, Washington, DC. 623-662.

78. Sala, O.E., Chapin, F.S., Armesto, J.J., Berlow, E., Bloomfield, J., Dirzo, R., Huber-Sanwald, E., Huenneke, L.F., Jackson, R.B., Kinzig, A., Leemans, R., Lodge, D.M., Mooney, H.A., Oesterheld, M., Poff, N.L., Sykes, M.T., Walker, B.H., Walker, M. and Wall, D.H. (2000). Biodiversity–global biodiversity scenarios for the year 2100. *Science.* 287: 1770-1774.

79. Samant, S.S., Dhar, U. and Palni, L.M.S. (1998). Medicinal Plants of Indian Himalaya, Himavikas Publ., GBPIHED, Kosi Katarmal, Almora.

80. Sawada, Y., Matsuda, A. and Kimura, H. (2010). On the role of technical cooperation in international technology transfers. *Journal of International Development.* 24: 316-340.

81. Scoones, I. (1995). Living with Uncertainty. International Institute for Environment and Development, ITP Ltd., London.

82. Scott, D., Malcom, J. and Lemieux, C.J. (2002). Climate change and biome representation in Canada's National Park system: implications for system planning and park mandates. *Global Ecology and Biogeography.* 11: 475-484.

83. Sharma, E., Tse-ring, K., Chettri, N. and Shrestha, A. (2008). Biodiversity in the Himalayas – Trends, Perception and Impacts of Climate Change. *IMBC-Plenary Session 1*: Climate Change and its Implications for Mountain. ICIMOD, Kathmandu, Nepal.

84. Shen, X., Lu, Z., Li, S. and Chen, N. (2012). Tibetan sacred sites: understanding the traditional management system and its role in modern conservation. *Ecology and Society.* 17: 04785-170213.

85. Simberloff, D., Farr, J.A., Cox, J. and Mehlman, D.W. (1992). Movement corridors—conservation bargains or poor investments. *Conservation Biology.* 6: 493-504.

86. Singh, G.S. (1999). Utility of non-timber forest products in a small watershed in the Indian Himalayas: The threat of its degradation. *Natural Resources forum.* 23: 65-77.

87. Singh, G.S. (2000). Dynamics of sacred groves in western Himalaya: A paradigm of community based practices. *J. Indian Anthropological Society.* 35: 101-107.

88. Singh, G.S. and Saxena, K.G. (1998). *Sacred groves in the rural landscape: A case study of Shekhala village in Jodhpur district of desert Rajasthan.* In PS Ramakrishnan, KG Saxena and UM Chandrashekra (Eds.) Conserving the Sacred for Biodiversity Management, Oxford & IBH Publishing Co. Pvt. Ltd, New Delhi. 227-288.

89. Singh, G.S., Ram, S.C. and Kuniyal, J.C. (1997). Energy and economic efficiency of the mountain farming system: A case study from Lahul valley. *Journal of Environmental Systems.* 25: 195-211.

90. Singh, J.S. (2002). The Biodiversity Crisis: A multifaceted review. *Current Science.* 82(6).

91. Singh, D.K. and Hajara, P.K. (1996). Changing Perspectives of Biodiversity Status in the Himalaya (eds. Gujral, G. S. and Sharma, V.), British Council Division, New Delhi. 23-38.

92. Smith, J., Mapendembe, A., Vega, A., Hernandez Morcillo, M., Walpole, M. and Herkenrath, P. (2010). Linking the thematic Programmes of Work of the Convention on Biological Diversity (CBD) to Poverty Reduction. Biodiversity for Development: New Approaches for National Biodiversity Strategies. CBD Secretariat, Montreal.

93. Spehn, E.M., Liberman, M. and Korner, C. (eds) (2006). *Land use change and mountain biodiversity*. Boca Raton: CRC Publishers.

94. Spehn, E, and Korner, C. (eds) (2009). *Data mining for global trends in mountain biodiversity*. Boca Raton: CRC Publishers.

95. Stern, N. (2007). STERN Review – The Economics of Climate Change. Cambridge University Press.

96. Thomas, C.D., Cameron, A., Green, R.E., Bakkenes, M., Beaumont, L.J., Collingham, Y.C., Erasmus, B.F.N., de Siqueira, M.F., Grainger, A., Hannah, L., Hughes, L., Huntley, B., van Jaarsveld, A.S., Midgley, G.F., Miles, L., Ortega-Huerta, M.A., Peterson, A.T., Phillips, O.L. and Williams, S.E. (2004). Extinction risk from climate change. *Nature*. 427: 145-148.

97. Thuiller, W. (2007). Biodiversity: climate change and the ecologist. *Nature*. 448: 550-552.

98. Tripathi, A., Singh, G.S. (2013). Perception, anticipation and responses of people to changing climate in the Gangetic Plain of India. *Current Science*. 105(12).

99. UNCCD, UNDP and UNEP. (2009). Climate Change in the African Drylands: Options and Opportunities for Adaptation and Mitigation. UNCCD, UNDP & UNEP, Bonn, New York and Nairobi.

100. UNEP-WCMC. (2011). Saryarka – Steppe and Lakes of Northern Kazakhstan.

101. Uniyal, R.C., Uniyal, M.R. and Jain, P. (2000). *Cultivation of Medicinal Plants in India*: A Reference Book. 802 (TRAFFIC India and WWF India, New Delhi).

102. Xu, J., Grumbine, R.E., Shrestha, A.B., Eriksson, M., Yang, X., Wang Y. and Wilkes, A. (2009). The melting Himalayas; Cascading effects of climate change on water, biodiversity and livelihoods. *Conservation Biology*. 24(3): 520-530.

18

The Legal Foundations of Biodiversity Conservation

Ankita Yadav[1] and Abhay Singh Yadav[2*]

*[1]Department of Law,
University of Rajasthan, Jaipur-3002004, Rajasthan, India
[2]Department of Zoology, Kurukshetra University,
Kurukshetra-136119, Haryana, India
E-mail: abyzkuk@gmail.com*

Abstract

The biodiversity conservation aimed at its preservation, protection of endangered species and environment. These are progressively established in and enabled by laws globally. Biodiversity conservation has gained legal, political, scientific and cultural incentives and reinforcements. It was not the conservation biologists who inspired the laws to protect biodiversity, but it was conservation law that came first, in manifestations like the Convention on the International Trade in Endangered Species, 1973. The impact of modern environmental laws on biodiversity conservation is three-fold: First, giving legal incentives and approval for biodiversity preservation. Second, affirming the aims of biodiversity conservation and influencing the public to value conservation. Third, providing an environment that requires and sustains scientific research, management and monitoring. Laws not only represent current social values but laws also shape values for future generations. Laws empower action, providing political resources and social force to achieve specific goals. Because laws are difficult to repeal, they give sense of permanence to the values they establish. The United Nations and its Environmental Programmes (UNEP) are responsible for international conservation initiative and multinational agreements. Though it is beyond the scope of this chapter to discuss all the international treaties, important ones have been included here.

Keywords: Conservation law, Biodiversity, Sustainable environment, CBD, MDGs

Introduction

Environmental Law and Policy may be referred as "the use of government authority to protect the natural environment and human health from the impacts of pollution and development" (Salzman and Thompson, 2003). Policy is distinct from law as being the necessary outcome of all laws that are actually enforced. Legal scholars acknowledge two general views of international law, including international conservation laws: (a) 'positivist view'-it holds that international law consists of neutral rules. In this, the goal of national governments and international agencies is to enforce rules; (b) 'process view'-it holds that international law provides the standard framework and procedures for coordinating behaviour, controlling conflict, facilitating cooperation and achieving values (Weiss, 1999). In the realm of international conservation law, it is the process view that seems to best explain actual behaviour, particularly in democratic nations. In democracies, laws originate with issues that gain the attention of politicians and bureaucrats. But issues do not become law and laws are not translated into policies without going through lengthy examination and development by all concerned parties. To be effective, law must be supplemented by attendant policies that support and clarify its intentions. The environmental policy may be defined as "a set of principles and intentions used to guide decision making about human management of natural capital and environmental services" (Roberts, 2004). However, conservation law and environmental policy are inseparable.

Environmental and conservation laws are rooted in three conceptual frameworks: ethical rights, utilitarian interests and equitable distribution of risks (Salzman and Thompson, 2003). The earliest laws addressing the use or treatment of plants and animals were embedded in concepts of ethical rights. In ancient Roman, Chinese, Jewish and Indian legal traditions, animals, plants, rivers (water) and the land itself were protected from certain forms of abuse and exploitation. Although the intention of such laws was not directed toward conservation as we understand today, but rather towards providing justice. These traditions, however, did establish a basis for treating non-human creatures and environment as moral subjects. A second category of laws were prohibitions against the use of plants or animals found on private property. In this case the protection was for the right of the landowner to enjoy a productive, healthy and peaceful environment. But these laws cannot be considered as expressions of conservation. Such laws were rooted in utilitarian interests of the landowner. Conservation laws arising from concepts of rights, grounded in moral values, tend to advocate full protection for the entity to be conserved, regardless of the cost. In contrast, laws rooted in utilitarian interest use cost-benefit analyses as the primary guide to making the right decision. Historically, there were notable and commendable exceptions to the pattern of making conservation serve only as an expression of privilege for the fortunate few. Ashoka, the Emperor of India, proclaimed and enforced an edict for the protection of forests, mammals, birds and fishes in 252 BC. Much of international conservation law has been crafted from laws that were first formulated in individual nations. Even today, with a strong and growing body of international conservation law designed to empower the world conservation efforts, international laws and treaties invariably suffer constraints that cannot be overcome at international levels. Without sustainable environment approach, biodiversity conservation will remain a distant dream (Yadav, 2012).

The Process of International Conservation Law

The behaviour of modern global community has shown the increasing importance of soft law (nonbinding agreements that eventually define the norms and standards for international behaviour). This is evident from the ways, the international laws on environmental conservation created (Yadav, 2014). A Working Group of Experts on Environmental Law was established in 1977 by UNEP. Then, its recommendations were endorsed by the UNEP Governing Council and later on by the UN General Assembly in 1982. Although individual nations were not legally bound to use these guidelines, most of the soft law recommendations, over time, have become an increasingly recognized international standard (Sand, 1988). If international conservation law consisted solely of value-neutral rules, its most important element would be hard law (formal conventions and treaties adopted by nations with explicit mechanisms for enforcement). Even soft laws in conservation must have a catalyst. Every international convention, treaty or protocol is a consequence of unique circumstances. The development of international instrument in conservation follows a four step process: (i) issue definition; (ii) fact finding; (iii) creation of an international body to address the problem and (iv) consolidation and strengthening it. An ideal international conservation agreement is one in which there are clear and feasible mechanisms of implementation, high levels of compliance and workable methods of enforcement, all leading to accomplishing the goals for which the agreement was made (Dyke, 2008). A conceptual model, based on the actual success of a host of international environmental agreements, to show how various factors affect implementation, compliance and effectiveness of international conservation treaties has been developed by Weiss and Jacobson (1999). They have suggested three strategies for international compliance. (1) Sunshine approach-it focuses on mechanisms to bring the behaviour of main parties into open for public scrutiny. Here NGOs in conservation play an important role. According to international legal scholars Faure and Lefevere (1999), "The stronger and more active NGOs are with respect to the issue area of the treaty, the larger the probability of compliance." (2) Compliance information systems – that are built into government structures whose aim is to ensure compliance and report non-compliance. At the international level, the primary coordinating body for such compliance system is the Global Environmental Facility (GEF), established in 1991 by the World Bank in association with UNEP and UNDP. The UNESCO has instituted a number of programs to provide positive incentives toward compliance, such as World Heritage List (WHL) of sites of cultural and natural heritage. Administered by the World Heritage Centre in Paris, France, the WHL, a program created by the Convention Concerning the Protection of the World Cultural and Natural Heritage of 1972, is designed to identify and protect sites of outstanding cultural and natural value in every nation. (3) Coercive measures – if the above mechanisms fail, measures like penalties, sanctions, loss of membership or of privileges in international organizations can be useful in motivating unwilling parties to comply with agreements (Noss, 2001).

Nature of International Legal Interdependence

Both international conservation law and the national laws of modern nation states have grown progressively in breadth and matured in application. The factors affecting compliance are complex and national responses to international conservation efforts are not uniform. National laws and international conventions are usually aiming at the same goal, but in many cases they run afoul of one another. The conservation efforts at world level are guided, as well as constrained, by two overriding principles that often pull in opposite directions. However, both of these have significant applications at the national level. The first is the growing awareness and consensus that every nation has a responsibility to conserve its natural resources and must protect them for use by future generations. This first principle is rooted in the postulate that has become even more foundational to international conservation: 'the commitment to intergenerational equity.' Intergenerational equity gain support by the following core ideas:

1. Each generation should be required to conserve the natural and cultural resource base of its own nation so that it does not restrict the options available to future generations in solving their problems and attaining their goals.

2. Each generation should maintain their environmental quality in such a state that it is in no bad condition than that which they received.

4. Members of every generation should have comparable rights of access to the legacy of past generations and should conserve this access for the future generations.

The second fundamental principle of modern international environment law is that every nation has sovereign rights over its own natural resources, and these rights are not to be infringed by other nations.

Environmental policy scholar Weiss (1999) has noted, "In international environment law, the most important development for the next century may be the emerging interaction of intergovernmental environment law with transnational law." He further states, "International law has always been linked with national law, for it is implemented through national, provincial and local laws. National laws independent of any treaty, provide protection to other countries or their citizens for harm that occurs within the country but injures those outside." Thus there is an unavoidable complexity of and connection between the claims of environmental protection and national sovereignty, national and international conservation law, and governments and non-governmental organizations.

Following two instances will justify this complexity.

Marine Mammal Protection Act (MMPA) was enacted in 1972 by the US Congress. The Act`s aim was 'to protect certain species and population stocks of marine mammals that are, or may be, in danger of extinction or depletion as a result of man`s activities.' One of the MMPA`s mechanisms to attain the goal was to reduce 'incidental kill or serious injury of marine mammals.' The 'incidental killing' of dolphins by tuna fishermen had become a cause for scandal and condemnation by the public and the press. This problem had been cropping up since 1950s when tuna fishermen began to use purse-seine nets in capturing tuna fish. Such nets captured tuna in large schools when they fed near the surface. After surrounding the tuna by purse-seine net, the bottom of the net was pulled together, trapping the tuna and all other animals inside (Joyner and Tyler, 2000). According to an estimate, about six million dolphins have disappeared in this manner.

Then in late 1980s US environmental and conservation NGOs successfully forced the US Congress to add an amendment to the MMPA which established strict guidelines for US tuna fishermen and all tuna fishing in its waters to assure protection for dolphins and other species. Obviously, other countries, including those harvesting the majority of tuna, were not following guidelines set by the MMPA. Such countries were required to reduce incidental kill of non-tuna species to the level of US fishing fleets. They were prohibited from using large scale drift nets, encircling marine mammals without direct evidence of the presence of tuna, or using purse-seine nets after sundown. The amendments in MMPA also specified that failure to comply would lead the ban on imports of tuna from countries violating the regulation or from countries they sold tuna to prevent trans-national shipment as a way of getting around the regulation. In effect, this placed a US embargo on the tuna products of the offending country (Salzman and Thompson, 2003).

The US Congress, in 1989, added a provision (Section 609) to Public Law 101-162 that became to known as the Sea Turtle Act (Olson, 1990). This Act was motivated by major concern over worldwide declines in the populations of all seven species of sea turtles and by scientific findings that implicated shrimp nets in sea turtle's high mortality. The US was one of the first nations to use the 'turtle excluder device' (TED). TED is a grid trapdoor installed inside a trawling net that keeps shrimp in the net but directs other larger objects or animals out. Earlier legislation had already required TEDs for all shrimp trawlers operating in the Gulf of Mexico and in the Atlantic Ocean off the Southeast coast of the US. The Sea Turtle Act prohibited fish imports from any nation that failed to adopt sea turtle conservation measures comparable to those in the US. Initially such sanctions were applied only to western Atlantic and Caribbean nations which eventually complied. However, the largest shrimp importers were Asian nations that did not use TEDs. As these events were taking place, the US was engaged in negotiations to ratify the General Agreement on Tariffs and Free Trade (GATT). The US administration was reluctant to create controversy with Asian nations over sea turtles that could delay or halt ratification of GATT. So, the US officials delayed enforcement of the Act against its most important shrimp suppliers. Such silence led to a federal lawsuit by the Earth Island Institute, a US NGO. Earth Island Institute pleaded that the provisions of the Sea Turtle Act be enforced uniformly against all nations exporting shrimp to the US. The Earth Island Institute, after a series of appeals, won the case in the US Court of International Trade, thus forcing the US to ban imports from nations that had not complied with the Sea Turtle Act.

The tuna and shrimp embargoes led to the legal challenges by the sanctioned nations before the World Trade Organization (WTO). In separate but similar cases, the tuna and shrimp-exporting nations argued that the MMPA and Sea Turtle Act were violations of the free trade provisions guaranteed by GATT (Miller and Croston, 1998). In the case of tuna and dolphins, the European Community also joined in challenging the MMPA, for the embargoes prevented them from selling tuna they had purchased from Asian nations that did not comply with the MMPA to the US. The plaintiffs argued that under the terms of GATT an individual nation could not impose restrictions on imports from other nations, that those nations had not been party to developing these. Also, the US could not impose sanctions based on the 'processing and production' of a product, but only on the product. That is what mattered was the tuna in the can, not how the tuna got in the can (WCED, 1987). Further, the bottle-nosed dolphin (*Tursiops truncatus*), the main species affected by the tuna-fishing methods in question, was not an endangered species, and as such not subject to international protection. The plaintiff nations, finally charged that the entire embargo was only a deceit to protect US tuna fishers to give them an unfair competitive advantage in US markets (WCED, 1987). The WTO finally agreed and ruled against the US in the case of both dolphins and sea turtles, agreeing with the plaintiffs that the US laws constituted unfair barriers to free trade. The US appealed against the decisions, but its appeals were not fruitful (Olson, 1990).

The above discussed cases provide insight into a world of complex interactions between national and international conservation law, public interest, private industry, NGOs and government bureaucracies. Conservation laws of individual countries can no longer be enforced or enacted without first considering the interests of other nations. According to Salzman and Thompson (2003), an individual country cannot be confident that it will win in the international courts unless:

(1) the measure is not unilaterally imposed, and

(2) the harm done is local i.e. within the jurisdiction of the country imposing the sanctions.

International conservation is basically driven by the utilitarian interests and by the equalization of risks, usually in the form of mutual international interdependence and increased concern for trans-generational equity. Recent efforts in international conservation law arise from one or more of the following sources:

(a) Bilateral or multilateral treaties among nations;

(b) Binding acts of international organizations;

(c) Rules of customary international law and

(d) Judgements of an international court or tribunal.

The United Nations and Environmental Programs

Largely through the efforts of UNEP and other UN environmental programs, modern nation states have entered into over 250 treaties, conventions and agreements focusing on conservation during the last three decades, and more than 1,000 international legal instruments, most of them binding, contain at least one section or provision that addresses environmental conservation (Sands, 1999). In addition to stimulating the formation of new regional international organizations and encouraging their work in conservation legislation, UNEP and other UN programs have directly propagated the development of conservation agreements among nations thereby serving as a catalyst for more coordinated regional action for environmental conservation. Perhaps the UN's greatest initial contribution to the world conservation effort has been that it provided a forum for the discussion of international conservation issues and a means to allow adoption of international conservation agreements (Sands, 1999). A turning point in international conservation came in 1972, with the convening of the UN Conference on the Global Environment in Stockholm, Sweden, known as Stockholm Conference.

Stockholm Conference

Most legal scholars mark the beginnings of modern international environmental and conservation law with the holding of the Stockholm Conference of 1972 (Di Mento, 2003). Its most significant achievement was the production of the Declaration on the Human Environment, a document containing 26 principles and 109 recommendations related to environmental protection and conservation. UN's first environmental agency, UNEP was also created during this conference. UNEP was given the responsibility for making both new international conventions to foster conservation and protect the environment as well as their enforcement. UNEP made environmental concerns and programs a permanent fixture of the UN agenda. The Stockholm Conference was significant in that the UN became involved in world conservation in systematic and comprehensive way.

The UNEP, in 1973, had an almost immediate impact on world conservation as it declared regional seas to be an important conservation priority. This resulted directly in the development of the **Barcelona Convention of 1976** for the **Protection of the Mediterranean Sea against Pollution** and to preserve native sea species (Di Mento, 2003). The Barcelona Convention provided the incentive and model for regional environmental and conservation treaties that would follow during the next 30 years. By 1988, over 100 nations and 50 international organizations were cooperating in regional sea programs (Sand, 1988). Also, the 1982 **Montego Bay Convention,** developed in association with the Third UN Conference on the Law of the Sea, addressed major issues of ocean conservation worldwide. Consequently, the most important trend has been a shift from use-oriented approach to resource-oriented approaches. The use-oriented approach emphasized navigation and fishing. While then resource-oriented approach emphasizes sustainable development and harvest of ocean resources, focusing on defining and enforcing standards of protection, conservation, management and development (Sand, 1988).

The CITES

To deal directly with the problems of endangered species or the conservation of world biodiversity, the most important international agreement, The Convention on International Trade in Endangered Species of Wild Flora and Fauna of 1973 (CITES), was created out of the combined efforts of the International Union for the Conservation of Nature (IUCN) and UNEP.

The CITES of 1973 is the most important international conservation agreement functional today, for it specifically regulates or prohibits commercial trade in globally endangered species or their products. The Treaty has currently been ratified by more than 150 countries. CITES, headquartered in Switzerland, create lists (known as Appendices) of species for which international trade is to be controlled or monitored. International treaties like CITES are implemented when a country signing the treaty passes laws to enforce it. Countries may also establish Red Data Books of endangered species, which are national versions of the international Red Lists prepared by the IUCN. Laws may protect both species lists by CITES and the national Red Data Books. Once species protection laws are passed within a country, police, customs inspectors, wildlife officers and other government agents can arrest and prosecute individuals possessing or trading in protected species and seize the products or organisms involved. The CITES Secretariat periodically sends out bulletins aimed at publicising specific illegal activities. Member countries are required to establish their own management and scientific authorities to implement their CITES obligations. Technical advice is provided by nongovernment organizations such as International Union for the Conservation of Nature (IUCN) Wildlife Trade Specific Group, the World Wildlife Fund (WWF) Traffic Network and the World Conservation Monitoring Centre (WCMC) Wildlife Trade Monitoring Unit, which is part of the UN. The Treaty has been instrumental in restricting the trade in certain endangered wildlife species. Its most notable success has been a global ban on the ivory trade when poaching was causing severe decline in African elephant populations. A difficulty in enforcing CITES is that shipments of both living plants and animals and preserved parts of plants and animals are often mislabelled, due to either an ignorance of species names or a deliberate attempt to avoid the restrictions of the Treaty.

Three appendices of CITES are more relevant as these list categories of species regulated under the terms of the treaty. Red Data Books are the most important source and most respected authority for identifying the world`s endangered species and their status. Parties to the treaty meet every two years to make amendments to the appendices and develop new species and animal products lists and identification manuals to improve enforcement (Slocombe, 1989).

Another key treaty is the **Convention on Conservation of Migratory Species of Wild Animals,** often referred to as the **Bonn Convention**, signed in 1979, with a primary focus on bird species. This convention serves as an important complement to CITES by encouraging international efforts to conserve bird species that migrate across international borders and by emphasizing regional approaches to research, management, and hunting regulations. The Convention now includes protection of bats and their habitats and cetaceans in the Baltic and North Seas.

International Agreements to Protect Habitat

Habitat conventions at the international level complement species conventions by emphasizing unique biological communities and ecosystem features that need to be protected. The most significant are the Convention on Wetlands of International Importance Especially as Waterfowl Habitat or the **Ramsar Convention on Wetlands;** the Convention Concerning the Protection of the World Cultural and Natural Heritage or the **World Heritage Convention;** and the UNESCO Man and the Biosphere Program or the **Biosphere Reserves Program.** These three conventions establish an overarching consensus regarding appropriate conservation of protected areas and certain habitat types.

As the effectiveness of international cooperation to form agreements for promoting conservation was increasing, so the next step was to integrate conservation with the problems of human poverty and development. In fact, this was the subject of the UN Conference on Environment and Development in Rio de Janeiro, Brazil in 1992, popularly known as Rio Summit or Earth Summit.

Rio Summit and Post-summit Efforts

The main aim of the Rio Summit was to integrate efforts to protect ecosystems with economic development of the poor nations of the world. To that end, it created five documents. Of these, the best known is the Rio Declaration or *Earth Charter* that summarized international consensus on environmental policy and development. The most comprehensive document signed at the Rio Summit was *Agenda 21*. It is an 800 page work plan document addressing social and economic dimensions of environment and development, conservation and management of resources and means of implementation. Its structure was based on key environmental and conservation issues including the problems of desertification, protecting the atmosphere and managing toxic waste (Greene, 1994).

In its final section on means of implementation, Agenda 21 supported promoting public awareness, establishing a new UN body, the **Sustainable Development Commission (SDC)**, to coordinate pursuit of sustainable development among international organizations and monitor progress toward reaching the goals set out in the Agenda (Hails, 1996).

However, Agenda 21 failed to reach agreement on issues of fish stocks, targets and dead-lines for increases in development assistance and the governance of the Global Environmental Facility (GEF). Another significant shortcoming of the Agenda has been its failure to establish new regimes of international development, especially to benefit poorer countries. Despite all these shortcomings, Agenda 21 has deeply influenced international conservation law and policy. Although not all are legally binding, the principles of the Agenda have already found their way into many UN resolutions; the conventions on climate change and biodiversity have set the standard of international policy, practice and expectation on the issues they address.

The Convention on Biological Diversity (CBD) addressed conservation and sustainable use of biodiversity along with fair sharing of genetic resources.

Initially 153 nations signed it and took pledge to develop plans to protect habitats and species, provide fund and technology to assist developing countries, ensure commercial access to biological resources for development, establish safety regulations and accept liability for risks associated with biotechnology development. Entering into force only 18 months later on December 29, 1993, over 175 nations had signed on by 2001 and most of the provisions of the CBD are now being implemented (Hails, 1996).

Bangalore Declaration on Biodiversity: A conference of Asian region countries was held in Bangalore on 23-24 August, 1994. The conference was organized by the Ministry of Environment and Forest (MoEF) with the assistance of UNEP in which the subject for deliberation was preservation of biological hereditary resources and their sustainable use and dissimilation of related knowledge for studies and research.

United Nations Commission on Sustainable Development: For the first time the word "sustainable development" was used commonly in the **Brundtland Commission**. It was used in the World Commission on Environment and Development (WCED) in its seminal 1987 report, *Our Common Future*. This Report has given a very important definition of sustainable development: "Sustainable development is development that meets the needs of the present without compromising the ability of the future generations to meet their own needs."

Various principles for biodiversity protection and sustainable development adopted by the Brundtland Commission are as under:

1. Environmental Standards and Monitoring: States should have duty to make standards for environment protection and publish relevant data on environmental quality and resource usage.

2. Sustainable Development and Assistance: States shall ensure conservation is treated as an integral part of the planning and implementation of development activities and provide assistance to developing countries in support of preserving flora and fauna.

3. Fundamental Human Rights: According to this Commission all human beings have fundamental right to get good environmental and good health and well being.

4. Inter-generational Equity: It is the duty of state to preserve and conserve natural resources for the benefit of present and future generations.

5. Prior Environmental Assessment: States shall make or require prior environmental assessment of proposed activities which may significantly affect the environment and use of natural resources.

6. General Obligation to Co-operate: States shall co-operate each other for implementing the preceding rights and obligations.

In 1991, World Conservation Union, UNEP and World Wild Fund, jointly produced a document called *"Caring for the Earth: A Strategy for Sustainable Living"* which defined the "sustainability" as a characteristics or state that can be maintained indefinitely whereas "development" is defined as the increasing capacity to meet human needs and improve the quality of human life.

The success of Agenda 21 was evaluated in "Earth Summit Plus Five" in June, 1997 at New York at Special session of United Nations General Assembly.

Earth Summit Plus Five: This is known as "Earth Summit Plus Five" because it was after five years from the historic "Earth Summit" in Rio de-Janeiro in 1992. In this conference it was observed that the planet's oceans, forests and atmosphere are still in trouble and its population of poor people is growing, another area of concern was Global Warming. According to Razali Ismail, Malaysian diplomat and Summit Chairman, "the skin results were pretty sobering (Hanley, 1997a)." The Third World Network's Martin Khor put it simply, "Five years later we are extremely disappointed (Hanley, 1997b)." The Biodiversity Convention was meant to ensure that the destruction of species was slowed down and eventually halted but latest estimates suggest a rising toll owed to deforestation rate that rose from 11,000 km^2 a year in 1991 to 15,000 km^2 a year in 1995 (Leake and Wavell, 1997).

The delegates agreed on few concrete remedies in such critical areas as global warming. Environmentalists also gave a call to create a "World Environment Court" to solve the international environmental disputes. In the end the Conference concluded, "We are deeply concerned that overall trends are worse today than they were in 1992". "Statement of Commitment" kept Agenda 21 alive and recorded the willingness of the member states to hold a session in 2002 A.D. "We commit ourselves to work together in the spirit of global partnership to reinforce our joint efforts to meet equitably the needs of the present and future generations" (The Tribune, 1997).

Global Environment Facility (GEF): UN sponsored GEF in April 1998, told the industrialised nations to enforce stringent emission norms and limit pollution levels to save the world from an impending ecological disaster as their high levels of economic activity and consumption were causing sever environmental degradation. Shri Atal Bihari Vajpayee, former Prime Minister of India, while opening the Assembly warned the industrialised nations that global warming and climate change could damage the ecology severely (The Tribune, 1998).

Biodiversity Target, 2010: In 2001, in Gothenburg, the European Union heads of State had decided in EU Summit that biodiversity loss should be halted by 2010 with the aim of reaching its objectives. The decision said, "Parties commit themselves to a more effective and coherent implementation of the three objectives of the CBD, to achieve by 2010, a significant reduction of the current rate of biodiversity loss at the global, regional and national level as a contribution to poverty alleviation and to the benefit of all life on earth.

Second Earth Summit, 2002: This has been called as Earth Summit +10 because it was held ten years after the First Earth Summit of 1992. The summit was held from 16th August to 4th September, 2002 in Johannesburg (South Africa). It was the Second World Summit on Sustainable Development.

The UN Environmental World Summit on Sustainable Development, Rio Earth Summit, 2012: The UN Conference on Sustainable Development was held on June 20-22, 2012 in Rio-de-Janeiro (Brazil) which is also known as Rio or *Earth Summit* +20 because it was held 20 years after the first Rio Summit of 1992. Many documents were open for discussion: (1) Rio-declaration on environment and development; (2) Agenda 21; and (3) Forest principles set out for effective protection of forest resources.

Two more important legally binding documents were opened for signature: (a) Convention on Biological Diversity, and (b) UN Framework Convention on Climate Change.

Intergovernmental Panel on Climate Change (IPCC), 2014: IPCC is a specialised body jointly established by the UNEP and the World Meteorological Organization to prepare scientific assessments on various aspects of Climate Change. The UN`s IPCC report "Climate Change 2014 Impacts, Adaptation and Vulnerability" was released on April 1, 2014 at a meeting in Yokohama, Japan. It was by far the most detailed investigation of the global impacts of climate change-extending from oceans to mountains and from the poles to the equator.

The Editorial column of The Tribune dated April 2, 2014 described that " this report brings home for the fifth time since 1990 the point that global warming is for real and a threat to life as we know it. Only this time the warning has been worded rather alarmingly. The reason for that is the apparent deafness the previous reports have been met with around the world, India included. The science is simple: Life – plant, animal or human-evolved under a given set of climatic and environmental conditions. Change the temperature by 2^{0}C, all the weather systems-rain, seasons, glaciers, polar ice, oceanic currents-go haywire. The very foundation on which life is based changes. We may debate the disaster projections of the report, but the changes are already upon us. Glaciers which are banks of water for our rivers are disappearing across the entire Himalayan range. The extreme weather events this winter in India as well as Europe and North America were most likely related to climate change. Practically, everything from food security to transport and economy is set to be hit. It is time to learn how we live".

The key environmental challenges in India have been sharper in past two decades. Climate Change is impacting our natural ecosystems and is projected to have substantial adverse effects mainly on agriculture, melting of Himalayan glaciers, sea level rise, and threats to a long coast-line and habitation. Climate Change will also induce increased frequency of extreme events such as droughts, floods and diseases. These will have an impact on our food security and water security (Economic Survey, 2012-13).

The 1970's gave rise to a new concept of development where the environmental issue is not seen in a separate sector but as an inherent aspect of development. Now it is fully realized in national governments and multilateral institutions that it is impossible to separate economic development issues from environmental issues as many forms of development erode the environmental resources upon which they must be based and environmental degradation can undermine economic development (Yadav, 2012). Many present development trends leave increasing numbers of people poor and vulnerable, while at the same time degrading environment. How can such development serve next century's world of twice as many people relying on the same environment? This realization broadened the concept of development. Thus, development came to be seen not in its restricted context of economic growth in developing countries but as one that sustained human progress (Malyuja, 1997).

Environment Impact Assessment is another widely accepted norm of International Environmental Law. Typically, such an assessment balances economic benefits must substantially exceed its environmental costs. India has adopted this norm for select projects which are covered under the Environmental Impact Assessment (EIA) regulations introduced in January, 1994. Precautionary Principle and Polluter Pay Principle are part of the environmental law of India. There is constitutional mandate to protect and improve the environment. Polluter pay principle basically means that the producer of goods or other items should be responsible for the cost of preventing of dealing with any pollution that the process causes. This include environmental cost as well as direct cost incurred in avoiding pollution and not just those related to remedying any damage. The Polluter Pay Principle does not mean that the polluter can pollute and pay for it. Precautionary Principle provide for taking protection against specific environmental hazards by avoiding or reducing environmental risks before specific harms are experienced (Yadav, 2013).

The Millennium Development Goals (MDGs)

"Respect for nature. Prudence must be shown in the management of all living species and natural resources in accordance with the precepts of sustainable development. Only in this way can the immeasurable riches provided to us by nature be preserved and passed on to our descendants. The current unsustainable patterns of production and consumption must be changed in the interest of our future welfare and that of our descendants (UNMD, 2000)."

The MDGs originated from the Millennium Declaration adopted by the General Assembly of the UN in September 2000. The MDGs have helped in bringing out a much needed focus and pressure on basic developmental issues, which subsequently led the governments at national and sub-national levels to do better planning and implement more intensive policies and programmes. The MDGs consists of eight goals. These address myriad development issues (*www.mospi.nic.in*). The eight goals are as follows:

- Goal 1: Eradicate Extreme Poverty and Hunger
- Goal 2: Achieve Universal Primary Education
- Goal 3: Promote Gender Equality and Empower women
- Goal 4: Reduce Child Mortality
- Goal 5: Improve Maternal Health
- Goal 6: Combat HIV/AIDS, Malaria and TB
- Goal 7: Ensure Environmental Sustainability

- Goal 8: Develop Global Partnership for Development

For attaining the goals 18 targets were set as quantitative benchmarks. The United Nations Development Group (UNDG) in 2003 devised a framework of 53 indicators for measuring the progress towards individual targets. A revised indicator framework drawn up by the Inter-Agency and Expert Group (IAEG) on MDGs came into effect in 2008. This framework had 8 Goals, 21 targets and 60 indicators. India has not endorsed this.

MDGs – India Country Report 2015: India's MDGs framework is based on UNDG's MDGs 2003 framework. It includes all the 8 Goals, 12 out of 18 targets which are relevant for India and related 35 indicators (*www.mospi.nic.in*).

In India there has been considerable emphasis and consequent progress on all the MDGs. It will be pertinent to discuss the Goal 7 pertaining to Ensuring Environment Sustainability. The MDG Goal 7 addresses the concern for sustainable development to reverse environment degradation and loss with focus on improving / monitoring indicators associated with it.

Target: To Reverse the Loss to Environmental Resources:

Indicator: Proportion of land area covered by forests: Improving forest cover and protected areas are measures towards ensuring sustainable environment and biodiversity. Forest cover includes all lands which have a tree canopy density of 10 per cent and above and have a minimum area of one hectare. Thus all tree species along with bamboos, fruit bearing trees, palm, coconut etc. and all trees including forest, private and community of institutional lands meeting these criteria have been termed as forest cover. As per assessment in 2013, the total forest cover of the country is 697898 sq. km which is 21.23 per cent of the geographic area of the country. During 2011-2013, there is an increase of 5871 sq. km in forest cover (Annual Report, MoEF, 2012-13). This positive change can be attributed to the conservation measures or management interventions such as afforestation programmes, active involvement of local people for better protection measures in plantation areas as well as in traditional forest areas. During 2011-2013, there was also a decline of 1991 sq. km of moderate forests while an increase of 31 sq. km and 7831 sq. km was recorded for the categories of very dense forest and open forests, respectively. The decline in forest cover is due to developmental activities, mining, harvesting of short rotational plantation, clearance of encroached areas, biotic pressures and shifting cultivation practices etc.

Indicator: Ratio of area protected to maintain biological diversity to surface area: According to IUCN, "Protected Areas (PAs) are those in which human occupation or at least the exploitation of resources is limited." The term "Protected areas" also includes Marine Protected Areas, the boundaries of which will include some area of ocean, and Trans-boundary Protected Areas that overlap multiple countries which remove the borders inside the area for conservation and economic purpose.

India has taken significant steps in inventorying her vast and diverse biological heritage. In 2014, there are 692 Protected Areas (103 National Parks, 525 wildlife

Sanctuaries, 4 Community Reserves and 60 Conservation Reserves) which is 4.83 per cent of the total geographic area of the country. India has 23 Marine Protected Areas in peninsular India and 106 in the islands. India, a mega diverse country with only 2.4 per cent of the world's area, harbours 7-8 per cent of all recorded species, including over 45,000 species of plants and 91,000 species of animals. It is also amongst the few countries that have developed a bio-geographic classification for conservation planning and has mapped biodiversity-rich areas in the country. Of the 34 global biodiversity hotspots, four are present in India, represented by the Himalaya, the Western Ghats, the North-East and the Nicobar Islands (*www.moef. nic.in*).

The UN is working with governments, civil society and other partners to build on its MDGs and craft a post-2015 agenda. Significant progress has been made, but a lot remains to be done before the deadline. The MDGs have been the most successful global anti-poverty push in the history. The UN MDGs Report 2013 looked at the areas where action is needed most. The resource base is in serious decline, with continuing losses of forests, species and fish stocks. The achievement of MDGs has been uneven among and within countries. As per this report, global emissions of carbon dioxide have increased by over 46 per cent since 1990. About one third of marine fish stocks have been overexploited. Many species are at risk of extinction, despite increase in protected areas. Estimated 863 million people reside in slums in the developing world. Forests are safety net for the poor, but they are disappearing at an alarming rate.

Programmes and Policies Pertaining to Development of Sustainable Environment in India

Green India Mission (GIM): The Ministry of Environment, Forests and Climate Change, Government of India, has launched the National Mission for Green India as part of NAPCC through a consultative process involving relevant stakeholders. The GIM acknowledges the influences that the forestry sector has on environmental amelioration through climate mitigation, water security, food security, biodiversity conservation and livelihood security of forest dependent communities. The Mission proposes a holistic view of greening and focuses on multiple ecosystem services, particularly biodiversity, water, biomass and other natural resources along with carbon sequestration as a co-benefit. Establishment of **National Green Tribunal** in 2010 is a quite significant step.

National Afforestation Programme (NAP): The NAP is the flagship scheme of National Afforestation and Eco-Development Board (NAEB). It gives support, both in physical and capacity building terms, to the Forest Development Agencies (FDAs). The FDA has been conceived and created as a federation of Joint Forest Management Committees (JFMCs) to carry out holistic development in the forestry sector with people's participation. This is a paradigm shift from the earlier programmes wherein funds were routed through the State Governments. The village is reckoned as a unit of planning and implementation. All activities under the programme are conceptualized at the village level. This significantly empowers the local people to take part in the decision making process.

Conservation of biodiversity: India has established six National Bureaus dealing with genetic resources of plants, animals, insects, microorganisms, fish and soil sciences.

1. National Bureau of Plant Genetic Resources (NBPGR): It has a total of 4,08,186 plant genetic resource accessions.

2. National Bureau of Animal Genetic Resources (NBAGR): It has a total holding of 1,23,483 frozen semen doses from 276 breeding males representing various breeds of cattle, buffalo, sheep, goat, camel, yak and horse for *ex situ* conservation.

3. National Bureau of Agriculturally Important Microorganisms (NBAIM): It has a repository of 4668 cultures, including 4644 indigenous and 24 exotic accessions.

4. National Bureau of Fish Genetic Resources (NBFGR): It has a repository of 2553 native fin-fishes and fish DNA Barcode Information System.

5. National Bureau of Agriculturally Important Insects (NBAII): It has 593 insect germplasm holdings.

6. National Bureau of Soil Survey and Land Use Planning (NBSS & LUP): It is engaged in land resource inventorization of India.

The Government of India in 1999 prepared the National Policy and Macro level Action Strategy on biodiversity through a consultative process. The MoEF implemented an externally aided project, the National Biodiversity Strategy and Action Plan (NBSAP) from 2000 to 2004. India was one of the first countries to have a proactive legislation and enacted a comprehensive Biological Diversity Act in 2002 to implement the provisions of CBD. India is among the select countries in the world that have developed their own National Biodiversity Targets.

Legal provisions

A variety of policy measures has been developed over a period which provides opportunities for strengthening documentation of data collection; empowering local communities; ownerships, rights; concessions and creating suitable institutions. The mandates of National Forest Policy, 1988 and National Environment Policy 2006 recognize the need to address the conservation of areas of biodiversity importance, restoring degraded areas and enhancing forest productivity. The legislative provisions created as a follow-up to such national policies are as under:

1. Indian Forest Act, 1927

2. Wildlife (Protection) Act, 1972

3. Environment Protection Act, 1986

4. Protection of Plant Varieties and Farmer's Rights Act, 2001

5. Biological Diversity Act, 2002

6. The Schedule Tribes and Other Traditional Forest Dwellers Act, also known as Forest Rights Act (FRA), 2006

7. State-level legislations pertaining to various aspects of biodiversity conservation and ecosystem services are also addressing the relevant issues.

Conclusion

Environmental issues on a global scale may require a greater level of coordination. Law and policy require an integrated, interdisciplinary approach. Earlier, environmental laws have been based on individual scientific premises and have the application of those logics regardless of what new findings discovered. Today, such legal certainties are inconsistent with the state of our knowledge of ecosystems. Modern conservation law and policy must mature to the point that they can deal with such uncertainties, rather than simply ignore or reject it and thereby better manage risk to threatened species. The development of conservation law and policy demonstrates repeated themes:

1. The scrutiny of a free press and the involvement of an educated public enables private organizations and citizens to make a difference in how things turn out.

2. Even failed attempts at international legislation, such as the Rio Summit, may produce positive results, and should be pursued toward the ultimate goal of a comprehensive and coordinated system of international conservation legislation.

3. Programs of lasting effectiveness in conservation are strongly influenced by economic incentives, as evidenced by the efforts to save dolphins and sea turtles.

The development activities undertaken to improve the living conditions of people sometimes affect the natural environment adversely in many ways and cause severe threats to human health and biodiversity. After perusal of all the conventions and conferences, it is submitted that many nations showed their concern about depleting environmental conditions but still much needs to be done in this direction. Climate Change is a serious global environmental concern. Its effects particularly on developing countries are adverse as their capacity and resources to deal with the challenges is limited. The Government is implementing NAPCC to enhance the ecological sustainability of India's development path and to address Climate Change. Modern conservation law and policy must mature to the point that they can deal with the uncertainties of climate and ecosystem estimates, rather than simply reject or ignore these, and thereby better manage risks to threatened biodiversity. At the international level, the developed nations have more responsibility towards the protection of environment, biodiversity and sustainable development. More bold steps are needed in the area of environmental sustainability at the global level. The actions really needed to conserve biodiversity must be resolved at national and local levels because only in national and local communities one can hope to get a consensus of shared values that can support more effective actions which are required to achieve real conservation goals. International conservation laws become meaningless without national and local enforcement and monitoring. The global community has cumulatively consumed or damaged a large part of the planet's natural resources, including clean air, clean water, forests, biodiversity and healthy soil. In addition, unrestricted and unregulated economic development and growth has also polluted the environment beyond acceptable limits. The mandate of the time is loud and clear. Preserve the Plant's resources to save the life form on the Planet. The natural resources of the earth, including the air, water, lands, flora and fauna and especially natural ecosystems, must be safeguarded for the benefit of present and future generations through careful planning or management.

References

1. Annual Report. (2012-13). Ministry of Environment and Forest, Govt. of India. 350-352.

2. Annual Report. (2012-13). Ministry of Environment and Forests, Govt. of India. 349.

3. DiMento, J.F. (2003). *The Global Environment and International Law.* University of Texas Press, Austin, TX.

4. Dyke, F. V. (2008). Conservation Biology. 2nd edn. Springer Science, USA.

5. Economic Survey. (2012-13). Ministry of Finance, Govt. of India. 256-257.

6. Faure, M. and Lefevere, J. (1999). Compliance with international environment agreements. In: Vig, N. J. & Axelford, R. S. (eds) *The global environment: institutions, law and policy.* Congressional Quarterly, Washington DC. 116-137.

7. Greene, G. (1994). Caring for the Earth: the World Conservation Union, the UNEP and WWF for Nature. *Environment.* 36: 25-28.

8. Hails, A.J. (ed.) (1996). Wetland, biodiversity and the Ramsar Convention: The role of the Convention on wetlands in the conservation and wise use of biodiversity. Ramsar Convention Bureau, Gland, Switzerland.

9. Hanley, C. J. (1997a). Rio spirit is dying, say environmentalists. *The Times of India.* 12: 27 June, New Delhi.

10. Hanley, C. J. (1997b). Earth Summit lends few answers to troubled planet. *The Sunday Times of India,* 29 June, New Delhi. 14.

11. http://unfcc.int/essential_background/convention

12. IUCN. (1990). Caring for the Earth. 16.

13. Joyner, C.C. and Tyler, Z. (2000). Marine conservation versus international free trade: reconciling dolphins with tuna and sea turtles with shrimp. *Ocean Development and International Law.* 31: 127-150.

14. Leake, J. and Wavell, S. (1997). Rio plus 5 finds green hopes withering, *The Times of India,* 25 June, New Delhi. 13.

15. Malvuja, R.A. (1997). Sustainable Development and Environment: Emerging Trends and Issues. *Indian Journal of International Law.* 58-59.

16. Millennium Development Goals Report. (2007). 23.

17. Miller, C.J. and Croston, J. L. (1998). WTO scrutiny v. environmental objectives: assessment of the international dolphin conservation program. *American Business Law Journal.* 37: 73-125.

18. Nomani, Md. Mahfooz, Z. (1996). Law Relating to Environmental Liability and Dispute Redressal: Emergence and Dimension. *Indian Bar Review.* 23: 139.

19. Noss, R.F. (2001). Beyond Kyoto: forest management in a time of rapid climate change. *Conservation Biology.* 15: 578-590.

20. Olson, J.M. (1990). Shifting the burden of Proof. *Environmental Law.* 20: 112.

21. Roberts, J. (2004). Environmental policy. Routledge, London.

22. Salzman, J. and Thompson, B.H. (2003). *Environmental law and policy.* Foundation New York.

23. Sand, P.H. (1988). *Marine environment law in the United Nations Environment Program: an emergent eco-regime.* Tycooly, London.

24. Sands, P. (1999). Environmental protection in the 21st century: Sustainable development and international law. In: Vig, N.J. & Axelford, R.S. (eds.). *The global environment: institutions, law and policy.* Congressional Quarterly, Washington DC. 116-137.

25. Slocombe, D.S. (1989). CITES, the wildlife trade and sustainable development. *Alternatives.* 16: 20-29.

26. Supreme Court Case. (2005). *Research Foundation for Science Technology and National Resource Policy v. Union of India & others,* 2005(I) SCJ 407.

27. The Indian Express. (1997). Editorial, "Environmental diplomacy – Empty promises pollute the air.", 1 July, Chandigarh. 6.

28. The Tribune. (2013). Excerpted from the "UN millennium Development Goals Report 2013", Chandigarh. 133: 9.

29. The Tribune. (1997). Editorial, "Earth Summit fails", 1 July, Chandigarh. 6.

30. The Tribune. (1998). South must check pollution: PM, 2 April, Chandigarh. 1.

31. UNEP (United Nations Environment Program). (1995). *Global Biodiversity Assessment.* Cambridge University Press, Cambridge.

32. UNMD (United Nations Millennium Declaration). (2000). UNGA Res. A/RES/55/2. Item 5, paragraph 6 of Part 1.

33. Upadhyay, Sanjay. (2000). Environmental Protection. *Land and Energy Law.* 3: 111-112.

34. WCED. (1987). *Our Common Future.* The World Commission on Environment and Development (Brundtland Report). 114: 43.

35. Weiss, E.B. (1999). The emerging structure of international environmental law. In: Vig, N.J. & Axelford, R.S. (eds) *The global environment: institutions, law, policy.* CQ, Washington, DC. 98-115.

36. Weiss, E.B. and Jacobson, H.K. (1999). Getting countries to comply with international agreements. *Environment.* 41: 16-31.

37. www.moef.nic.in

38. www.mospi.nic.in

39. Yadav, A. (2012). Conserving biodiversity through sustainable development. *Orient Journal of Law and Social Sciences.* 4: 84-93.

40. Yadav, A. (2013). Need of judicial review in management of bio-medical waste. *Journal of Legal Studies.* 11(4): 143-147.

41. Yadav, A. (2014). International aspects of biodiversity protection. *Journal of Legal Studies. 11(5):* 174-176.

19

Biological Diversity Conservation and its Legal Paradigm

Himangini Singh

*University Institute of Management and Commerce,
Rani Durgavati University, Jabalapur-482001, Madhya Pradesh, India
E-mail: speakhima@gmail.com*

Abstract

Biological diversity celebrates the difference in cross section of various species of flora and fauna including within its fold micro-organism to the larger variety of living organism spread across the geographical, climatic and ecological sphere. The various constituent of diversity are so intertwined with each other that on a number of occasions their distinctiveness may not be seen. However if one of the links goes missing the entire system becomes vulnerable. Thus from this point of view protection and conservation of biodiversity in its intactness becomes vital for sustenance all the life forms. The fragility of biodiversity in the current scenario cannot be protected by a mere intent to do so. Though the awareness about importance of protection of biodiversity, it can go a long way. However, till the awareness of protection of biodiversity percolates to the masses a strong and impregnable legal mechanism to facilitate the conservation of biodiversity becomes vital.

Introduction

Biological diversity (biodiversity) comprises of all living organisms (plants, animals, microbes, etc.) and the genetic differences among them. It exists at species, community, ecosystem and landscape. Biodiversity means the species of life on planet earth and the ecological pattern that support the vast cycle of diversity of ecosystems.

Genetic Diversity is the difference in the genes of plants, animals and other organisms. With the exception of plants used in cultivation, there is not much known about the genetic diversity of most rangeland species. Fortunately, new and modern techniques are becoming available to facilitate the acquaintance of genetic diversity.

Species Diversity is the variety of species (plants, animals and other organisms) on Earth. The perception that rangelands have low biodiversity is false. A wide variety of plants, animals and other organisms can be found on rangelands. The problem is, there is not an accurate accounting of many of the components of diversity except for selected plants, birds and mammals

Community Diversity is the variation observed in forms of communities of species. It involves the richness (population) and dominance of species. The protection includes appreciable and striking a balance between actions which includes protecting and compromising between various species.

Ecosystem Diversity is the amalgamation of biological communities and adjoining physical environment of ecosystems. The diversity of species, life forms, lifecycles and blue print and structured plans for growing and reproducing, affects the micro-environment energy flow and nutrient cycles for each and specific ecosystem.

Landscape Diversity involves the diversity occasioned due to various changing landscape like hillside, dessert or such other regions. The reason for this diversity to vary according to the landscape arises on account of the impact of landscape on the weather, availability of water, food supply and other environmental factors.

The Importance of Biodiversity

Biodiversity plays a vital role in individual's life and health of whole ecosystem.

Following reasons showcase the importance of Biodiversity:

- Biodiversity contributes majorly in living our lives in a healthy and happy way. The supply of food is dependent on Biodiversity, the health of the economy also relies upon the fruits of the Biodiversity for its inputs. The agricultural production is totally dependent on the diversity of range of pollinators, plants and seeds.

- Nature to make the earth livable had evolved a number of complex systems to facilitate the life like the plants for releasing oxygen, the trees to increase precipitation etc, When a link in the bio diversity is broken the delicate system to facilitate life is broken which leads to overall tampering with the working of the life sustaining macro system.

- Biodiversity allows for ecosystems to adjust to disturbances and natural calamities.

- Genetic diversity plays preventive role to check various diseases.

- The various forms of life respond differently when they come in contact with different diseases, this is the key to fighting the disease in the form of medicine and vaccines. Every time a form of life becomes extinct we lose the knowhow about its response to the threat caused by the disease.

Legal Paradigm of Bio Diversity

India has an evolved constitutional and the legal framework for protection of biodiversity. The subject "Protection of wild animals and birds" falls under List III, Entry 17B of Seventh Schedule. The Parliament passed The Wild Life (Protection) Act 53 of 1972 to provide for the protection of wild animals and birds with a view to ensuring the ecological and environmental security of the country. The Parliament vide Constitution (42nd Amendment) Act, 1976 inserted Article 48A w.e.f. 03.01.1977 in Part IV of the Constitution placing responsibility on the State "to endeavor to protect and improve the environment and to safeguard the forests and wild life of the country." Article 51A was also introduced in Part IVA by the above-mentioned amendment stating that

"it shall be the duty of every citizen of India to protect and improve the natural environment including forests, lakes, rivers and wildlife and to have compassion for living creatures".

Further Section 5A authorizes the Central Government to constitute the National Board for Wild Life (in short 'NBWL'). The NBWL is, therefore, the top most scientific body established to frame policies and advise the Central and State Governments on the ways and means of promoting wild life conservation and to review the progress in the field of wild life conservation in the country and suggesting measures for improvement thereto. The Central and the State Governments cannot brush aside its opinion without any cogent or acceptable reasons. Legislation in its wisdom has conferred a duty on NBWL to provide conservation and development of wild life and forests.

The Supreme Court of India Court in Sansar Chand v. State of Rajasthan, (2010) 10 SCC 604 held that all efforts must be made to implement the spirit and provisions of the Wild Life (Protection) Act, 1972; the provisions of which are salutary and are necessary to be implemented to maintain ecological chain and balance.

The Stockholm Declaration, the Declaration of United Nations, Conventions on Human Environment signed in the year 1972, to which India is the signatory, have laid down the foundation of sustainable development and urged the nations to work together for the protection of the environment. Conventions on Biological Diversity, signed in the year 1962 at Rio Summit, recognized for the first time in International Law that the conservation of biological diversity is "a common concern of human kind" and is an integral part of the development process.

The Parliament enacted the Biological Diversity Act in the year 2002 followed by the National Biodiversity Rules in the year 2004. The main objective of the Act is the conservation of biological diversity, sustainable use of its components and fair and equitable sharing of the benefits arising out of the utilization of genetic resources. Bio-diversity and biological diversity includes all the organisms found on our planet i.e. plants, animals and micro-organisms, the genes they contain and the different eco-systems of which they form a part. The rapid deterioration of the ecology due to human interference is aiding the rapid disappearance of several wild animal species. Poaching and the wildlife trade, habitat loss, human-animal conflict, epidemic etc. are also some of the reasons which threaten and endanger some of the species.

India is known for its rich heritage of biological diversity and has so far documented over 91,200 species of animals. In India's bio-graphic regions, 45,500 species of plants are documented as per IUCN Red List 2008. India has many critically threatened animal species. The IUCN adopted a resolution of 1963 by which a multi-lateral treaty was drafted as the Washington Convention also known as the Convention on International Trade in Endangered Species of Wild Fauna and Flora (CITES), 1973. CITES entered into force on 1st July, 1975, which aims to ensure that international trade in specimens of wild animals and plants does not threaten the survival of the species in the wild, and it accords varying degrees of protection to more than 33,000 species of animals and plants. Appendix 1 of CITES refers to 1200 species which are threatened with extinction.

For achieving the objectives of various conventions including Convention on Biological Diversity (CBD) and also for proper implementation of IUCN, CITES etc., and the provisions of the Wild Life (Protection) Act, Bio-diversity Act, Forest Conservation Act etc. in the light of Articles 48A and 51A(g), the Government of India has laid down various policies and action plans such as the National Forest Policy (NFP) 1988, National Environment Policy (NEP) 2006, National Bio-diversity Action Plan (NBAP) 2008, National Action Plan on Climate Change (NAPCC) 2008 and the Integrated development of wild life habitats and centrally sponsored scheme framed in the year 2009 and integrated development of National Wild-life Action Plan (NWAP) 2002-2016.

The National Biodiversity Action Plan approved in November 2008 to augment natural resource base and its sustainable utilization. The Plan draws from the principles of National Environment Policy, incorporates suggestions made by a consultative committee and proposes to design actions based on the assessment of current and future needs of conservation and sustainable utilization. The key features of the National Bio Diversity Action plan are as follows:-

i.   Strengthening and integration of in situ, on-farm and ex situ conservation.

ii. Augmentation of natural resource base and its sustainable utilization; Ensuring inter and intra-generational equity.

iii. Regulation of introduction of invasive alien species and their management.

iv. Assessment of vulnerability and adaptation to climate change and desertification Integration of biodiversity concerns in economic and social development.

v.   To prevent, minimize and abate impacts of pollution Development and integration of biodiversity databases.

vi. Strengthening implementation of policy, legislative and administrative measures for biodiversity conservation and management.

vii. Building of national capacities for biodiversity conservation and appropriate use of new technologies.

viii. Valuation of goods and services provided by biodiversity and use of economic instruments in the decision-making processes International cooperation to consolidate and strengthen bilateral, regional and multilateral cooperation on issues related to biodiversity

Another aspect of Bio diversity is protection of bio diversity for promotion of individual or corporate financial interests. The financial aspects of bio diversity are further advanced by the exclusivity in access to it and the right and knowledge of the exclusive use of it. Intellectual Property Rights refer to new ideas and knowledge which has evolved as a result of research and innovation, these rights leads to the exclusivity of the usage of the knowledge by the inventor to the exclusion of the others. The main purpose the legal framework of it is regulated by various intellectual property laws. However the aspects of genetic mutation and modification of the fauna and flora so that they can be used most productively for commercial interests is also a key issue. Though the exploration of bio diversity for commercial and scientific activities has enormous monetary and social advantages but the same bring with them there share of ethical and environmental concerns. The threat of the genetically modified species becoming invasive species and leading to the extinction of the pure and original specie is always there. Another aspect of the regulation of the use of bio diversity by intellectual property rights legislations is to distance the traditional know how from the common people who have been traditionally the beneficiaries of the knowledge, however not possessing the financial capability to obtain the right to use it under the Intellectual property rights legal framework.

References

1. Basu, D.D. Shorter Constitution of India 13th Edition.
2. Dutfield, G. (2000). "The public and private domains: Intellectual property rights in traditional knowledge".
3. Malik, S. and Malik, S. Supreme Court on Environmental Law 2nd Edition.
4. Supreme Court Cases 104. 2010(10).
5. www.rangeland.org

20

Traditional Knowledge among Indian Tribes and Conservation of Phytodiversity through Faith, Taboos and Sacred Groves

Ashok K. Jain

Institute of Ethnobiology,
Jiwaji University, Gwalior-474011, Madhya Pradesh, India
E-mail: asokjain2003@yahoo.co.in

Abstract

Indian tribes are the treasure of traditional knowledge, which they have inherited through generation to generation without any written script. Nature has been the main source of their learning, where they accepted and discarded plants and other natural resources on the basis of trial and error. Properties of a large number of plants including medicinal, edible, toxic etc. are known to them. Due to faith, taboos and beliefs a large number of plant species are conserved by them. A good number of sacred groves shelter several rare and threatened species. Such groves are sustaining even today because they are protected by tribes. Several species of nutritional value are used during scarcity of food. Tribes have unique knowledge of climatic forecasting by observing various natural phenomena.

Key Words: Traditional Knowledge, Ethnobotany, Tribes, Conservation

Introduction

Along with the gradual progress of the human civilization, his dependence on plants and vegetation increased day by day. Beginning with the use of plants for food and medicines man gradually depended on his surrounding vegetation for house building, clothing, gum, resin, oil, construction of household as well as agricultural implements, musical instruments and for other numerous purposes. But any recorded account of this man-plant relationship and history of man's gradual dependence on plant world from the initial stage is not available as the method of writing or recording was not developed then. In later stages when man was able to develop the method of writing, these ethnobotanical accounts were recorded, although not in a systematic way. Now scientists from various disciplines and especially ethnobotanists, are recording such information scientifically with systematic and authentic manner.

Indian subcontinent is inhabited by over 53 million tribals belonging to over 550 different communities under 227 ethnic groups (Anonymous, 1994), residing in about 5000 villages (Sikarwar, 2002). Folk-lore is an integrated part of tribal life. It includes myths, legends, tales, proverbs, riddles, the texts of ballads and other songs (Dash And Mohapatra, 1979). Traditional knowledge is commonly preserved and transmitted orally from generation to generation in the form of folk lore. Ethnobotanical studies indicate that a good number of folk-lorescomprise valuable information on ecology, agriculture and weather forecasting. Several folk-lores mention about the conservation of several threatened taxa.

Based on the activities of birds, animals; shape, size, color, speed and direction of clouds; phenological events of plants, the ancient Indian folk people developed the knowledge of forecasting weather of upcoming months, growth of plants and yield of grains, vegetables and other edible parts. Even today in several parts of the country, on the basis of height; body shape and size; color; types of horns, teats, ears, tail etc. folk people tell about the quality of animal and production of milk. A large number of beliefs, faith, taboos prevailing amongtribals in Indiahave helped in conserving a large variety of plants. As per the view of Schultes (1996) there is ample proof from recent surveys that, folklore is indeed a valuable tool when properly analyzed.

The sacred groves are considered to be rich source of medicinal, rare and endemic plants as refuge for relic flora of a region and as center of seed dispersal (Whittaker, 1975; Jeevaet al., 2006). Plant wealth and self-conservation potentials of sacred groves are impressive enough for them to be acknowledging as mini biosphere reserve (Gadgil and Vartak, 1975). Several plant species conserved in sacred groves in India have been documented by a good number of workers (Vartak *et al.,* 1987; Bhakat and Pandit, 2003; Bhandari and Chandrasekhar, 2003). With this realization, the recent upsurge of interests in studying sacred groves vis-a-vis medicinal plants has not only established the topic as one of ecological significance but this tradition of nature conservation based on socio-cultural grounds has got a new found value as well.

Different regions of India with some important tribes

1. **Central Region:**
    a. **South Central** (Andhra Pradesh, Chhatisgarh, South Orissa) Gonds, Oraon, Konda, Bhil, Reddis, Koyan, Kodagu, Pardhan, Saravas, Khonda
    b. **North Central** (Bihar, West Bengal, North Orissa, East M.P.) Santhal, Oraon, Munda, Kolho, Mal Pahariya, Bhag, Sahariya.

2. **Western Region:**
    Rajasthan, Gujrat, Maharashtra, Dadra & Nagar Haveli, Goa,Daman & Diu:

Bhills, Dhodia. Gamit.Warli, Naika, Rathwa, Kolidher.

3. **North Eastern Region:**

    (Meghalaya, Arunachal Predesh, Nagaland, Manipur, Mizoram, Tripura, Assam) Nagal, Khasi, Mizo, Miri, Tripuri, Adi, Aka, Apatan, Maripa, Mishmi, Garos, Tagin, Naga, Singhpho, Sherdukpen.

4. **North Western Region:**

    (U.P., Uttaranchal, H.P.) Jaunsaries, Laholies, Kinnoras, Tibetans, Battis, Bhotai.

5. **Southern Region:**

    Karnataka, Kerala, Tamil Nadu Malayali, Kani. Kurumba, Muthuvan, Solugu.

6. **Island Region:**

    Andaman &Nicibar, Lakshadweep Onges, Great Andamanese, Sentenelese, Jarwas, Nicobarese, Shompen.

Some other facts about Indian tribes:

*   Tribal lands all over India yield the major part of the country's mineral wealth, including iron ore, mica, copper, chromium and coal.
*   Due to various laws, majority of Indian tribes had to leave their home land and are under debt.
*   Yet, 75 per cent of the tribes live below the poverty line.
*   Tribal hamlets like **Amlasole** in West Bengal and **Kalahandi** in Orissa are synonymous with starvation deaths and misery.
*   Many important plant species have been devasted from tribal inhabited localities.
*   Whenever dams have come up, it has been on tribal land. And so have wildlife sanctuaries and biosphere reserves.

Uses of plant species in different systems of Medicine

Under different systems of medicines,a large number of plant species are used, but maximum number of species are used under folk or traditional system of medicines.

Folk Medicine	:	*44%*
Ayurveda	:	19%
Siddha	:	12%
Unani	:	10%
Homoeopathy	:	8%
Tibetan	:	5%

Methodology

The observations were made in the following manner:

1. Extensive survey of tribal inhabited localities.

2. Life styles, socio-cultural and economic aspects of tribes were studied.

3. Plants concerned with tribes were collected, identified and preserved in the form of herbarium.

4. Records on uses, faith, taboos, beliefs, offerings, totems related to plants were made with the help of standard questionnaire (Jain, 1987).

5. Information on folklores, folk-proverbs, folktales in reference to ecological and other climatic aspects prevail among tribals, were recorded.

6. Information on sacred groves was collected by personal observation and published literature.

7. Emphasis was made on tribe's perception on climatic forecasting, agriculture, soil testing, cattle, varieties of crops, conservation and threatened plant species.

The data generated after the detailed ethnobotanical studies in tribal localities are presented here, however the emphasis has been made on following aspects:

(A). Climatic forecasting for agricultural purposes

Folk proverbs are very common among folk people. Many natural phenomena are observed vary carefully and then forecasting are made. Some folk proverbs related to the forecasting of climate and having a bearing on agriculture (Jain, 1996). The original proverbs were expressed in folk "Hindi' language, however their meanings are presented here in English language:

1. *"Ookhgodiketuratdababe, to phirookhbahut such pabe"*

2. To get good yield of sugarcane (*Saccharumspontaneum*) the farmer should bury its stump immediately after harvesting.

3. *"Pug pugdooribajra, mendhakkudnijwar Aisebobejo koi, gharkobharebhandar"*

4. It means, to get high yield, millets must be sown at a foot step distance and *Sorghum vulgare* needs at least a frog's leap distance between its seeds.

5. *"Aamasamabadrapuravpashchimjaye, panchmilawamaghjidus din jhadilagaye"*

6. It means in the tenth lunar month of Hindu calendar (December-January) if clouds move from two ends face to face and move towards east and west, they are sure to bring heavy rains, which can last for 10 days at a stretch.

7. *"Sravanakerepratham din ugatnadeekheBhan, Char mahinapani pare janoisepraman"*

8. Invisibility of Sun rise on the first day of Shravana (Fifth lunar month as per Hindu calendar ensures good rains throughout the rainy season.

9. *"SudiashadmeinBudhkoudaybhayojodekh Shukraastasavanlakhomahakalabrekh"*

10. It means Visibility of Mercury during the lighted month 'Ashad' (June/July) and the return of Venus during the month of Shravana (July/August) depicts that the famine is the destiny.

In reference to climatic forecasting, Mohanty (2008) has made some interesting observations from Orissa.The meaning of some of the folk proverbs collected by him from Orissa are as follows:

1. Appearance of stars within the orb of the Moon indicates heavy rains in future.

2. The appearance of 'Ardraballi' and 'Roi' stars in the eastern and western segments of the sky is a sure indication of heavy rain and resultant flood in coming future.

3. Rainfall in the month of 'Chaitra' (March-April) accelerates the growth of grass, cows get sufficient food and milk production enhances.

4. A hot 'Pausa' (December-January), a stormy 'Baisakha' (April-May), moderate rainfall in the beginning of 'Asadha' (June-July) indicates no rains in Sravana (July-August) and 'Bhadraba' (August-September) leading to drought and famine due to failure of crop in that year.

The farmers of Orissa state of India have been cultivating rice since time immemorial. A good number of folklores are concerned with germ plasm, cultivation techniques and yield of the crop (Mohanty, 2008). For example the meaning of some of the folklores are:

1. Paddy cultivation through transplantation method never fails to give desired return.

2. Paddy crop developed by broadcasting method requires less water, whereas the same by the transplantation method needs plenty of water for required growth.

3. 'Nauri' (Local name of an insect) is a beneficial insect which distracts the pathogens, whereas 'Giria' is a destructive pest in the rice field.

4. One proverb says that for agricultural purposes a bullock with thin and narrow horn as well as thick tail is not fit for hard work and should not be purchased.

5. Tamarind (*Tamarindusindica*) tree in front of the house and Palmyra palm (*Borassusflabellifer*) at the backyard are harmful for the health of the members residing there.

6. Common jujub in lunar month 'Magha' (Jan. – Feb.) and radish in solar month 'Makara' (Feb.-March) are prohibited for consumption.

(B). Tribal way of conservation

Many practices used by indigenous peoples serve to manage species diversity, create habitat heterogeneity on the landscape scale, and manage intensity of use, thereby enhancing the diversity of biological resources available.

We assume that most part, *biodiversity conservation is the indirect outcome, rather than the objective, of these practices.* Nevertheless, conscious conservation of biodiversity in some situations should not be ruled out (Ruddle*et al.*, 1992). Many crops such as *Panicummiliaceum, Echinocloacolona, Paspalumscrobiculatum* and

Setariaitalica are now cultivated and conserved only by the tribal people in many parts of southern India.It is believed that in 1730, the king of Jodhpur ordered his men to cut timber from Bishnoi land. The local people, led by Attri Devi, hugged the trees to save them. The king's men hacked down 263 children, women and men before they gave up. The Bishnoi religion was initiated by Guru Jambeshwar about 500 years ago. Followers believe in a set of 29 principles, hence are called the 'Bishnoi'. These principles include a ban on felling treesand a ban on killing animals. In particular, they consider the kejari (*Prosopiscinereria*) and blackbuck sacred species.

Among Santal tribes of West Bengal several following taboos are prevalent –

- *Emblicamyrobalan:* Cutting of trees or branches are prohibited, Only peeling of bark allowed, that too by rotation
- Restriction on plucking fruits of *Zizyphus*spp.
- Pulling out *Evolvulusalsinoides*is prohibited.
- Not to cut tree of *S. robusta*in october& on full moon days

Among Gond tribe of Central India following taboos are prevalent:

- Do not cut trees. Fallen branches of *Buchnanialanzana*are used only for making musical instruments.
- Do not eat fruits of *Dilleniaindica*&*Mangifera* before worship Lodha tribes of West Bengal do not cut trees of *Madhucalongifolia,Buteasuperb, H. antidysenterica, Aeglemarmelos, Alstoniascholaris, Lygodiumflexuosum, Cocosnucifera*etc. Tribes of Andaman and Nicobar do not dig *Dioscoreaglabra* in rainy season, as it is considered the food of 'Rain God'. *Ocimum sanctum and Ficusreligiosa*are considered sactred all over India.
- **"Kadars"** of Tamil Nadu for example select only mature plants of the yam *Dioscorea* for harvesting the tubers. They first examine the vine and choose only those whose leaves are yellow which is an indication of maturity. Tubers of young green vines are never dug out. After harvesting the mature yams they cut off the upper portion of the tuber along with the vine and replant it in the pit. They cover the pit with loose soil for the tuber to grow again in the coming season for whoever may harvest it in the future.
- *Irulas, Malayalis* **and** *Muthuvas* inhabiting Tamil Nadu have been cultivating the traditional cultivars *viz.* paddy, millets, pulses and vegetable crops. Their subsistence life style, local diet habits and dependence on rain fed irrigation have influenced them to cultivate and conserve the traditional cultivars or land races.

(C). Conservation through clans:

Many of the clans of tribes are related to plants and named after concerned plant species. A marriage relation between persons of the same clan is strictly prohibited (Jain, 1988). During the study it was observed that the tribes do not harm the species of their concerned clan. Similar observations have been made in

Jhabua district by Kadel (2007). Some clans named after plants and related taboos are mentioned below (Table 1):

Table 1: Clans of Bheel tribe named after plant species in Jhabua district, MadhyaPradesh

S. No.	Clan	Name of Plant	Local Name	Taboo
1	Amalia	*Mangiferaindica*	Aam	Stem & leaves used in worship.
2	Awalia	*Emblicaofficinalis*	Anwala	Never cut the tree & worship it.
3	Banskela	*Bombaxceiba*	Bans	They make musical instruments, used in worship.
4	Barela	*Ficusbenghalensis*	Bargad	Never cut the tree & Worship it.
5	Amalaharya	*Curcuma amada*	Ambahaldi	Use for worship purpose.
6	Semaliya	*Bombaxceiba*	Semal	Never cut the tree.
7	Singadia	*Trapabispinosa*	Singada	Never use the fruit.
8	Bakhla	*Viciafaba*	Sukla	Use for worship.
9	Maoli	*Madhucalongifloia*	Mahua	The birth of child under tree.
10	Sarpota	*Achyranthesaspera*	Chirchira	Plant use in worship.
11	Jamania	*Syzygiumcuminii*	Jamun	Never cut the tree.
12	Rohini	*Soymidafebrifuga*	Rohan	Never cut the tree & Worship it.
13	Barbaria	*Ziziphus spp.*	Ber	Fruits used in worship.
14	Bilwaliya	*Aeglemarmelos*	Bel	Leaves and fruits are used in worship.
15	Dodiyar	*Zea mays*	Makka	Grains used in ceremonies.
16	Tad	*Borassusflabellifer*	Tadi	Drinks taken.

(D). Sacred groves:

In India, there are more than 100,000 sacred groves alone (Malhotra*et al.*, 2001). There are about more than 100 protected areas around the world that are important to one or more faiths (Dudley *et al.*, 2005). The sacred groves, in India, are known by different names at different places, such as '*Devray*' in Maharastra, '*Devarkand*' and '*Siddarvanam*' in Karnataka, '*Oraans*', '*Kenkari*', '*Malvan*' and '*Yogmaya*' in Rajasthan and '*Saranya*' in Bihar [9,10]. Overall, 120 plant species belonging to 51 families were documented from Sacred Jungles of Kwand and Bund of Pakistan (Shah *et al.*, 2112).

In tribal region of Jharkhand and Orissa sacred groves are popularly known as *Jaher* [11]. Some facts about these groves are following:

- Mostly the head of the village or Padihar collected the fruits from the sacred grove, particularly *Madhucaindica*and *Tamarindusindica*.

- There were some customs and cultural practices associated with collection and generally there was no restriction on the collection of fruits, however the local people were afraid of collection from the sacred groves.

- There was a general belief that since sacred groves were pure, the gatherers might be punished by spirits and deities for unauthorized collection of natural resources, though all deities were not considered unsafe.

The study conducted in Jhabua district of Madhya Pradesh indicates the presence of about 10 sacred groves in the region. These Sacred groves are lying at and around temples. A good number of medicinal and rare plant species are growing over there. The tribes of Jhabua believe that their deities live inside these sacred groves. They also believe that these deities would be offended if any damage is caused to the plants and animals residing in these groves. Hence, nobody collects anything from these localities, not even fallen branches of trees. Similar observations have been made from Gwalior, Shivpuri and Guna districts of Madhya Pradesh. Some important plant species in such sacred groves are mentioned below in Table 2:

(E). Species of nutritional value:

Observations in several tribal inhabited localities indicate that a large number of plant species are collected, stored and used during adverse conditions viz. drought, famine, floods etc. Biochemical analysis of such species has shown the richness of important compounds in them (Jain and Tiwari, 2012). Table 3 depicts about the amino acids concentration in 10 selected species, used by tribals during emergency.

Table 2: Some important plant species in sacred groves of Jhabua district

1. *Vitex negundo*	19. *Enicostema axillare*
2. *Withani asomnifera*	20. *Diospyros melanoxylon*
3. *Woodfordia fruticosa*	21. *Dioscorea bulbifera*
4. *Vanda tessellate*	22. *Datura metel*
5. *Urginea indica*	23. *Curcuma pseudomontana*
6. *Tinospora cordifolia*	24. *Curculigo orchioides*
7. *Syzigium heyneanum*	25. *Cissampelospareira.* var. *hirsute*
8. *Sterculia urens*	26. *Chlorophytum borivilianum*
9. *Spigelia anthelmia.*	27. *Ceropegia bulbosa*
10. *Sphaeranthus indicus*	28. *Cassine glauca*
11. *Santalum album*	29. *Buchanania lanzan*
12. *Mucuna pruriens*	30. *Bacopa monnieri*
13. *Ichnocarpus frutescens*	31. *Asparagus racemosus*
14. *Hemidesmus indicus*	32. *Arisaema tortuosum*
15. *Helicteres isora*	33. *Aristolochia indica*
16. *Hardwickia binata*	34. *Amorphophallus paeoniifolius*
17. *Gloriosa superb*	35. *Aegle marmelos*
18. *Gardenia latifolia*	36. *Abrus precatorius*

Table 3: Qualitative estimation of amino acids in different parts of selected plant species.

S. N.	Edible Parts	Plant Name	Essential amino acids	Non-essential amino acid
1.	Leaves			
a.		O. cornicu-lata	Phenylalanine, Isoleucine, Valin`e.	Asparagine, Alanine, Cystine.
b.		M. oleifera	Phenylalanine, Isoleucine, Leucine, Valine, Methionine, Threonine,Glutamine.	Serine, Ornithine.
c.		C. obtusi-folia	Isoleucine, valine.	Asparagine, Ornithine, Alanine,CysteinAspartate.
d.		B. diffusa	Phenylalanine, Isoleucine, Va-line, Methionine,Leucine.	Asparagine, Serine, Ornithine,
e.		A. virdis	Phenylalanine, Isoleucine, Methionine, Valine.	Alanine.
2.	Fruits			
a.		R. parvi-flora	Phenylalanine, Isoleucine, Valine.	Asparagine, Aspartate, Serine, Ornithine.
b.		P. cineraria	Phenylalanine, Isoleucine, Leucine, Valine.	Aspartate, Ornithine.
3.	Flowers			
a.		M. oleifera	Phenylalanine, Isoleucine, Methionine, Valine, Leucine, Threonine.	Asparagine, Serine, Ornithine.
b.		C. fistula	Phenylalanine, Isoleucine, Valine.	Serine, Ornithine,Proline.
4.	Seeds		Phenylalanine, Isoleucine, Valine,	Alanine, Asparagine, Serine, Aspartate, Cystine, Ornithine.
a.		B. vahlii	Histidine,Leucine.	Semi Essential
b.		A.aspera	Phenylalanine, Isoleucine, Valine,	Tyrosine.
5.	Bark			
a.		P. cineraria	Glycine.	Alanine, Asparagine, Serine, Ornithine,Proline.
			Phenylalanine, Isoleucine.	Serine, Ornithine. Asparagine, Serine, Ornithine. Alanine, Proline.

Discussion

Living very close to nature, folk people certainly possess a vast knowledge of properties and activities of various living and non-living components of the nature. They do not have any modern scientific laboratory for analysis or proof but their long experience has provided them this wisdom. The nature happens to be their laboratory. From the above observations it is clear that Indian folk people are the treasure of knowledge of various disciplines of nature. Folklores are the best means to learn or explain any incidence or event (Agarwal, 1981). Thousands of such folklores relating to various aspects of natural, cultural, mythological, political, traditional and other activities of people prevail in Indian folk life (Jain, 1996). Similarly these folklores concerning the season, weather and their consequent effects are a part of numerous poetic versions of ancient wisdoms gained through man's long experiences and observations. They were composed by some anonymous creative genius in the country side and transferred orally from generation to generation in form of folk songs, proverbs or as sayings.

With the advancement of the modern civilization the traditional knowledge about the uses of plants for various purposes, among the aboriginal cultures is rapidly disappearing. There is urgent need for ethnobotanical observations to find out the secrets of nature of such rapidly disappearing primitive culture.

References

1.   Agarwal, S.R. (1981). Trees, flowers and fruits in Indian folksongs, folkproverbs and folk tales.In   Glimpses of Indian Ethnobotany, Jain, S.K. (Ed.) Oxford & IBH Pub. New Delhi.

2.   Anonymous. (1994). Ethnobotany in India: A Status Report.All India Coordinated Research Project in Ethnobiology. Min. of Env. & Forests, Govt. of India, New Delhi.

3.   Bhakat, R.K. and Pandit, P.K. (2003). Role of a sacred grove in conservation of medicinal plants. *Indian Forester.* 129: 224-232.

4.   Bhandari, M.J. and Chandrasekhar, K.R. (2003). Sacred groves of DakhinaKanada and Udupi districts of Karnataka.*Current Science.* 85: 1655-1656.

5.   Dash, K.B. and Mohapatra, L.K. (1979). Folklores of Orissa. National Book Trust, New Delhi.

6.   Gadgil, M. and Vartak, V.D. (1975). Sacred groves of India: a plea for continued conservation. *J. Bombay Nat. His. Soc.*72: 314-320.

7.   Jain, A.K. (1988). "Tribal clans in Central India and their role in conservation" *Environ. Conservation, Sweden.* 15(1): 368.

8.   Jain, A.K. (1996). "Climatic Forecasting for Agricultural use in Ancient Indian FolkProverbs". *Ethnobotany.* 88(1,2): 85-87.

9.   Jain, A.K. and Tiwari, P. (2012). Nutritional value of some traditional edible plants used by tribal communities during emergency with reference to Central India. Ind. *J. Traditional Knowledge.* 11(1): 51-57.

10. Jain, S.K. (1987). A Manual of Ethnobotany. Sci. Pub., Jodhpur.

11. Jeeva, S., Mishra, B.P., Venugopal, N., Kharlukhi, L. and Laloo, R.C. (2006). Traditional Knowledge and biodiversity conservation in the sacred groves of Meghalaya.*Indian J. Trad. Know.*5: 563-568.

12. Kadel, ChitraSoni. (2007). "Ethnobotanical Studies on tribal communities of Jhabua district (M.P.)". Ph.D. Thesis, Jiwaji University, Gwalior.

13. Mohanty, R. (2008). Ethnobotanical study of the folklores of Orissa.D.Sc. Thesis, F.M. University, Balasore.

14. Schultes, R.E. (1996). The Plant Kingdom-A thesaurus of biodynamic constituents. *Ethnobotany.* 8: 2-13.

15. Sikarwar, R.L.S. (2002). Ethnogynaecological uses of plants new to India. *Ethnobotany.*14: 112-115.

16. Whittaker, R.H. (1975). *Communities and Ecosystems.* Macmillan Publishing Company. New York.

21

Conservation of Species under International Environmental Law

Jagdish Khobragade, P.P. Singh and Vikash Agrawal*

*School of Law, Dr.Harisingh Gour Vishwavidhyala,
Sagar-470003, Madhya Pradesh, India
E-mail: jkhobragade@gmail.com*

Abstract

The write up aims to analyze and trace out the detailed provisions regarding conservation of species and ecosystem with current international environmental law. The environment consists of millions of living species, which are to be found in the ecosystem. The importance of conservation of these species along with the ecosystem has been increasing consistently due to the shocking impact of the advancements in science and technology. The relation of these species with the ecosystem creates a biotic enterprise of which human beings are also a part.[1] The conservation of biology consists of conservation of genetics, biogeography, wildlife, forestry and fisheries etc. There are many instruments and norms in international law which directly deal with conservation of species and ecosystem. However, in this paper conservation of species mainly focused on protection of seals, whales and migratory species. This study makes an attempt to analyse the nexus between the conservation and protection of species under ecosystem with the help of existing legal mechanism in international environmental law.

Key Words: Conservation of Species, Ecosystem, Migratory Species, Biodiversity, Environmental Law

Introduction

Ecosystem has always been under threat because of habitat destruction, global warming and other human induced environmental degradation

---

[1]B. NATH, L. HENS, P. COMPTON AND D. DEVUYST, "ENVIRONMENTAL MANAGEMENT IN PRACTICE" (New York: Routledge Publication, Reprinted 2002), pp.4-5

activities. In beginning in the 1970s, the international community started addressing environmental concerns through the adoption of international treaties, agreements, and declarations designed to curb these problems.[2] Rio Declaration on Environment and Development, the United Nations Conference on Environment and Development, having met at Rio de Janeiro from 3 to 14 June 1992, Reaffirming the Declaration of the United Nations Conference on the Human Environment, adopted at Stockholm on 16 June 1972, and seeking to build upon it a proper implementation mechanism of the environmental issues. Following the Rio Declaration of 1992, a large majority of States undertook to develop strategies to reverse the trend and protect the global environment.[3] The Endangered Species Act, 1973[4] as well as the Convention on Biological Diversity, 1992 defines the concept of species. It is accepted fact that we could not protect the endangered species unless we have some basic understanding of the concept of the species, their importance and role in the ecosystem. Though, it is difficult to understand the term 'species' and 'ecosystem' but in general species is the one of the part of ecosystem and whole ecosystem is mixture of living species. In addition conservation of species is not only an environmental issue but the part of life cycle of living animals including human being. Here it is worth to mention that the primary object of this Convention on Biological Diversity is to protect and conserve the biological diversity therefore it came under the conservation of species and ecosystem. With the development of various legislations on conservation of species many NGO's also come forward to work in the field of environment. Today many big NGO's and institutions are working in this field like IUCN,[5] Conservation of Biological Diversity Society and WWF etc.

I: What is Biodiversity?

Generally Biodiversity means 'the variety of life on earth'[6] and this variety can be measured on several different levels. It includes Genetic, Species and Ecosystem as a whole. It is worth noting that Biodiversity and ecosystems sustain each other.

The Convention on Biodiversity conservation defines biodiversity "as the variety and differences among living organisms from all sources, including terrestrial, marine, and other aquatic ecosystems and ecological complexes of which

---

[2]JASON GRAY, *"Indigenous Communities and Biodiversity conservation: Protected areas and the right to consultation"* GONZAGA JOURNAL OF INTERNATIONAL LAW, Vol.2-Issue 2 (2008-09) P.1
[3]*Supra* note 2
[4]See, The Endangered Species Act of 1973 (16 U.S.C. 1531-1544, 87 Stat. 884)
[5]See, The International Union for Conservation of Nature (IUCN) helps the world find pragmatic solutions to our most pressing environment and development challenges. It supports scientific research, manages field projects all over the world and brings governments, non-government organizations, United Nations agencies, companies and local communities together to develop and implement policy, laws and best practice.
[6]DR. BARBARA CORKER, *Biodiversity and conservation*, Off well Woodland and Wildlife Trust, see also www.countrysideinfo.co.uk

they are a part."[7] This includes genetic diversity within and between species and of ecosystems. Thus, the biodiversity represents all life and its conservation is the primary obligation of all human beings. However, biodiversity is currently being lost due to human activities and this trend can only be reversed if the awareness among the people about conservation of species and ecosystem is spread. The Convention under Article 8 provides for promoting the conservation of species.[8] It imposes an obligation to:

(d) Promote the protection of ecosystems, natural habitats and the maintenance of viable populations of species in natural surroundings;

(e) Promote the wide implementation and further development of the **ecosystem approach,**[9] as being elaborated in the ongoing work of the Convention;

(f) Promote concrete international support and partnership for the conservation and sustainable use of biodiversity, including in ecosystems, at World Heritage sites or the protection of **endangered red species,** in particular through the appropriate channelling of financial resources and technology to developing countries and countries with economies in transition;

(i) Strengthen national, regional and international efforts to control invasive alien species,[10] which are one of the main causes of biodiversity loss, and encourage the development of effective work programme on **invasive alien species** at all levels;

II: What is Species?

A common definition species is that of "a group of organisms capable of interbreeding and producing fertile offspring of both genders, and separated from other such groups with which interbreeding does not normally happen."[11]

In addition, diversity is the variety of species in a given region or area. This can either be determined by counting the number of different species present, or by determining taxonomic diversity. Taxonomic diversity is more precise and considers the relationship of species to each other. It can be measured by counting the number of different taxa (the main categories of classification) present.

For example, a pond containing three species of snails and two fish is more diverse than a pond containing five species of snails, even though they both contain the same number of species. High species biodiversity is not always necessarily a good thing. For example, a habitat may have high species biodiversity because many common and widespread species are invading it at the expense of species restricted to that habitat.[12]

---

[7]The Convention on Biological Diversity, 1992

[8]Ibid

[9]Ibid

[10]"Invasive species" means an alien species whose introduction does or is likely to cause economic or environmental harm or harm to human health.

[11]http://en.wikipedia.org/wiki/Species

[12]*Supra* note 5

Importance of Conservation of Species

All the aspects of human life are dependent on ecosystem services like the food from the oceans, the ability of forests to absorb carbon and to provide oxygen and water etc. While focusing on conservation of species and ecosystem we have to consider various factors because it is a part of human living and ecosystem. Although conservation of species have been protected by the various Convention but there are many reasons for the conservation of species. For example; Species are the **building blocks** of ecosystem; Species for **recreation; Medicinal** plants and animal; the **connection** between wild and **domestic animals;** Inherent **cultural** and **spiritual** values of species, etc.

Protection of Species

Since, the beginning of international environmental law there has been long debate on the protection of species. Although Protection of species is important but there are some of the areas where we need to focus more and that area is protection of migratory species, protection seals and protection of whales. The one of finest example of protection of environment is Antarctica treaty which specifically focuses on Antarctica region only and species living therein. Whereas there are other conventions which need to be discuss at length. In the ongoing explanation all the issues are covered for the better understanding of the conservation of species and ecosystem.

Protection of Migratory Species

The Bonn Convention on the conservation of Migratory Species of Wild Animal (CMS) was adopted in 1979. The conventions entered into force in 1983. Today the Convention includes seventy-nine parties: Africa (25), America and Caribbean (6), Asia (9), Oceania (3), Europe (36)[13] and the purpose of the convention is the conservation of migratory species including birds, mammals, reptiles and fish.[14] There are many types of migratory species, stock that breeds on the territory of a state and then migrates into the sea like seals, sea turtles, anadromous fish etc. and highly migratory species that travel between Exclusive Economic Zones (EEZ), and between the EEZ and the high seas (tuna, whales); and territorial species that live in border areas and usually cross jurisdictional boundaries e.g gorillas and elephants etc.

The convention requires the immediate protection of endangered migratory species included in Appendix I[15] and the conclusion of additional agreement for the protection of migratory species included in Appendix II.[16] Overall, the Convention on Migratory Species has not enjoyed the attention devoted to other international conventions, such as the Convention on Biological Diversity. This is because the

[13]See Guide to Convention on the conservation of Migratory species of Wild Animals (Secretariat of the UNEP, Jan 2002)
[14]ELLI LOUKA, INTERNATIONAL ENVIRONMENTAL LAW-FAIRNESS, EFFECTIVENESS AND WORLD ORDER, (New York: Cambridge University Press 2006, 1st edn.) p.335
[15]Convention on the Conservation of Migratory Species of Wild Animals Article III Endangered Migratory Species: Appendix I
[16]Ibid

convention is a highly specialised tool and many of the issues it addresses are not politically charged.

Protection of Whales

The second most important controversial agreement for the protection of mammals is the 1946 International Convention for the Regulation of Whaling.[17] The purpose of the whaling Convention is to regulate the development of the whaling industry. As explicitly provided for in the preamble, the purpose of the convention is to 'make possible the orderly development of the whaling industry' through the proper conservation of whale stocks.[18] The convention has a quite large jurisdictional coverage as it applies to factory ships, land stations, and whale catchers, and it is effective in all waters in which whale hunting is exercised.[19]

The Whaling Convention established the International Whaling Commission. The regulatory powers of the IWC are extensive.[20] An IWC regulation is binding on state parties unless the parties object within the time limit provided for in the convention. In case state parties object, regulations adopted are not binding on the states that have objected, states can be granted exemptions for scientific purposes. In that case state party has to give report of research accomplished. This commission had notable development and in 2003, the commission noted the increase in the number of whale stocks, especially of **Antarctic minke whales** and of **Antarctic blue whales**, but has not undertaken decisive action to life the moratorium with regard to these stocks. The commission similarly noted its intention to reduce the anthropogenic mortality of North Atlantic whales to zero.[21]

Protection of Seals

The protection of seals is the most important issue in the mid 80's because the decision of the Behring Sea Arbitrators,[22] rendered at Paris on the fifteenth day of August, marked the close of one of the most important international controversies. In this case the original subject of complaint was the seizure by the United States of three Canadian sealing vessels in Behring Sea, in the summer of 1886, for the alleged violation of an act of Congress which forbade the taking of fur seals in Alaskan waters. The British Foreign Office at once protested that these seizures were illegal because they had been made in the open sea at a greater distance from land than three miles and therefore outside of the territorial jurisdiction of the United States. A prolonged diplomatic controversy between the two governments ensued. The Behring Sea decision also furnishes one more precedent in support of a proposition of international law which was laid down nearly a century ago by

---

[17]*Supra* note 13 p.337

[18]Ibid

[19]Ibid

[20]www.iwcoffice.org

[21]*Supra* note 19

[22]RUSSELL DUANE, *"The Decision of the Behring Sea Arbitrators,"* THE AMERICAN LAW REGISTER AND REVIEW, Vol. 41, No. 10, (First Series) Volume 32 (Second Series, Volume 6) (Oct., 1893), pp. 901-921 Published by: The University of Pennsylvania Law Review

Lord Stowell in the celebrated case of 'Le Louis,' and which has since been affirmed by several text-writers,[23] viz: that the right of 'visitation and search' cannot lawfully be exercised upon the high seas by any nation in time of peace. Lastly, the Behring Sea Arbitration is likely to prove of immeasurable benefit to the world in that it has furnished one more instance of the peaceful settlement of a most serious international dispute by the submission of the questions at issue to an impartial judicial authority.

This case also noted in the history of protection of seals. The Behring sea controversy includes two questions. First question was that whether the suppression of unauthorized sealing in Behring Sea was necessary for the preservation of seal fishery? Second was that, if such suppression was necessary, whether the United States had the right to suppress it at points more than three miles from land? This dispute was solved by the international arbitration tribunal which affirmed the freedom of fishing in the high seas but, at the same time, favoured the adoption of regulatory measures for the protection of seals. The first treaty to regulate the management of seals was adopted in 1911 among the states of Japan, Russia, the United Kingdom, and the United States. The treaty banned pelagic sealing and established a certification system to prevent illegal trade in seal skins. In case of Antarctica the commercial sealing is prohibited. A convention was adopted in 1972 to regulate commercial sealing in the Antarctica amid fears that the resumption of commercial sealing was eminent. In addition to the Antarctica seals convention, seals in the Antarctica are protected by the protocol to the Antarctica Treaty on Environmental protection.[24]

Bird Extinction

Bird species all over the world are becoming extinct far more rapidly than earlier. As many as 12 percent of all existing bird species and about 1,250 could disappear by the end of this century, according to a study published in the journal proceeding of the National Academy of Sciences on July 4, 2006.[25] The study, conducted by researchers from Stanford and Duke Universities and the Missouri Botanical Garden in St. Louis, is the most thorough and comprehensive analysis of global bird species ever conducted.

Prior to the finding of the new study, scientists had documented the extinction of about 130 bird species since the year 1500, but the authors of the study say it is more likely that 500 of the 10,000 known bird species become extinct during that period, about one extinction per year over the past 500 years. The new figures extinct bird species identified only from fossil records and others never officially declared extinct.

Who is responsible for majority of bird extinctions?

The rate of extinction among bird species is 100 times higher than what scientists considered natural before human beings began hunting birds for food

---

[23]*Supra* note, 13 pp.339-342

[24]Protocol on Environmental protection to the Antarctica Treaty

[25]LARRY WEST, "*Bird Extinctions Occurring Faster than Previously Believed 12 Percent of All Bird Species Could Be Extinct By End of 21st Century,*" see also, www.about. com Guide.

and sport, destroying bird habitat by clearing land for agriculture and other uses, and introducing disease and non-native species of birds, rats and snakes that either prey on native bird species or compete aggressively with them for food and habitat.[26]

"The extinctions all have to do with people in one way or another," said Peter Raven, president of the Missouri Botanical Garden, according to an article in the *San Francisco Chronicle*.[27] Raven, who has been studying plant and animal extinctions for 40 years, and his co-authors said habitat destruction from human activity and global warming is expected to be the leading cause of bird species extinction throughout the 21st century.

Species Threatened By the Global Warming

There are estimates among scientists that one million species are threatened with extinction by climate change. In a research work titled as, *Extinction Risk from Climate Change*,[28] it was concluded that from 15 to 37% of all the species in the regions studied could be driven to extinction by the climate changes between now and 2050. The study's lead author, Professor Chris Thomas, of the University of Leeds, UK, says: "If the projections can be extrapolated globally, and to other groups of land animals and plants, our analyses suggest that well over a million species could be threatened with extinction.

Global Warming's Effect on Whales and Seals

The report "Whales in hot water?" examines the impacts on cetaceans including.[29]Due to Changes in sea temperature there has been declining salinity because of the melting of ice and increased rainfall. Sea level rise and it causes loss of icy polar habitats. And decline of krill populations in key areas. Krill is a tiny shrimp-like marine animal that is dependent on sea ice and is the main source of food for many of the great whales.

Mark Simmonds, International Director of Science at WCDS, Said 'Whales, dolphins and porpoises have some capacity to adapt to their changing environment,' "But the climate is now changing at such a fast pace that it is unclear to what extent whales and dolphins will be able to adjust, and we believe many populations to be very vulnerable to predicted changes."[30]

Climate change impacts are currently greatest in the Arctic and the Antarctic. According to the report, cetaceans that rely on polar, icy waters for their habitat and food resources, such as belugas, narwhal, and bowhead whales are likely to be dramatically affected by the reduction of sea ice cover. As sea ice cover decreases there will be more human activities such as commercial shipping, oil,

---

[26]*Supra* note 24

[27]Ibid

[28]http://www.nature.com/nature/journal/v427/n6970/abs/nature02121.html

[29]Whales, dolphins and porpoises (cetaceans) are facing increasing threats from climate change, according to a new report published by WWF and the Whale and Dolphin Conservation Society, SCIENCE DAILY (May 23, 2007)

[30]See *Supra* note 29

gas and mining exploration and development, and military activities in previously untouched areas of the Arctic.

Trade and Species

Organisations such as WWF, founded in 1961 by Sir Peter Scott, the eminent naturalist, are highly effective in publicising the plight of endangered species world-wide.[31] They also play a large role in raising charitable funds towards projects concerned with saving wildlife in various areas of the globe. Many such conservation organisations pay for the basic resources needed by under-developed countries to enforce their laws. This can be as basic as providing a means of transport and salaries for enforcement officers. However, how effective these campaigns and projects are in the long run remains to be seen. Loss of habitat is still the most pressing problem.

In some areas, biodiversity is seriously threatened as a result of trade in endangered species. The international trade in wildlife is estimated to be worth £12 billion a year. Up to a quarter of that trade is almost certainly illegal. The main piece of legislation limiting trade in endangered species is CITES (the Convention on International Trade in Endangered Species). This is a UN convention which came into effect in 1975. CITES prohibits commercial trade in endangered species of plants and animals. Legitimate international trade in species which are not now threatened, but which may become so if trade is not controlled, is allowed via a permit system. Responsibility for implementing it lies with signatory nations.

Many of the problems involved in protecting habitats and species arise because local people either need to use the resources available in sensitive habitats to provide the necessities for subsistence or survival, or traditionally have always done so. UNESCO (United Nations Educational, Scientific and Cultural Organisation), through its 'Man and the Biosphere' programme, has set up a number of Internationally recognised biosphere reserves in an attempt to address this problem.[32]

Most of the initial treaties for the protection of biodiversity dealt with specific species and particular ecosystems. Most treaties also were regional treaties. The biodiversity convention adopted during the 1992 UNCED conference was the first attempt to address biodiversity as a global issue. Treaties on species may concern the protection of individual species for nature conservation purposes, or may have as their objective the organisation of the rational exploitation of species harvested by more than one country. As the purposes and mechanisms of such treaties are different, they will be considered separately in the sections dealing respectively with conservation and exploitation treaties

III: THE International Convention on Conservation of Species

Species and ecosystems are seldom neatly confined within national boundaries. Many species roam across the national borders and ocean are owned by none. Trade

---

[31]DE KLEMM, C. AND SHINE, C. "THE BIOLOGICAL DIVERSITY CONSERVATION AND THE LAW," (IUCN, Gland, Switzerland and Cambridge,UK, 1996),pp.111-118

[32]*Supra* note 31

in endangered species is international and pollution produced on one side of the world may wind up affecting regions on the other side of the globe. Biodiversity conservation is thus an international problem requiring international solutions.

International conservation organisations play an important role in the wide publicising of environmental information. IUCN was responsible for the idea of compiling lists of threatened species as a means of drawing attention to the plight of species faced with extinction. These lists became known as Red Data Books (RDBs). In these, species are placed into one of several categories which range from 'extinct' to 'vulnerable' or 'rare', depending on the degree of threat to their existence. The first internationally applicable RDB was published in 1996. The 'red' stands for 'danger' and the concept has since been adopted by many different countries, including Britain. RDBs point the way for government agencies charged with environmental protection, as well as for non-governmental organisations concerned about maintaining diversity.

Convention on International Trade in Endangered Species of Wild Fauna and Flora (Cites)

CITES (Washington Convention) governs international trade in endangered plants and animals. It Aim is to ensure that international trade in specimens of wild animals and plants does not threaten their survival.[33]It involves voluntary participation and was signed in Washington D.C. on March 3, 1973.CITES primary concern is the conservation of species. Its preamble lists the economic value among species' values, and the convention does not generally prohibit but merely strives to coordinate trade in species that may become endangered.

IUCN General Assembly of 1963 held in Nairobi called for the creation of an international convention to regulate export, transit and import of rare and threatened wild species or skins and trophies, the first draft of which was made in 1964.

In 1972, UN Conference on human Environment held in Stockholm adopted Action Plan for Human Environment. Plenipotentiary conference to be convened as soon as possible, under appropriate governmental or intergovernmental auspices, to prepare and adopt a convention on export, import and transit of certain species of wild animals and plants.

The *Plenipotentiary Conference to Conclude an International Convention on Trade in Certain Species of Wildlife,*[34]hosted by the United States of America in Washington D.C. from 12 February to 2 March 1973 which ultimately led to the adoption of CITES and entered into force on 1 July 1975 and Switzerland agreed to be the depositary State.

---

[33]HUTTON AND DICKINSON, *"Endangered Species Threatened Convention: The Past, Present and Future of CITES,"* London: Africa Resources Trust, 2000.

[34]International Plenipotentiary Conference to Conclude an International Convention on Trade in Certain Species of Wildlife: Convention on International Trade in Endangered Species of Wild Fauna and Flora Source: THE AMERICAN JOURNAL OF INTERNATIONAL LAW, Vol. 68, No. 1 (Jan., 1974), pp. 197-211(Published by: American Society of International Law)

A Critical Analysis

An inherent bias in favour of trade over protection and conservation of species seems to be exists. The Focus on trade at species level without addressing the issue of habitat loss is also another issue of concern. The most important lacunae of this convention are lack of eco-system approach to conservation of species. It seeks to prevent use which is unsustainable rather than promoting *"sustainable use"*, in resemblance with Convention on Biological Diversity. There is lack of effective enforcement by member nation's egregious shortage of personnel, inadequate training, and reliance on importer's certificate. Mostly dependent on signatories for determinations on whether trade in a given species is *"non-detrimental."*[35] It regulates and monitors trade in the manner of *"negative list"*. Soft law mechanisms & inadequate national laws therefore CITES is incapable of addressing Domestic Trade in listed species. There are mainly two mechanisms for the species conservation treaties i.e technique and implementation of treaties.[36]

The Technique of Species conservation Treaties

Most treaties dealing with the protection of species impose obligations on contracting parties as to the means of such protection. Species are generally listed in appendices, a separate appendix being used for each different legal category *e.g.* fully protected and partly protected attributed to the species in question.

The implementation of Species conservation Treaties

Nature conservation treaties are law making treaties setting forth rules of general application, unlike the contract making treaties such as trade agreements, which contain reciprocal obligations. Whereas non compliance by one of the parties to a contact treaty may be sanctioned by the withdrawal of an advantage conceded by another party and no such sanctions exist under law making treaties. If one party allows a protected species to be destroyed, other parties cannot obtain satisfaction by doing the same thing. Most of conventions do not have effective retaliatory measures under the treaty, where retaliation is attempted by other means; this raises difficult issues and may even be of questionable legality, particularly with reference to the GATT. In recent and well-publicised case, the United States imposed restrictions upon imports of tuna from Mexico,[37] in a move intended to stop the incidental taking and drowning of dolphins in the nets of Mexican tuna fishing boats. Mexico complained to GATT on the ground that this constituted an un-permissible trade barrier and this position was upheld by the GATT panel to which the case was submitted. If trade restrictions in the interest of the conservation of biological diversity are to be made possible, the GATT rules would have to be amended accordingly.[38]

---

[35]See, Convention on International Trade in Endangered Species of Wild Fauna and Flora, Articles *III, IV & V.*
[36]*Supra* note 33 pp. 44-45
[37]*Mexico etc. V. U.S.,(* tuna dolphin case)
[38]*Supra* note 31

Regional Treaties and Other Instruments

The regional treaties has laid the foundations of the international protection of species like in **Africa** with respect to faunal resources, the African convention on the conservation of Nature and Natural resources of 1968 requires contracting states to ensure their conservation, wise use and development with the framework of land-use planning and economic and social development.[39] Outside protected areas, contracting states shall manage exploitable wildlife populations for an optimum sustainable yield. There are so many substantive provisions on protected species, hunting and protected areas, which broadly reproduce those of the London convention of 1993.[40]

In **America** the only treaties in force are the Western Hemisphere convention of 1940 and convection for the conservation of Biological Diversity and the protection of priority Wild Areas in Central America of 1992. The latter contains general provisions for the protection of species, which include the obligation for each party to encourage the development of national legislation for the conservation and sustainable use of the components of biological diversity; to promote species recovery plans; to establish machinery to strengthen controls on illegal traffic in specimens of wild fauna and flora between the countries of the region; to control the collection of biological resources in natural habitats; and to regulate domestic trade in such resources by national legislation.

Europe is much developed in case of convention and the important convention are as follows; The Berne convention of 1979, European community legislation and The Alpine convention etc.[41]

So far as **Asia** is concern the only treaty covering the protection of species in Asia is the ASEAN Agreement on the conservation of Nature and Natural Resources of 1985. Parties are required to give special protection to threatened and endemic species and to preserve those areas which constitute the critical habitats of endangered or rare species, of species that are endemic to a small area and of migratory species. In addition, an Appendix listing endangered species deserving special attention will be adopted by a meeting of the contracting parties. The taking of and trade in these species is prohibited.[42]

V: Comparative Study

International development with regard to conservation of species and ecosystem has shown the importance of environmental protection. However, there are many developing countries like china doing better in the field of environment. As far as developed countries by using science and technology are much better position as compare to developing countries.

China

China has promulgated a series of laws and regulations concerning biodiversity conservation, especially for the protection of wild animals and

[39]See, *Supra* note 31 pp.29-43
[40]*Supra* note 13 pp. 323-334
[41]Ibid
[42]Ibid

other natural resources. The enforcement of the statutes, has led to progress for biodiversity conservation. However, some gaps in legislation still exist. Based on the present status of conservation legislation in China, and in accordance with the Convention on Biological Diversity, there is a need to perfect and enhance conservation legislation at ecosystem, species and genetic levels of biodiversity with more attention to conservation of genetic resources, wild plant species, and some natural ecosystems.[43]

Moreover, Effective biodiversity conservation needs sound legal systems that include legislation at international, national and local levels. The Convention on Biological Diversity (CBD) that entered into force on 29 Dec. 1993 is an international legal instrument. As a frame convention, it has formed a legal system, together with the other inter-national Agreements such as the Convention on Wetlands of International Importance Especially as Waterfowl Habitat, Ramsar, 1971; the Convention Concerning the Protection of the World Cultural and Natural Heritage, Paris, 1972; the Convention on International Trade in Endangered Species of Wild Fauna and Flora, Washington, 1973; the Convention on the Conservation of Migratory Species of Wild Animals, Bonn, 1979.[44]

Although the effective implementation of the international legal agreements is largely dependent on the actions taken by the parties to the Agreement, but the most importantly is establishment of legal systems for biodiversity conservation at the state level. Each country needs a legal system with a series of statutes to conserve natural ecosystems, wild species, and genetic resources. Since the late of 1 970s, the Chinese government has paid more attention to protecting natural resources and the environment.

Legislation in China

Legislation has greatly promoted the conservation of natural ecosystems. Based on the variability and large differences in the landscape, Nature Reserves of eco-system category can be sorted into 5 types, i.e. (i) forest ecosystems; (ii) grassland ecosystems; (iii) desert ecosystems; (iv) terrestrial wetland and water ecosystems; (v) ocean and coastal ecosystems (1). By the end of 1993, China had set up a total of 433 nature reserves oriented to the protection of various natural ecosystems. Existing Legislation for Wild Species Conservation the Wild Animal Conservation Law issued in 1988 is the first special law to protect wildlife in Chinese history.[45]

United Kingdom

Several international conventions exist for the preservation of biodiversity. These include such conventions as the Ramsar Convention (1976) which provides for the conservation of internationally important wetlands and the Bern Convention (1979) which requires the protection of endangered and vulnerable species of flora

---

[43]XUE DAYUAN,"*The Legislation, Enforcement, and Further Needs for Biodiversity Conservation in China,*"AMBIO, Vol. 27, No. 6 (Sep., 1998), pp. 489-491 (Published by: Allen Press on behalf of Royal Swedish Academy of Sciences)
[44]Ibid
[45]See *Supra* note 41

and fauna in Europe and their habitats.[46] There are many others. Signatory nations to these conventions must ratify national laws to ensure compliance with the conventions.

In Britain, the main piece of legislation covering conservation is the Wildlife and Countryside Act 1981 and 1985, which implements preceding EU conventions. It protects both species and sites of UK importance. Enforcement of conservation directives is the responsibility of the Environment Agency, a government organisation. English Nature, a government funded watchdog, is also responsible for the promotion of the conservation of England's wildlife.

In addition to the enforcement of laws, the Environment Agency is also responsible for data collection and monitoring. Environmental monitoring and biodiversity surveys are important because they provide information on the condition of ecosystems and the changes that are taking place within them. They therefore provide the scientific information on which to base environmental policy decisions. Similarly, assessments of the environmental impact of large development projects are vital before relevant authorities can either grant permission to proceed, or require that changes be made to development designs.

Despite that voluntary organizations both in UK and other dependencies, coordinated by the UK Dependent territories conservation forum, are providing much of the driving force for biodiversity research and conservation action but are limited in their capacity to undertake research. There is a need for further collaborative research in many subject areas, including those indicated, and for researchers to ensure that their studies are planned to include local consultation and to ensure that their results are made available to local interests.

Legislation in the United Kingdom

The legislation in the United Kingdom provides for the protection of certain species of wild plants, birds and animals at all times; some species of bird are protected at certain times of the year only, while certain methods of taking or killing wild animals and birds are prohibited.[47]

The legislative provisions in Great Britain for the protection of wild animals are contained primarily in the Wildlife and Countryside Act, 1981, Sections 9-12, the wild animals which are protected are listed in Schedules 5-7 of the Act and the provisions for the granting of licenses and enforcement are set out in Sections 16-27.

In **England** and Wales, enforcement provisions were extended and some amendments for protection made by the Countryside Rights of Access Act 2000 (CROW ACT) Section 81 and Schedule 12.

In **Scotland**, enforcement provisions were extended and some amendments for protection made by Section 50 and Schedule 6 of the Nature Conservation (Scotland) Act 2004. Specific legislation for protecting badgers is provided by the

---

[46]DIRK S. SCHMELLER, *"European species and habitat monitoring: where are we now?"* Published online: 31 October 2008 © Springer Science+Business Media B.V. 2008

[47]http://www.jncc.gov.uk/page-1747

Badgers Act, 1992 (amended, for Scotland, by the Nature Conservation (Scotland) Act 2004), for wild deer in the Deer Act, 1963 (Amended By The Deer Act, 1991), and for seals in the Conservation of Seals Act 1970. The close season for seals in (some areas of) Scotland was extended by the Conservation of Seals (Scotland) Order 2002.

In **Northern Ireland,** the legal provisions are set out in the Wildlife (Northern Ireland) Order, 1985 (amended 1995), Articles 10-13 and 16-29, and the protected wild animals are listed on Schedules 5-7.

The protection of European animal species in Great Britain is covered by the Conservation (Natural Habitats, & c.) Regulations, 1994, Part II, Regulations 38-41 and Schedules 2-3 and in Northern Ireland provisions for European species are laid down in the Conservation (Natural Habitats, etc.) Regulations (NI) 1995, Part II, Regulations 33-36 and Schedules 2-3. Whaling in UK waters is prohibited by the Whaling Industry (Regulations) Act 1934, as amended by the Fishing Limits Act, 1981.

United States of America

After several amendment to the Endangered Species Conservation Act of 1969, the Endangered Species Act 1973 (ESA) come into existence. It is perhaps the most far-ranging and important federal legislation specifically concerned with the preservation of wildlife. The Act establishes a scheme under which any species which is deemed 'endangered' or 'threatened' is protected from various private and governmental activities which will or may jeopardize its continued existence. The Act is aimed at protecting endangered and threatened species and the ecosystems on which they depend.

The interesting thing is that CITES is implemented in the US through the ESA. A "species "is considered endangered if it is in danger of extinction throughout all or a significant portion of its range. A species is considered threatened if it is likely to become an endangered species within the foreseeable future.

There are approximately 1,895 total species listed under the ESA. The listing is done under Article 4 of the Act.[48]The Act requires the Federal government to designate "critical habitat" for any species it lists under the ESA. Section 4(f) of the Act directs NOAA's (National Oceanic and Atmospheric Administration), National Marine Fisheries Service (NMFS) to develop and implement recovery plans for threatened and endangered species, unless such a plan would not promote conservation of the species. Section 6 of the ESA provides a mechanism for cooperation between NOAA's National Marine Fisheries Service (NMFS) and States in the conservation of threatened, endangered, and candidate species. Federal agencies are directed, under section 7(a) (1) of the Act to utilize their authorities to carry out programs for the conservation of threatened and endangered species. Thus, Federal agencies must consult with NMFS, under section 7(a) (2) of the ESA, on activities that may affect a listed species. When non-Federal entities such as states, counties, local governments, and private landowners wish to conduct an otherwise lawful activity that might incidentally, but not intentionally, 'take'

---
[48]The Endangered Species Act 1973

a listed species, an incidental take permit[49] must first be obtained from NOAA Fisheries. To receive a permit, the applicant must submit a Conservation Plan (CP) that meets the criteria included in the ESA and its implementing regulations. Because many CPs are habitat-based, the term Habitat Conservation Plan (HCP) is often used interchangeably with CP.

Since passage of the Endangered Species Act in 1973, over 1,300 endangered and threatened species have been protected in the United States and its territories. In the case of New Jersey appellate court regarding alleged 'endangered species.'[50] Rejecting a plea from the New Jersey Builders Association, the court ruled, in essence, that it was more important to save the Pine Barrens tree frog or the bog turtle than to provide housing for those who live in the most densely populated state in the Union.

India

India is a Party to the Convention on Biological Diversity, 1992. This convention by recognizing the sovereign rights of States to use their own biological resources, the Convention expects the parties to facilitate access to genetic resources by other Parties subject to national legislation and on mutually agreed upon terms.[51]

India is one of the developing countries who are the party to the convention on Biological diversity. With only 2.5% of the land area, India already accounts for 7.8% of the global recorded species. India is also rich in traditional and indigenous knowledge, both coded and informal. India is also a signatory of Convention on International Trade in Endangered Species of Flora and Fauna since 1976.Despite of this fact, in India the illegal trade includes diverse products including mongoose hair; snake skins; Rhino horn; Tiger and Leopard claws, bones, skins, whiskers; Elephant tusks; deer antlers; shahtoosh shawl; turtle shells; musk pods; bear bile; medicinal plants; timber and caged birds such as parakeets, mynas, munias etc.

Legislation in India

These are the following **Legislation in India** which plays a vital role in protection of species and ecosystem.

- The Wild Bird Protection Act, 1887
- The Wild Bird and Animal Protection Act, 1912
- Indian Forest Act, 1927
- The Prevention of Cruelty to Animals Act, 1960
- The Wildlife Protection Act, 1972
- CITES, 1975 (its regulations and notifications)

---

[49]*Supra* note 42 Section 10(a)(1)(B)
[50]ALAN CARUBA, *"The Endangered Species Act: An Environmental Law That Deserves Extinction, The Progressive Conservative's,* U.S.A AN ONLINE JOURNAL OF POLITICAL COMMENTARY & ANALYSIS, Volume VI, Issue 8, January 12, 2004
[51]Article 3 And 15 Of Convention On Biological Diversity,1992

- The Forest (Conservation) Act, 1981.
- The Environment Protection Act, 1986

After an extensive and intensive consultation the Central Government has brought **Biological Diversity Act, 2002** with the following salient features:-

- to regulate access to biological resources of the country with the purpose of securing equitable share in benefits arising out of the use of biological resources;
- and associated knowledge relating to biological resources;
- to conserve and sustainably use biological diversity;
- to respect and protect knowledge of local communities related to biodiversity;
- to secure sharing of benefits with local people as conservers of biological resources and holders of knowledge and information relating to the use of biological resources;
- conservation and development of areas of importance from the standpoint of biological diversity by declaring them as biological diversity heritage sites;
- protection and rehabilitation of threatened species;
- Involvement of institutions of state governments in the broad scheme of the implementation of the Biological Diversity Act through constitution of committees.

VI: Judicial Decisions

There are some of the most popular cases under international environmental law which requires attention to understand the future of environmental law. The following few cases are very important from the point of conservation of species under environmental law.

Tuna Dolphin case

The United States banned imports of Mexican tuna because Mexico had not taken steps to reduce the number of Eastern Pacific Tropical dolphins killed each year due to tuna fishing. Mexico appealed the case to the General Agreement on Tariffs and Trade (GATT), where the panel ruled in favour of Mexico. The ruling was in part due to the discriminatory manner in which the United States implemented the measure and in part due to the GATT resistance to cases where the process of production is a major factor. Furthermore, the panel found that the U.S. labelling of "Dolphin Free" tuna did not conform to GATT standards. The case was, however, solved bilaterally between the United States and Mexico.

Rapanos v. United States, 547 U.S. 715 (2006),

This is the United States Supreme Court case challenging the Clean Water Act. It was the first major environmental case heard by the newly appointed Chief Justice, John Roberts and Associate Justice, Samuel Alito. The Supreme Court heard the case on February 21, 2006 and issued a decision on June 19, 2006.

While five justices agreed to void rulings against the plaintiffs, who wanted to fill their wetlands to build a shopping mall and condos, the court was split over further details, with the four more conservative justices arguing in favour of a more restrictive reading of the term "navigable waters "than the four more liberal justices.

There has been much discussion, mostly in the way of Congressional Committee testimony, discussing the plurality nature of the *Rapanos* decision. To summarize, four justices wanted to limit EPA, and therefore federal jurisdiction, under the previously described Scalia standard. One justice, Kennedy, wanted a determination of federal jurisdiction to be based on *ecological* connections, the described "substantial nexus" test in cases of water bodies that are not traditionally considered "navigable." Finally, four dissenters wanted to allow substantial deference to the administrative agency's determination of federal jurisdiction. In light of this distribution, some commentators have suggested the Kennedy decision is more restrictive than the dissenters, and therefore a step-back in the extension of federal jurisdiction to non-navigable water bodies.

National wildlife federation v. Gale Norton, secretary, U.S. Department of the Interior, NO. CIV.A. 03-1393 (JR). AUG. 20, 2004.

Brief facts of the case

In this case three conservation groups, the National Wildlife Federation, the Florida Wildlife Federation, and the Florida Panther Society, challenge the U.S. Army Corps of Engineers' (Corps) issuance to Florida Rock Industries of a Clean Water Act 'dredge and fill' permit for the operation of a limestone mine on a 6000 acre site near Ft. Myers, Florida. Plaintiffs assert that mining operations on that site will unacceptably reduce the habitat of the endangered Florida panther. They have sued the Corps of Engineers and the U.S. Fish and Wildlife Service (FWS), which issued the Biological Opinion upon which the permit is based, alleging violations of the Endangered Species Act, the National Environmental Policy Act, the Clean Water Act, and the Administrative Procedure Act.

Issues Discussed In the Case

The Endangered Species Act is the most comprehensive legislation for the preservation of endangered species ever enacted by any nation.[52] Congress enacted the ESA to provide a means whereby the ecosystems upon which endangered species and threatened species depend may be conserved, and to provide a programme for the conservation of such endangered species.

The plain intent of Congress in enacting the statute was to halt and reverse the trend toward species extinction, whatever the cost. The ESA requires the Secretary to protect "species" defined to include any subspecies of fish, wildlife, or plants, and any distinct population segment of any species of vertebrate fish or wildlife.

An endangered species is in danger of extinction throughout all or a significant portion of its range. A threatened species is a likely to become an endangered species within the foreseeable future throughout all or a significant portion of its range.

---

[52]*Tenn. Valley Auth. vs. Hill,* 437 U.S. 153, 180 (1978)

The Secretary is charged with determining whether a species should be listed as threatened or endangered based upon five statutorily prescribed factors. (collectively referred to as listing factors). Each factor is equally important and a finding by the Secretary that a species is negatively affected by just one of the factors warrants a nondiscretionary listing as either endangered or threatened.

The same five factors are used to determined whether threats to the species have been diminished or removed to the point that down listing or delisting is appropriate. The FWS shall make listing determinations solely on the basis of the best scientific and commercial data available.

The ESA makes it unlawful for any person to take any endangered species within the United States. The term "take" means to kill, harass, hunt, wound, trap, capture, collect, or harm a species. Under ESA section 4(d), the (Fish and Wildlife Service) FWS can adopt rules that allow the taking of threatened species under certain circumstances. To fulfill its goals of species preservation, the ESA requires the Secretary to develop and implement recovery plans under its duty to conserve.

Although the Act does not define "recovery," FWS has essentially defined the term to mean conservation, the use of all methods and procedures which are necessary to bring any endangered species or threatened species to the point at which the measures provided pursuant to this Act are no longer necessary.

Decision

The US court after seeing all the facts and figures come to the conclusion that the rule made by the Secretary for enlisting the Species were inconsistent with the existing laws. And no such authority can go against the ESA and therefore dismisses the arguments of the defendant and affirmed the motion the plaintiff.

VII: Conclusion

Since the Conservation of species and biological diversity both are very much important in the study of ecosystem. The importance of conservation of these species along with the ecosystem has been increasing consistently due to the devastating impact of the advancements in science and technology. The relation of these species with the ecosystem can never be ignored because human beings are depending on them. However, the man-made Conventions are trying to solve the environmental issues by imposing liability on the polluters through the *"polluter pays principle"*. There has not been much success in the field of conservation of species.

Humans are responsible for the bird's extinction and today the wildlife and other species are in danger because of unprecedented activities of human life. Moreover, the international environmental law focused on conservation of species due to the Convention on International Trade in Endangered Species of Wild Fauna and Flora (CITES). The most important lacunae of this convention are lack of eco-system approach to conservation of species. It seeks to prevent use which is unsustainable rather than promoting *"sustainable use"*, in resemblance with Convention on Biological Diversity. There is lack of effective enforcement by member Nations, inadequate training, and reliance on importer's certificate. As far as regional measures are concern every state is trying to do the best for the conservation of species and ecosystem. The judicial decision of USA and other countries shows the importance of the environmental issues and given priority to the environment than the trade and business. The WTO has funding to the most of international organisation for the protection of environment. But it is not sure that whether this funding really for the protection of environment? As a result the importance of conservation of species has been increased and international law giving priority to the environment but there is lacunae in some of Convention and regional treaties. In this way even though there are sufficient legislation for the conservation of species it requires support from all over the world and proper implementation of the existing laws.

References

1. Bird Extinctions Occurring Faster than Previously Believed 12 Percent of All Bird Species Could Be Extinct By End of 21st Century," see also, www. about.com Guide.

2. Caruba, A. (2004). *"The Endangered Species Act: An Environmental Law That Deserves Extinction, The Progressive Conservative's,* U.S.A AN ONLINE JOURNAL OF POLITICAL COMMENTARY & ANALYSIS. 6(8).

3. Corker, C. (1996). Biodiversity and conservation, Off well Woodland and Wildlife Trust, see also www.countrysideinfo.co.uk

4. Dayuan, X. (1998). *"The Legislation, Enforcement, and Further Needs for Biodiversity Conservation in China,"*AMBIO. (Published by: Allen Press on behalf of Royal Swedish Academy of Sciences). 27(6): 489-491.

5. Deklemm, C. and Shine, C. "THE BIOLOGICAL DIVERSITY CONSERVATION AND THE LAW," (IUCN, Gland, Switzerland and Cambridge,UK. 111-118.

6. Duane, R. (1893). *"The Decision of the Behring Sea Arbitrators,"* THE AMERICAN LAW REGISTER AND REVIEW, Vol. 41, No. 10, (First Series) Volume 32 (Second Series, Volume 6) (Oct., 1893), pp. 901-921 Published by: The University of Pennsylvania Law Review.

7. ELLI LOUKA, INTERNATIONAL ENVIRONMENTAL LAW-FAIRNESS, EFFECTIVENESS AND WORLD ORDER, (New York: Cambridge University Press 2006, 1st edn.). 335.

8. Gray, J. (2008). "Indigenous Communities and Biodiversity conservation: Protected areas and the right to consultation" *GONZAGA JOURNAL OF INTERNATIONAL LAW.* 2(2).

9. HUTTON AND DICKINSON, *"Endangered Species Threatened Convention: The Past, Present and Future of CITES,"* London: Africa Resources Trust, 2000.

10. International Plenipotentiary Conference to Conclude an International Convention on Trade in Certain Species of Wildlife: Convention on International Trade in Endangered Species of Wild Fauna and Flora Source: THE AMERICAN JOURNAL OF INTERNATIONAL LAW, Vol. 68, No. 1 (Jan., 1974), pp. 197-211(Published by: American Society of International Law).

11. *Mexico etc. V. U.S.,*(tuna dolphin case).

12. Nath, B., Hens, L., Compton, P. and Devuyst, D. (2002). "Environmental management in practice" (New York: Routledge Publication, Reprinted). 4-5.

13. *National wildlife federation v. Gale Norton, secretary, U.S. Department of the Interior,* NO. CIV.A. 03-1393 (JR). AUG. 20, 2004.

14. Protocol on Environmental protection to the Antarctica Treaty.

15. *Rapanos v. United States.*

16. Schmeller, D.S. (2008).*"European species and habitat monitoring: where are we now?"* Published online: 31 October 2008 © Springer Science+Business Media B.V.

17. *Tenn. Valley Auth. vs. Hill,* 437 U.S. 153, 180 (1978).

18. The Convention on Biological Diversity, 1992.

19. The Convention on International Trade in Endangered Species of Wild Fauna and Flora, Articles *III, IV & V.*

20. The Convention on the conservation of Migratory species of Wild Animals (Secretariat of the UNEP, Jan 2002).

21. The Convention on the Conservation of Migratory Species of Wild Animals Article III Endangered Migratory Species: Appendix I.

22. The Endangered Species Act 1973.

23. Whales, dolphins and porpoises (cetaceans) are facing increasing threats from climate change, according to a new report published by WWF and the Whale and Dolphin Conservation Society, SCIENCE DAILY (May 23, 2007).

Biodiversity Conservation and International Conventions

Nair Bindu Vijay

Gujarat National Law University,
Koba, Gandhinagar-382007, Gujarat, India
E-mail: bvijay@gnlu.ac.in

Abstract

Biodiversity is the diverse forms of life. It includes not only the various species but also various genes contained in those species. There are several definitions for biodiversity. However the most accepted definition is the one given by CBD. Biodiversity is considered as an important valuable resource and a lot of effort is being made each day to preserve and conserve this valuable resource as it is the sole resource which can cater human needs viz; food feed fiber and farma. Besides this biological diversity plays a major role in poverty reduction and ensuring food security. This paper aims to explore these important uses, the need for conservation and the various biodiversity related conventions.

Key Words: Biodiversity, Biodiversity hot spots, goods and services, CBD, traditional knowledge.

Introduction

The planet Earth houses millions of varied kind of species. Out of these human beings are considered to be the most intelligent species that is supported by these million species. These various species is unique in it and existence of each of them is essential for the survival of other species. This varied species is what we call "biodiversity" or "biological diversity". Bio means life and diversity means varied kinds of species. These diverse kinds of species are existing in earth since its origin. However it's only since past 25 years or so that the word biodiversity has gained global attention and importance.. There was a time when biodiversity was considered as a common heritage. In the recent years we see that biodiversity is being discussed, debated in academics, conferences and in policy issues. As a result several definitions are used. However the most accepted definition internationally is the one given by Convention on Biological Diversity. According to CBD "Biological diversity" is the variability among living organisms from all sources including, inter alia, terrestrial, marine and other aquatic ecosystems and the ecological complexes of which they are a part; this includes diversity within species, between species and of ecosystems.

Thus conservation of ecosystem ensures conservation of biodiversity which in turn ensures the productivity of ecosystem, which is why biological diversity is rightly termed nature's basket of goods and services by CBD. In spite of its economic importance recent years have witnessed the significant loss of biodiversity. This loss is two-three times higher than the loss happened in the geological past. This loss is a concern for the world at large as the loss of these recourses would ultimately affect the very existence of humans.

This chapter aim to explore these important uses, the need for conservation and the various biodiversity related conventions.

Importance of Biological diversity

Around the word the rural communities, totally depend on the biological resources for their food, medicine, feed etc. in fact they even make their livelihood from these resources. For them these source of income and they even worship them. Indians specially in rural areas worship the forest. For instance, the Bishnois tribe of Western Rajasthan worship nature conserve trees and medicinal plants; provide food and water to animals[1]. Many households in developing countries, especially in Asia, derive as much as 50–80 per cent of annual household income from non-timber forest products[2].

Biodiversity is considered as a valuable asset as it is natures' basket of goods and services. Mankind depends on biological resources for basic needs and beyond. It provides us with food fiber, medicines.According to CBD secretariat 'goods and services' provided by ecosystem include: Provision for food, fuel, and fibre, Provision of shelter and building materials, Purification of air water, detoxification and decomposition of wastes, stabilization and moderation of earth's climate, Moderation of floods, droughts, temperature extremes and forces of wind. Generation and renewal of soil fertility, including nutrient cycling, pollination of plants, including many crops, control of pests and diseases Maintenance of genetic resources,as a key inputs to crop varieties and livestock breeds, medicines, and other products, Cultural and aesthetic benefits, ability to adapt to change. This resource is especially valuable to rural poor because it's their source of food and livelihood. Thus biodiversity is important for the achievement of the Millennium Development Goals. Thus biological diversity is a valuable resource without which human life can come to stand still. Globalization and technological advancement especially modern biotechnology have made these bio-resources a more valuable asset. Interestingly it's these modern biotechnology which gives values to un explored species.

---

[1]Draft National Biodiversity Action Plan, Government Of India, Ministry Of Environment & Forests August, 2007
[2]Secretariat of the Convention on Biological Diversity (2014) *Global Biodiversity Outlook 4.*

The term 'ecosystem services' was defined in the Millennium Ecosystem Assessment (MEA 2005) as 'the benefits people obtain from ecosystems', both natural and managed. These services may be categorized as provisional, regulative, cultural or supporting services, also referred to as supporting processes, The first three categories have a direct impact on human well-being, whereas the latter has an indirect impact by supporting provisional, regulative, and cultural services. However, all these services, whether direct or indirect, are essential for human life and the well-being of humans[3] (Costanza *et al.*, 1997; Daily *et al.*, 1997; Wall, 2004; MEA, 2003, 2005).

Biodiversity threats

Extinction is a natural event and, from a geological perspective a routine factor. We now know that most species that have ever lived have gone extinct. There have also been occasional episodes of mass extinction, when many taxa representing a wide array of life forms have gone extinct in the same blink of geological time[4].

Global fossil record shows 5 mass extinctions in the geological past are estimated to have caused the loss of 50% to 90% of all species. All these mass extinctions happened before humans appeared on earth, and these extinctions are the results of natural calamities and asteroid collisions and the sixth mass extinction is mainly due to human activities. However the sixth extinction which is already undergoing can be a major threat to human existence, and if reports are to believed then this could be more disastrous than the fifth mass extinction which is reported to be responsible for the extinction of Dinosaurs. Hence it is important to understand the reasons for the threat and adopt appropriate measure to conserve the rich biological diversity which indeed is required for the existence of mankind. Otherwise future generations will blame us for the loss. The bio-resources belong to all and we do not have any right to harm it.

All most all the recorded data shows that biodiversity is threatened due to human activities. These include; habitat destruction/ fragmentation/loss, invasive alien species, pollution, and overexploitation of resources. Adding to these are climate change and global warming. As mentioned earlier the modern biotechnology has made biodiversity a valuable asset. This indirectly is a threat to biodiversity as this leads to bio piracy of valuable medicinal plants.

India's status

India, is a mega diverse country with only 2.4% of the world's land area, harbours 7-8% of all recorded species, including over 45,000 species of plants and 91,000 species of animals. Of the 34 global biodiversity hotspots, four are present in India, represented by the Himalaya, the Western Ghats, the North-east, and the Nicobar Islands[5]. India ranks among the top ten species-rich nations and shows high endemism. India is also rich in traditional and indigenous knowledge. For

[3]Wall, D. H. & Nielsen, U. N. (2012) Biodiversity and Ecosystem Services: Is It the Same Below Ground? Nature Education Knowledge 3(12):8

[4]www.Globalchange.Umich.Edu/Globalchange2/Current/Lectures/ Biodiversity/Biodiversity.html

[5]India's Fifth National Report To The Convention On Biological Diversity 2014

many diseases like jaundice people still prefer going to traditional healer. However these rich biological resources are under serious threat.

Sacred groves (India has over 19,000 sacred groves) are also getting eroded or getting converted to plantations.[6]

The demands of a growing human population for food, medicine, shelter and fuel, along with the need for economic development, are exerting the pressure on biodiversity and ecosystems. As a result large area of land area is cleared for various developmental projects. This lead to habitat loss and degradation causing threats to grass land depended species.

Mining and quarrying cause habitat loss and degradation, with severe consequences for the ecology of areas such as the Aravalli Range and the Western Ghats[7] (MoEF, 2013; Pillay *et al.*, 2011).

Next to habitat loss invasive alien species is considered as a major threat to biological diversity. They are spreading through both deliberate and accidental introductions as a result of increasing levels of global travel and trade[8]. Invasive alien species ranges from unicellular microbes to multi cellular plants and animals. India has an estimated 18,000 plants, 30 mammals, 4 birds, 300 freshwater fishes and 1100 arthropods that are invasive (Ali & Pelkey, 2013).

Biodiversity related convention

There are five major conventions which directly or indirectly deal with biodiversity conservation. These include: Convention on Biological Diversity (CBD), Convention on International Trade in Wild Species of Endangered Flora and Fauna (CITES), Ramsar Convention on Wetlands, World Heritage Convention, and the Convention on Conservation of Migratory Species[9].

The Convention on International Trade In Endangered Species Of Wild Flora And Fauna, 1975 (CITES)

One of the earliest global environmental instruments was the Convention on International Trade in Endangered Species of Wild Flora and Fauna (CITES), conceptualized in the 1960's and signed in 1973. Its aim is to ensure that international trade in specimens of wild animals and plants does not threaten their survival. Because the trade in wild animals and plants crosses borders between countries, the effort to regulate it requires international cooperation to safeguard certain species from over-exploitation. Today, it accords varying degrees of protection to more than 35,000 species of animals and plants, whether they are traded as live specimens, fur coats or dried herbs. CITES was drafted as a result of a resolution adopted in 1963 at a meeting of members of IUCN (The World Conservation Union) CITES now counts over 181 member states and India is signatory to CITES since 1976[10].

---

[6]Draft National Biodiversity Action Plan, Government Of India, Ministry Of Environment & Forests August, 2007

[7]ndia's Fifth National Report To The Convention On Biological Diversity 2014

[8]

[9]www.theebi.org/pdfs/conventions.pdf

[10]https://www.cites.org/

The Ramsar Convention, 1971

The Ramsar Convention – formally known as the Convention on Wetlands of International Importance especially as Waterfowl Habitat – provides a framework for national action and international cooperation for the conservation and wise use of wetlands and their resources. The Convention defines wetlands as "areas of marsh, fen, peatland or water, whether natural or artificial, permanent or temporary, with water that is static or flowing, fresh, brackish or salt, including areas of marine water the depth of which at low tide does not exceed 6 m." The Ramsar Convention is the only global convention dealing with a particular type of habitat. India is a signatory to the Ramsar Convention.. In India, only 26 sites are notified under the convention [11]

The World Heritage Convention 1972

The most significant feature of the 1972 World Heritage Convention is that it links together in a single document the concepts of nature conservation and the preservation of cultural properties. The Convention recognizes the way in which people interact with nature, and the fundamental need to preserve the balance between the two[12].The United Nations Educational, Scientific and Cultural Organization (UNESCO) houses the Convention's Secretariat, while IUCN– The World Conservation Union, the International Council of Monuments and Sites (ICOMOS) and the International Centre for the Study of the Preservation and Restoration of Cultural Property (ICCROM) act as the advisory bodies for, respectively, natural properties, cultural properties and the study of the preservation and restoration of cultural property.[13] India ratified the convention on November 1977.

Convention on Migratory Species, 1979 (Bonn Convention)

As an environmental treaty under the aegis of the United Nations Environment Programme, the Convention on the Conservation of Migratory Species of Wild Animals (also known as the Bonn Convention) aims to "conserve terrestrial, marine and avian migratory species throughout their range." The Convention has two appendices: Appendix I lists migratory species that are classified as endangered and where urgent international cooperation is necessary to address the issue. Appendix II lists other species that require or would benefit significantly from international agreements under the Convention. The Convention entered into force in 1983, and as of 1 October 2015 the Convention on Migratory Species has 122 Parties. India is party to Bonn convention since 1983[14].

Convention on Biological Diversity, 1992 (CBD)

The Convention on Biological Diversity is one of the three 'Rio Conventions', emerging from the UN Conference on Environment and Development, also known as the Earth Summit. The three main goals of the Convention on Biological Diversity (CBD) are the conservation of biological diversity, the sustainable use of its components, and the fair and equitable sharing of the benefits arising from

[11]http://www.frontline.in/environment/wetlands-in-peril/article8017664.ece
[12]http://whc.unesco.org/en/convention/
[13]*www.theebi.org/pdfs/conventions.pdf*
[14]http://www.cms.int/en/legalinstrument/cms

utilization of genetic resources including by appropriate access to genetic resources and by appropriate transfer of relevant technologies, taking into account all rights over those resources and to technologies, and by appropriate funding." There are currently 196 parties to the convention.

Thus India has India has participated in major international events on environment and biodiversity conservation since 1972. Other international agreements with a bearing on biodiversity, to which India is a Party include: UNFCCC, UNCCD, Commission on Sustainable Development, World Trade Organisation, and FAO facilitated International Treaty on Plant Genetic Resources and UN law of the Seas[15].

A new platform, the Intergovernmental Science-Policy Platform on Biodiversity and Ecosystem Services (IPBES), was established by the international community in 2012 and is open to all United Nations member countries. It is an independent intergovernmental body committed to providing scientifically-sound assessments on the state of the planet's biodiversity in order to support informed decision-making on biodiversity and ecosystem services conservation and use around the world[16]

Conclusion

Biodiversity is a foundation of developed and developing economies. Without healthy biodiversity, livelihoods, ecosystem services, natural habitats, and food security can be severely affected. It is essential to comply with the available national and international legislations to preserve and conserve these rich biological resources, otherwise the future generation will blame us for we belong to the century that is responsible for the destruction of these resources.

References

1. Ali, R. and Pelkey, N. (2013). *Satellite images indicate vegetation degradation due to invasive herbivores in the Andaman Islands. Current Science.* 105: 209-214.
2. Costanza, R. (1997). *The value of the world's ecosystem services and natural capital. Nature.* **387:** 253-260.
3. Daily, G.C. (1997). *Ecosystem services: Benefits supplied to human societies by natural ecosystems. Issues in Ecology.* **2:** 1-18.
4. MEA. (2003). *Millennium Ecosystem Assessment: A Framework for Assessment.* Washington, DC: Island Press.
5. MEA. (2005). *Millennium Ecosystem Assessment: Ecosystems and Human Well-being: Synthesis.* Washington, DC: Island Press.
6. MoEF. (2013). *Report of the High Level Working Group on Western Ghats,* Volume I. Ministry of Environment and Forests, Government of India.
7. Pillay, R., Johnsingh, A.J.T., Raghunath, R. and Madhusudan, M.D. (2011). *Patterns of spatiotemporal change in large mammal distribution and abundance in the Southern Western Ghats, India. Biological Conservation.* 144: 1567-1576.
8. Wall, D.H. (2004). ed. *Sustaining Biodiversity and Ecosystem Services in Soils and Sediments.* SCOPE 64 Washington, DC: Island Press.

[15]Draft Biodiversity Action plan, GOI, 2007
[16]Intergovernmental Platform on Biodiversity & Ecosystem Services (2013) About IPBES

23

Legal Implications of Biodiversity: Associated Policies, Laws, IPR and Patent

Manali Datta and S.L. Kothari*

*Amity Institute of Biotechnology, Amity University Rajasthan,
Jaipur-302001, Rajasthan, India
E-mail: mdatta@jpr.amity.edu*

Abstract

Asia's second largest nation with an area of 3,287,263 square km, India is known for its rich biological diversity. Conservation international has recognized India as one of the 17 mega diverse countries. With such a huge assortment of diversity, conservation and sustainability of biodiversity has become a necessity for our subsistence as well as maintenance of this precious inheritance. Economic and aesthetic implications of biodiversity are so evident that international and national bodies across the globe have taken initiatives to enable its conservation and preservation. Many fundamental laws and intellectual property rights have been laid by the monitoring bodies to prevent the obliteration of worthy and valuable asset of world.

Key Words: Biodiversity, Laws, IPR, Patent

Introduction

Biodiversity is fundamental to addressing some of the world's greatest challenges such as climate change, sustainable development and food security. Asia's second largest nation with an area of 3,287,263 square km, India is known for its rich biological diversity (Hoseitti, 2013). Conservation international has recognized India as one of the 17 mega diverse countries. The country has been broadly divided into four major demarcations namely the Gangetic plain, Deccan plateau the Lakshawadeep islands and Himalayan region. The major regions are further subdivided into ten bio-geographic regions namely Trans Himalayan zone, Desert zone, semiarid zone, eastern and western ghats to name some (Mast et al., 1997). Out of the total number of 7,63,812 species found in world, 2,10,210 species are found in India alone [UNEP-WCMC), 2004]. It constitutes almost 27% of the biodiversity found throughout the world. Subcontinent has more than 300 wild predecessors and close relatives of cultivated plants still budding and evolving under natural conditions. Conservation of biodiversity has become mandatory for maintenance of stable ecosystems. Its sustainability is necessary for our subsistence as well as precious in its own right. This is because it provides the fundamental building blocks for the many goods and services which provides a healthy environment to lead our life. Modern scientists are delving into our natural biological resources for generation of effective therapeutics. Nearly 65,000 native plants are still used prominently in indigenous health care systems (Ehlrich and Wilson, 1991). Considering the rich distribution of flora and fauna, it becomes a social and legal responsibility to maintain the rich legacy of the Indian subcontinent.

In this chapter we present an overview of the policies, patents, laws and IPR associated with biodiversity.

Table 1: Comparison between the Number of Species in India and the World

Group	Number of species in India (SI)	Number of species in the world	SI/SW [SW%]
Mammals	422	4,629	7.6
Bir ds	1,180	9,702	12.6
Reptiles	521	6,550	6.2
Amphibians	233	4,522	4.4
Vascular Plants	18,664	2,50,000	6.0

Source: World Conservation Monitoring Centre of the United Nations Environment Programme (UNEP-WCMC), 2004. Species Data (unpublished, September 2004).

Monitoring Bodies

India is signatory to several major international conventions related to conservation and management of wildlife. Some of these are Convention on International Trade in Endangered Species of Wild Fauna and Flora (CITES), Convention on Biological Diversity, Convention on the Conservation of Migratory Species of Wild Animals etc. These conventions provide the State and Union Territory Governments financial and technical assistance for protection and Management of Protected Areas as well as other forests under various Centrally Sponsored Schemes.

International trade in species of conservation concern is monitored by CITES. CITES is an international agreement between governments. In 1963, resolution of CITES was drafted and adopted at a meeting of members of IUCN (The World Conservation Union). The document of the convention was agreed at a meeting of representatives of 80 countries in Washington, D.C., the United States of America, on 3 March 1973, and on 1 July 1975 CITES was enforced. India is a party in CITES since 1976 (4). Currently the number of parties adhering to the convention is 181 in number out of which 147 countries have accepted the Convention adopted at Bonn (Germany), 22 June 1979 and 100 countries are signatories for the Convention adopted at Gaborone (Botswana), on 30 April 1983 (www.cites.org).

Its aim is to ensure that international transactions pertaining to certain specimens of wild animals and plants are restricted so that it does not threaten their survival. The species are enlisted in three appendices, appendix I enlists species susceptible to extinction and only exceptional circumstances permits trade of these species, appendix II involves species not necessarily endangered with extinction, but in which trade must be controlled in order to avoid exploitation incompatible with their existence, Appendix III contains species protected in at least one country and mutual assistance has been requested from other CITES Parties in controlling the trade. Around 35,000 species from different countries are protected by CITES by various degrees of intensity (CITES Convention) (Table 1). Approximately 1300 species native to India has been enlisted and protected from exploitation under this convention.

Not all trade is legal of course: between 2005 and 2009 EU enforcement authorities made over 12,000 seizures of illegal wildlife products in the EU.

Measures are continuously been implemented to ensure that illegal trade is restricted; Trans Pacific Partnership (TPP) in agreement with twelve countries in the Asia-Pacific region has further strengthened rules for combating wildlife trafficking and ensuring legal and sustainable trade (Summary of the Trans-Pacific Partnership Agreement (http://www.international.gc.ca)

Table 2: Table below shows the approximate numbers of species that are included in the CITES (as of 2 October 2013) [Data obtained from https://www.cites.org/sites/default/files/eng/disc/species_02.10.2013.pdf]

	Appendix I	*Appendix II*	*Appendix III*
Mammalia	300 spp. (incl. 11 popns) + 23 sspp. (incl. 3 popns)	501 spp. (incl. 16 popns) + 7 sspp. (incl. 2 popns)	45 spp. + 10 spp.
Aves	154 spp. (incl. 2 popns) + 10 sspp.	1278 spp. (incl. 1 popn) + 3 sspp.	25 spp.
Amphibia	17 spp.	126 spp	3 spp.
Fishes	16 spp.	126 spp	3 spp.
Arthropoda	3 spp.	69 spp.	17 spp. + 3 spp.
Annelida		2 spp.	
Mollusca	60 spp. + 5 sspp.	14 spp. + 1 ssp.	
Cnidaria		2077 spp.	4 spp.
Flora	301 spp. + 4 sspp.	29592 spp	12 spp. + 1 var.

TPP also provides a new international platform for enhanced regional and global cooperation among national and international authorities, such as CITES (7).

Additionally, an agreement has been signed by World Animal Health Organization (OIE) and CITES in Geneva, Paris to work together on animal health and welfare issues worldwideto safeguard biodiversity and protect animals (www.cites.org; Krugman, 2015).

The Convention on Biological Diversity (CBD) materialized on 29th December 1993 and is a comprehensive, binding multilateral agreement signed by 196 contracting parties.

CBD has primarily three objectives:-

1. Protection of biological diversity

2. Sustainable use of the diversity

Ensuring fair sharing of benefits arising due to utilization of biological resources (www.nbaindia.org).

CBD has further specified two important supplementary treaties namely Cartagena protocol and Nagoya protocol. The Cartagena Protocol on Biosafety to the Convention on Biological Diversity is an international treaty governing the trade and transactions of genetically modified organisms (GMOs). It was proposed on 29 January 2000 as a supplementary agreement to the Convention on Biological Diversity and entered into force on 11 September 2003.

The Nagoya Protocol on Access to Genetic Resources and the Fair and Equitable Sharing of benefits arising from their utilization to the Convention on Biological Diversity is an international agreement enforced on 12 October 2014. It aims at sharing the benefits arising from the utilization of genetic resources in a fair and impartial manner. Cartagena protocol has been signed by 170 parties and Nagoya protocol has been signed by 68 countries.

The objectives of the conventions are implemented by National Biodiversity Strategies and Action Plans (NBSAPs) at the national level. As per the NBSAP, preparation and implementation of national biodiversity strategy by each country is obligatory. Till date, 184 of 196 parties have formed NBSAP (www.cbd.int).

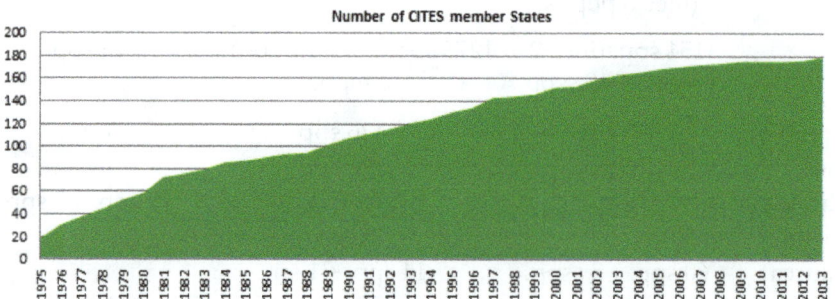

Figure 2: Increase in number of parties binding to CITES. State for which the Convention has entered into force is called a Party to CITES (https://www.cites.org/eng/disc/parties/index.php)

In India, implementation of the NBAP is mediated by Ministry of Environment and Forests (MoEF) and supported by 23 Ministries/Departments of the Government of India (GoI) along with the National Biodiversity Authority (NBA), State Biodiversity Boards (SBBs) and Biodiversity Management Committees (BMCs), which have been established under the National Biological Diversity Act, 2002 (Courierr, 1992; www.nbaindia.org).

Primary objectives of Biodiversity Act-2002 are conservation, sustainable use of biological resources in the country, issue related to access to genetic resources and associated knowledge and fair and equitable sharing of benefits arising from utilization of biological resources to the country and its people. The regulations of the Act are executed by a three tiered structure established under the Act at the national, state and local levels. At the local level, the Biodiversity Management Committees (BMCs) were established by institutions of local self-government for implementation of specific provisions of the Act and Rules. At the state level, the State Biodiversity Boards (SBBs) have been established to deal with all matters relating to implementation of the Act and the Rules (http://www.cms.int/) [Figure 4].

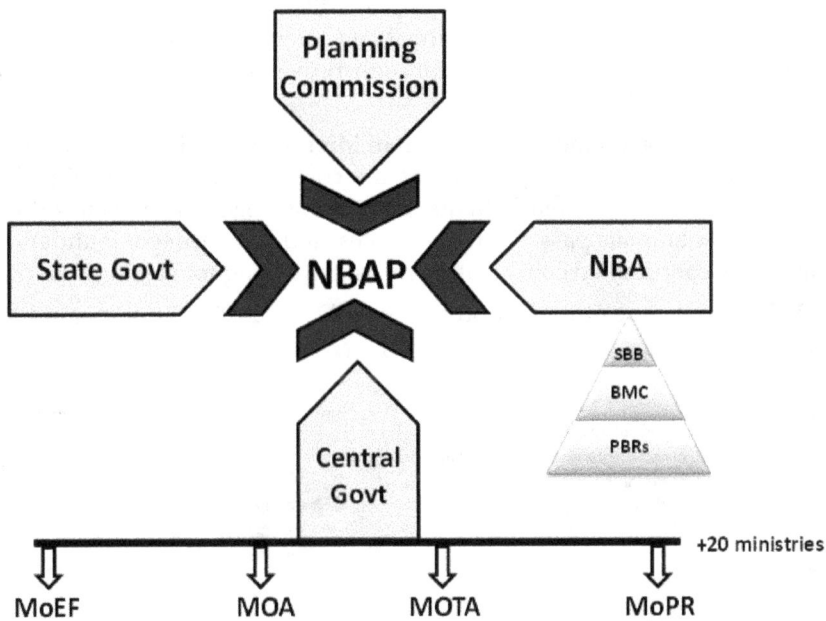

Figure 4: Hierarchy of the NBSAP prevalent in India (MOEF: Ministry of environment and forest; MOA: Ministry of agriculture; MOTA:Ministry of tribal affairs; MoPR:Ministry of Panchayati Raj; NBA: National biodiversity authority; SBB: State biodiversity board; BMC: Biodiversity Management committee; PBR: Panchayat biodiversity registers) (Fifth national report to the convention of national biodiversity 2014)

Around fourteen national repositories have been established allover India under the BD act 2002 and are involved in maintaining viable cultures of various species of microbes, flora and fauna (Table 2). At the national level, the National Biodiversity Authority (NBA) has been established to deal with all matters relating to implementation of the Act and the Rules. Each of these structure are required to be connected for decision making processes on various issues, including on issues of access and benefit sharing (ABS).

A centre, Centre for Biodiversity Policy and Law (CEBPOL) has been set up National Biodiversity Authority (NBA) to deal with governance related to biodiversity. The center is involved in:-

1. Preparation of trained experts in laws and policies related CBD

2. Providing professional support for negotiations related to biodiversity.

3. Creation of interactive information hub based on conferences, seminar and virtual sessions.

4. Establishment of capacity building programmes through multidisciplinary research and customise training programmes

5. Development of India as a regional and international resource Centre for Biodiversity Policy and Law through provision of training and human resource development.

CMS (Convention on the Conservation of Migratory Species of Wild Animals) provides a global platform (Figure 4) for the conservation and sustainable use of migratory animals and their habitats. CMS brings together the States through which migratory animals pass, the Range States, and lays the legal foundation for internationally coordinated conservation measures throughout a migratory range.

Table 3: National repositories approved by BD act 2002 ((http://nbaindia.org/uploaded/pdf/Repositorites_ of_BDAct.pdf)

1.	Botanical Survey of India, Kolkata
2.	National Bureau of Plant Genetic Resources, New Delhi
3.	National Botanical Research Institute, Lucknow
4.	Indian Council of forestry Research and Education, Deharadun
5.	Zoological Survey of India, Kolkata
6.	National Bureau of Animal Genetic Resources, Karnal, Haryana
7.	National Bureau of Fish Genetic Resources, Lucknow
8.	National Institute of Oceanography, Goa
9.	Wildlife Institute of India, Dehradun
10.	National Bureau of Agriculturally Important Micro-organisms, Mau Nathan Bhanjan, UP
11.	Institute of Microbial Technology, Chandigarh
12.	National Institute of virology, Pune
13.	Indian Agricultural Research Institute, New Delhi
14.	National Bureau of Agriculturally important insects , Bangalore

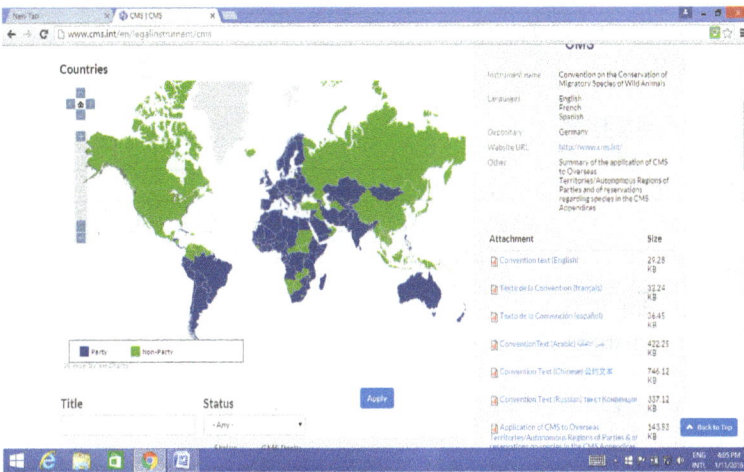

Figure 4: Contracting parties in CMS. Blue: Party; Green: Non Party (http://www.cms.int/en/legalinstrument /cms)

Another body set up by the Planning Commission/Government of India, National Natural Resource Management System (NNRMS) in early 1980s, aims towards optimal utilisation of remote sensing along with conventional data-for management and development of natural resources of the country. In order to guide the evolution of the system, a Preparatory Committee (later renamed Planning Committee) of NNRMS (PCNNRMS) chaired by Member (Science), Planning Commission, was constituted in 1982. Six standing committees (SCs) namely for Agriculture and Soils (SC-A), Bio-resources and Environment (SC-B), Geology and Mineral Resources (SC-G), Ocean Resources (SC-O), Remote Sensing Technology and Training (SC-T) and water resources (SC-W) were formed in 1984. These SCs have identified 56 priority areas (http://www.nnrms.gov.in).

Acts and Laws

Many other laws and acts have been formed which aids in maintenance of the biological diversity present in the subcontinent. At present, there are thirty six laws which aids and governs the conservation of biodiversity (Table 4). The laws are implemented by state, central and union government.

Table 4: Important Indian Acts passed related to Environment and Bio Diversity (http://www.nnrms.gov.in/)

1.	Fisheries Act, 1897.
2.	Destructive Insects and Pests Act, 1914
3.	The Indian Forest Act, 1927.
4.	Agricultural Produce (Grading and Marketing) Act,1937.
5.	Indian Coffee Act, 1942
6.	Import and Export (Control) Act, 1947.
7.	Rubber (Production and Marketing) Act, 1947.
8.	Tea Act, 1953.
9.	Mining and Mineral Development (Regulation) Act,1957

10.	Prevention of Cruelty to Animals Act, 1960.
11.	Customs Act, 1962.
12.	Cardamom Act, 1965.
13.	Seeds Act, 1966.
14.	The Patents Act, 1970.
15.	Wildlife (Protection) Act, 1972.
16.	Marine Products Export Development Authority Act,1972.
17.	Water (Prevention and Control of Pollution) Act, 1974.
18.	Tobacco Board Act, 1975.
19.	Territorial Water, Continental Shelf, Exclusive Economic Zone and other Maritime Zones Act, 1976.
20.	Water (Prevention and Control of Pollution) Cess Act, 1977.
21.	Maritime Zones of India (Regulation and Fishing by Foreign Vessels) Act. 1980.
22.	Forest (Conservation) Act, 1980. Air (Prevention and Control of Pollution) Act, 1981.
23.	Agricultural and Processed Food Products Export Development Authority Act, 1985/1986.
24.	Environment (Protection) Act, 1986
25.	Spices Board Act, 1986.
26.	National Dairy Development Board, 1987.
27.	Rules for the manufacture, use/import/export and storage of hazardous microorganisms/ genetically engineered organisms or cells, 1989
28.	Foreign Trade (Development and Regulation) Act, 1992.
29.	Protection of Plant Varieties and Farmers' Rights (PPVFR) Act, 2001
30.	Biological Diversity Act, 2002
31.	Plant Quarantine (Regulation of Import into India) Order, 2003
32.	Biological Diversity Rules, 2004
33.	The Food Safety and Standards Act, 2006
34.	Scheduled Tribes and Other Traditional Forest Dwellers (Recognition of Forest Rights) Act, 2006.

Fisheries Act, 1897

Also known as the Indian fisheries act, 1897, it is valid for the entire Indian subcontinent. The act details about the strict punishments to individuals using explosives and poison in water bodies to destruct species of fishes present. The punishment may range from two months imprisonment to payment of fines. State Government have been given the responsibility to make rules for protection of fish in selected waters i.e., all of such waters other then private waters where the state can define the status of private water bodies (ACT NO. 4 OF 1897)

Destructive Insects and Pests Act, 1914

This act prevents the transport and trade of any kind of infective agents of the agricultural and horticultural crops. It may be a fungus, insect or any kind of pests. The power of implementation and monitoring of this act lies with the central government. Violation of any rules laid by the act may lead to a fine of two thousand rupees (17). Procedure to be followed for declaring an area to be a Reserved Forest, a Protected Forest or Village Forest. It defines what is a forest offence, what are the acts prohibited inside a Reserved Forest, and penalties leviable on violation of the provisions of the Act (ACT NO. 2 OF 1914).

Wildlife (Protection) Act, 1972

The Act was passed to control habitat destruction and erosion due to agriculture, industries, urbanization and other human activities. The act enables setting up of National Parks, Wildlife Sanctuaries and constitution of Central Zoo Authority. The act also summarizes the provision for trade and transport of selected wild species. It also empowers the officers legally to punish the offenders in question. Several Conservation Projects for individual endangered species like Lion (1972), Tiger (1973), Crocodile (1974) and Brown antlered Deer (1981) were stated under this Act. The Act is adopted by all states in India except J & K, which has its own Act.

The act has been amended six times since 1972 till 2006. Amendment of the Act was done in January 2003 (Wild Life (Protection) Amendment Act, 2002) whereby punishment and penalty for offences under the Act have been made more stringent. Further amendments in the law have been proposed by Ministry to strengthen the Act. The Central Bureau of Investigation (CBI) has been empowered under the Wild Life (Protection) Act, 1972 to apprehend and prosecute wildlife offenders.One of the major decisions based on the Wildlife act is the business of Shahtoosh shawl as per the case of Cottage Industries Exposition Limited and Another v. Union of India and Others 2007 (143) DLT 477. The court decided that any component of the animal can be considered as a trophy and hence be preserved and protected under the law.Under Wild Life (Protection) act, transactions related to items like ivory, Pashmina shawls, tiger skin, leopard skin, shells of Nautilus, python skins have either been banned or is possible only under legit documentation.One of the most highlighted and mediacized case pertaining to this law was the infamous ' Black buck" case where Salman khan, a well known Bollywood actor, was penalized with Section 51 of the Wild Life Protection Act with the aid of Section 149 IPC (ACT No. 53 of 1972).

Marine Products Export Development Authority Act, 1972

The act is facilitated by the union government and responsibility is bestowed to a group named Marine Products Export Development Authority is constituted. The role is to keep a check on fisheries on all kinds, methods to increase export, marketing of marine products and processing of marine products. The act also requires the registration of processing and cold storage plants. The money generated from registration goes to Marine Products Export Development Fund. Additionally, money accumulated by the authority in term of taxes, fees, loans is directed into the fund (No. 13 OF 1972).

Water Prevention and Control of Pollution Act (1974) and Water (Prevention and Control of Pollution) Cess Act, 1977

The water act is comprehensive and applies to streams, inland waters, subterranean waters, sea or tidal waters. This act in 1974 to prevent the pollution of water by artificial and manmade activities that can contaminate water bodies. The Act mediates prevention, control and abatement of water pollution and the maintenance or restoration of the water quality. The act is implemented upon by central and state governments. The act was last modified in 1988.

Under this act, water (Prevention and Control of Pollution) Cess Act was enacted in 1977, which levies cess on water consumed by persons operating and carrying on certain types of industrial activities. The amount varies from 1.5-9.5 paise /kilolitre of water consumed depending on the type of usage. The act was last amended in 2003 (ACT No. 6 OF 1974; No.19 OF 2003).

The Patents Act, 1970

The patenting system in India is governed by Patent act 1970 and the latest amendment was implemented on April 2005. Patent Act explains on various inventions which are not qualified as patentable under Section 3 and 4 of Patent Act 1970. The law is enforced by central government. The patent act as per TRIPS agreement details about the possibility of numerous technology which can be patented except anything diagnostic, therapeutic and surgical method. The patent act grants the patentee the right to grant the freedom to use the patented technology. In the amendment enacted in 2005, any indigenous industry can use the patented technology by giving a nominal sum to the patent holder. Additionally pre-grant and post grant opposition avenues have been generated and time frame of patent granting has been reduced. In addition, term of every patent as on 20.5.2003 has now become 20 years from date of filing. Restoration time for a ceased patent, U/S 60 has now increased from 12 months to 18 months. For Example, an application for restoration of a patent ceased on or after 20th May, 2003 can be filed within 18 months from the date of cessation (No. 39 OF 1970).

Manufacture, use/import/export and storage of hazardous microorganisms/ genetically engineered organisms or cells, 1989

This act mainly pertains to microbes and genetically modified micro-organisms. Any product which utilizes or contains microbes comes under this act. Some monitoring committees have been formed which overlooks the stringency with which this act is imposed. These committees are mainly:-

- Recombinant DNA Advisory Committee (IXDAC)

   Committee reviews developments in Biotechnology at national and international levels and overlooks and suitably propose suitable safety regulations in India. The committee functions under the aegis of Department of Biotechnology.

- Review Committee on Genetic Manipulation (RCGM).

   Again a part of Department of Biotechnology, it functions to monitor the safety aspect of on-going research projects and activities involving GMOs/ hazardous microorganisms.

- Institutional Biosafety Committee (IBSC)
- Genetic Engineering Approval Committee (GEAC)

Department of Environment Forests and Wildlife governs this committee and monitors for approval of activities involving large scale use of hazardous microorganisms and recombinants in research and industrial production from the environmental angle.

- State Biotechnology Co-ordination Committee (SBCC)
- District Level Committee (DLC) (MoEF Notification, 1989)

The Food Safety and Standards Act, 2006

The Act aids in establishment of a single reference body, the Food Safety and Standards Authority of India Standards Authority of India with head office at Delhi. Food Safety and Standards Authority of India (FSSAI) along with State Food Safety Authorities are entrusted to enforce various provisions of the Act. The provisions of the act specify the standard of the food and edibles and enlist the guidelines for accreditation of various industries. FSSAI as per the guidelines of the act is also supposed to create an information network which will provide and collate data regarding food consumption, incidence and prevalence of biological risk. FSSAI is also entrusted with proving trainings (Act 34 of 2006).

Scheduled Tribes and Other Traditional Forest Dwellers (Recognition of Forest Rights) Act, 2006

The Ministry of Tribal Affairs [MOTA] implements the provisions of the Scheduled Tribes and Other Traditional Forest Dwellers (Recognition of Forest Rights) Act 2006.The Act recognizes forest rights and occupation of forest dwelling Scheduled Tribes and other traditional forest dwellers who are settled in forest for generations but whose rights could not be recorded. This Act permits forest inhabitants to cultivate for livelihood, controlled usage of minor forest produce, community rights such as nistar; The forest tribal groups and pre-agricultural communities may additionally use practices to enable regeneration and conservation of community forest for sustainable use. As per the Act, Gram Sabhas can divert forest land for public utility facilities managed by the Government, such as schools, dispensaries, fair price shops, electricity and telecommunication lines, water tanks, etc. with due recommendation

Ministry of Tribal Affairs has taken initiatives for the benefit of tribal people like:-

- ➢ Mechanism for marketing of Minor Forest Produce (MFP) through Minimum Support Price (MSP)
- ➢ Development of Value Chain for MFP.

As per the information provided by Minister of State for Tribal Affairs Ministry Shri Mansukhbhai Dhanjibhai Vasava in 2014, funds have been allocated from special central assistance fo Tribal infrastructure development (http://fra.org.in/).

Strategies and Associated Policies

Policy as per definition means a set of rules and regulation which has been agreed upon set of organizations and government. It also defines in case of an extreme situation a definite protocol would be followed to solve the situation.

Many policies have been proposed and are been implemented. One of the first policy to be followed was the global biodiversity strategy, developed jointly in agreement of world resources institute WRI, World conservation union (IUCN), United nations environment programme (UNEP) in collaboration with food and agriculture organization (FAO) united nations education, scientific and cultural organization (UNESCO) in 1992 (TRIPS: Annex 1c; Tansey and Mary 2000).

The strategy was conceptualized at around the same time when leaders across the world were forming CBD. The plan mainly details a three pronged approach in conservation of biodiversity which is study, saving and using biodiversity sustainably and equitably.

The five objectives of this approach:-

a. Framing of national and international policy for sustainable use of biological resources and preservation of biodiversity.

b. Creation of conditions and incentives for effectual conservation.

c. Strengthening and application of tools for conserving biodiversity in a broader respect.

d. Strengthening of human capacity particularly in developing country

e. International cooperation and national planning towards catalysis mediating conservation.

Another well known policy followed in Europe is, EU Biodiversity Strategy which aims to arrest global biodiversity loss and introduce restoration by 2020. The strategy has 6 targets and 20actions to attain the aim by 2020. European union adopted this strategy in May 2011. Some policies present in this strategy are Common Agricultural Policy, the Common Fisheries Policy and the Cohesion Policy (http://ec.europa.eu).

The six targets namely are:

1. Execution of EU nature legislation for protection of biodiversity

2. Utilization of green infrastructure

3. Sustainable agriculture and forestry

4. Efficient management of fish stocks

5. Stringent control of invasive alien species

6. EU contribution to averting global biodiversity loss

USA has also established its own policy named the USAID (www.usaid.gov) whereby the detailing present in the policy is categorized into goals and objectives.

The goals mentioned in USAID are as follows:-

a. Priority based conservation of biodiversity

b. Biodiversity to form an integral part of human development.

Objectives:

a. Supportive provisions for biodiversity conservation;

b. Conservation and development to go hand in hand to achieve outcomes;

c. Partnerships to activate resources

d. Promoting international policies to enhance biodiversity conservation

e. Integration of scientific technology and knowledge to enhance biodiversity conservation practice.

In the 10th meeting of CBD, known as COP-10, Aichi target was formulated (Fig 5). There were two time frames given. The short term is known "Strategic Plan for Biodiversity 2011-2020" aimed for 2020. There are 5 goals in the Aichi target composed of 20 targets. It intends to achieve 20 targets within this time frame.

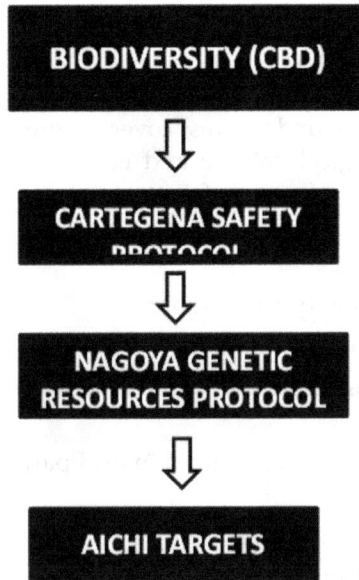

Figure 5: Hierarchy of global strategy for biodiversity conservation

The goals of AICHI target are (29):-

Strategic Goal A: Address the underlying causes of biodiversity loss by mainstreaming biodiversity across government and society

Strategic Goal B: Reduce the direct pressures on biodiversity and promote sustainable use

Strategic Goal C: To improve the status of biodiversity by safeguarding ecosystems, species and genetic diversity

Strategic Goal D: Enhance the benefits to all from biodiversity and ecosystem services

Strategic Goal E: Enhance implementation through participatory planning, knowledge management and capacity building (https://www.cbd.int/sp/targets/)

Talking about India, some of the policies applicable in national context are:-

1. National Forest Policy.

2. National Conservation Strategy and Policy statement on Environment and Development.

3. National Policy and macro-level action strategy on Biodiversity.

4. National Biodiversity Action Plan (2009).

5. National Agriculture Policy.

6. National Water Policy.

7. National Environment Policy (2006).

National Forest Policy

India has developed a forest policy since 1894. The policy was modified last in 1988. National Forest Policy (NFP), 1988 emphasized that 33% of the geographical area should be under forest cover as forest covers have a inevitable role in maintaining ecological balance and environmental stability. The forest policy details that method of saving forest is via protection, conservation and development of forests.

Aims:

1. Environmental stability Preservation

2. Natural heritage conservation

3. Control of soil erosion and denudation in catchment areas of rivers, lakes and reservoirs

4. Checking extension of sand dunes in arid parts of Rajasthan and coastal tracts

5. Aiding aforestation and forestry programs.

6. Attaining efficient utilization of forest produce and timber by rural and tribal populations.

7. Utilization of women as workforce.

By 2020, India's national forest policy aims to invest US$ 26.7 billion, to enable nationwide aforestation coupled with forest conservation, with the goal of increasing India's forest cover from 20% to 33% (MoEF report, 2009)

National Conservation Strategy and Policy Statement On Environment And Development, 1992

The policy basically lists out priorities and strategies for combating environmental problems. The key priorities are land and water, atmosphere, biodiversity and biomass. Support policies with an environmental perspective have been developed. The policy formulation and intervention have included areas like protecting watersheds, supporting forestry and plantations, protecting water bodies and sustaining fisheries, conserving biodiversity, increasing energy efficiency, developing and deploying renewable resources, preventing or decreasing pollution, managing urban wastes and preserving the cultural heritage.

This policy has also highlighted the importance of NGOs and women "to work on community involvement, providing information on environmental surveillance and monitoring, transmitting development in science and appropriate technology to the people at large" (Nath, 1992; Nair, 2011).

National Agriculture Policy [NAP]

NAP was announced on July 28, 2000 and broad objectives are to actualize the vast untapped growth potential of Indian agriculture. It aims to strengthen "rural infrastructure, promote value addition; accelerate the growth of agro business, create employment in rural areas; secure a fair standard of living for the agricultural population; discourage migration to urban areas; and face the challenges arising out of economic liberalization and globalization" (Fig 6). Although NAP promotes total utilization and expects an increased economic output, it safeguards the interests of the farmers on priority (Act No. 22 of 2005).

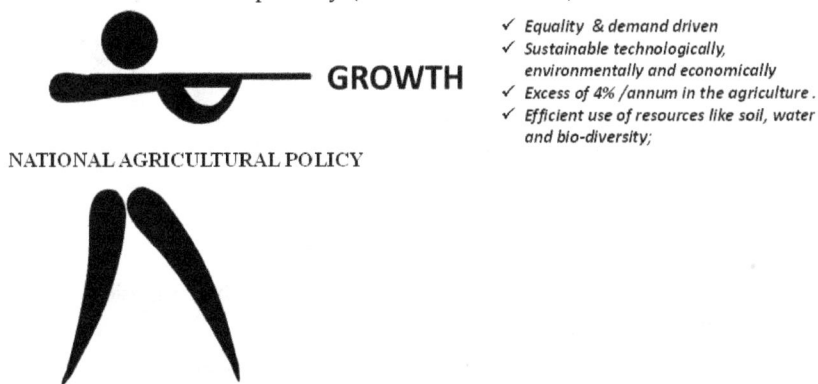

GROWTH

NATIONAL AGRICULTURAL POLICY

✓ Equality & demand driven
✓ Sustainable technologically, environmentally and economically
✓ Excess of 4% /annum in the agriculture .
✓ Efficient use of resources like soil, water and bio-diversity;

Figure 6: Aims of National agricultural policy

A nine pronged approach has been envisaged for the implementation of the agricultural policy which includes features like privatization of agriculture, private sector participation, strategy for livestock breeding, generation of location-specific, economically viable horticulture crops, livestock species and aquaculture to name a few. Stringency is trade of agricultural commodities and farm machinery and implements and fertilizers has also been reduced.

National Biodiversity Action Plan (NBAP)

Article 6 of CBD enjoins upon all Parties to prepare national strategies, plans or programmes for conservation and sustainable use of biodiversity, and to integrate conservation and sustainable use of biodiversity into relevant sectoral and cross-sectoral plans, programmes and policies.

The NBAP draws from the principle in the National Environment Policy that human beings are at the centre of concerns for sustainable development and they are entitled to a healthy and productive life in harmony with nature. This Action Plan identifies threats and constraints in biodiversity conservation taking into cognizance the existing legislations, implementation mechanisms, strategies, plans and programmes, based on which action points have been designed.

The implementation of the activities listed in NBAP would require substantial intersectoral coordination. This is on account of the cross-cutting issues in biodiversity. Apart from the Central Ministries/Departments and their agencies, and State Governments, the other actors involved are local bodies, research institutions, non-governmental organizations and civil society.

National Environmental Policy, 2006

The National Environmental Policy was documented on 2006 by MOEF. Its main aim targets towards conservation of biodiversity and quotes " conservation of environmental resources is necessary to secure livelihoods and well-being of all, the most secure basis for conservation is to ensure that people dependent on particular resources obtain better livelihoods from the fact of conservation, than from degradation of the resource".

Environment policy encompasses following objectives:-

1. Protection and conservation of critical ecological resources including man-made heritages.

2. Intra-generational Equity: secured access of resources to poor communities for their sustainable and additionally judicious utilization so that availability of resources is there for the future generations.

3. Integration of Environmental Concerns in policies plans, projects and programs for economic and social development

4. Efficiency in Environmental Resource Use for minimization of adverse effect on the environment.

5. Environmental Governance to be followed with "transparency, rationality, accountability, reduction in time and costs, participation, and regulatory independence".

6. Enrichment of Resources for Environmental Conservation (Five Year Plan 2002-2007,35)

Additionally, Indian Government has taken numerous protection steps. Important measures include:

1. Wetland (Conservation and Management) Rules 2010 have been framed for protection of wetlands and associated resources, a vital part of ecosystems. India being a signatory of Ramsar convention, framed in Iran 1971 and environment policy 2006, conservation of "ecological character" of wetlands is obligatory (Gazette of India, 2010). There are over 2000 Ramsar sites throughout the world, out of which, 26 sites have been identified in India by National Administrative authority.

2. The Centrally sponsored scheme of National Plan for Conservation of Aquatic Eco-System covers protection of aquatic ecosystems. Initially, National Wetlands Conservation Programme (NWCP) and the National Lake Conservation Plan (NLCP) were two separate plans but were merged in a cabinet (Cabinet Committee on Economic Affairs) meeting held in 2013. The plan aims to attain the quality standards of lakes and wetlands fit for survival for biological diversity.

3. Wildlife Crime Control Bureau, a statutory body established and its function is to complement the efforts of the state governments, primary enforcers of the Wildlife (Protection) Act, 1972 and other enforcement agencies of the country. Functionality is limited to control of illegal trade in wildlife, including endangered species.

4. The Centrally Sponsored Scheme 'Integrated Development of Wildlife Habitats has been modified by including a new component namely 'Recovery of Endangered Species and 16 species have been identified for recovery viz. "Snow Leopard, Bustard (including Floricans), Dolphin, Hangul, Nilgiri Tahr, Marine Turtles, Dugong, Edible Nest Swiftlet, Asian Wild Buffalo, Nicobar Megapode, Manipur Brow-antlered Deer, Vultures, Malabar Civet, Indian Rhinoceros, Asiatic Lion, Swamp Deer and Jerdon's Courser.

   Under the 'Recovery of Endangered Species' component of the Centrally Sponsored Scheme 'Integrated Development of Wildlife Habitats' for the recovery of endangered species viz. Hangul in Jammu and Kashmir, Snow Leopard in Jammu and Kashmir, Himachal Pradesh, Uttarakhand and Arunachal Pradesh, Vulture in Punjab, Haryana and Gujarat, Swiftlet in Andaman and Nicobar Islands, Nilgiri Tahr in Taml Nadu, Sanghai Deer in Manipur government has spend lakhs of rupees (http://moef.nic. in; Tansey, 1999).

IPR and Patenting

Intellectual property (IP) is defined as creation for which a monopoly is assigned to designated 'creators' by law. Lately IPR related issues have become important. IPR consists of copyrights, trademarks, patents, industrial design rights and trade secrets. In 1930, IPRs was extended to living beings and knowledge/ technologies related. In first of its kind, in 1930, the U.S. Plant Patent Act was passed and permitted patenting of new varieties of plants, excluding sexual and tuber-propagated plants. This led to increased awareness and implementation of IPR based protection of diversity. Two major international agreements, both legally binding, deal with IPR of biodiversity namely the Convention on Biological Diversity (CBD) and the Agreement on Trade-Related Aspects of Intellectual Property Rights (TRIPs) of the World Trade Organisation (WTO). World Intellectual Property Organization.

As per the report generated by CEAS Consultants (Wye) Ltd, Centre for European Agricultural Studies in association with Geoff Tansey and Queen Mary Intellectual Property Research Institute key, there are some international business associations involved with IPRs namely, UNICE, ICC, IFPMA, Europabio, AIPPI, ASSINSEL, and the principal non-business NGOs active in the area of TRIPS and biodiversity like GRAIN, RAFI, TWN, IPBN, WWF, ACTIONAID, GAIA.

One of the major frontrunner for IPR execution and protection is TRIPS. TRIPS Agreement is a legal text and comprises Annex 1C of the Marrakesh Agreement Establishing the World Trade Organization, conceptualized on 15 April 1994. Primary aspect of the TRIPS Agreement for food and farming is the requirement for WTO (World trade organization) members to make patents available for any inventions, whether products or processes, in all fields of technology without discrimination. Additionally, TRIPS also maintains patents can be granted for microorganisms for products, and microbiological and non-biological processes for plant and animal production, must be subject to patenting. Numerous times comparison is made between TRIPS and CBD, but clearly there is a clear cut difference between two. TRIPS protects IPR generated due to human resourcefulness whereas CBD protects resources already present in public domain (www.twn.my).

The laws laid down by TRIPS are mandatory for the WTO Members to follow. There are provision for different strata of countries like developed countries need to apply TRIPS within one year of entry into force of the Agreement, developing countries and economies in transition had the provision of extra four years for implementation whereas least developed countries have a 10 year transition period with a feasibility to apply for extensions (Art 66.1).

The International Union for the Protection of New Varieties of Plants (UPOV) is an intergovernmental organization with headquarters in Geneva (Switzerland) and is in complies with the requirements of Article 27.3(b) of the TRIPS Agreement. It has 51 members. The Convention was formed in Paris in 1961 and lastly revised on 1991. UPOV's mission is to provide and promote an effective system of plant variety protection sui generis, with the aim of encouraging the development of new varieties of plants, for the benefit of society. As per UPOV, the breeder or public breeding station, decides the conditions for authorizes the exploitation of his protected variety. UPOV in its agreement has favored mutual access to genetic resources and benefit sharing (www.upov.int).

Discussion

Many laws and conventions have been initiated to attain the strategic plans for conservation of biodiversity. As an aftereffect of Nagoya protocol a permit has been designed making feasibility of knowledge sharing available. The certificate is known as Access and Benefit-sharing (ABS). Based on this knowledge, University of Kent is using ethno-medicinal knowledge of the Siddi community from Gujarat for research and development (https://absch.cbd.int/). Implementation of any plan faces problems and these problems have been categorized in the report analysis. According to the national status, major hurdles are being faced in "integration of biodiversity knowledge in different sectors, loss of traditional knowledge, loss of biodiversity due to loss of habitat and lack of co-operation between stakeholders" (https://www.cbd.int/reports/analyzer.shtml)

Table 5: Non Governmental organization working for the cause of NGOs

---

WWF: World wildlife fund is one of the leading organization involved in wildlife and endangered species conservation. It also aims in stopping "degradation of the planet's natural environment and build a future in which humans live in harmony with nature" Objectives of WWF are:-

➢ Conservation of world's biological diversity

➢ Sustainable use of renewable natural resources

➢ Ensuring reduction in pollution and wasteful consumption

Currently there are 1.1 million supporters, WWF's partners, projects and experts are making a difference in creating a healthy future for our planet (www.worldwildlife.org).

GRAIN: GRAIN is a small international non-profit organization that aids farmers and social movements in their struggles for community-controlled and biodiversity-based food systems. The NGO monitors and analyzes trend that affect farmers' and rural communities' control over agricultural biodiversity. GRAIN's works initially by information gathering followed by movement building based on data (www.grain.org).

RAFI: RAFI (Rural Advancement Foundation International), an NGO-based in Canada devoted to the cause of biopiracy. RAFI now ETC Group is the first civil society organization (nationally or internationally) to draw attention to the socioeconomic and scientific issues related to the conservation and use of plant genetic resources, intellectual property and biotechnology. Their main interests lies in study of ecological erosion and development of agro based technologies. They also oversee corporate concentration and technology trade in plant genetic resources, biotechnologies and biological diversity. Their main zones of working are Africa, Asia and Latin America.

Their current interests pertain to biodiversity, climate and geoengineering, corporate monopolicies, sustainable development, synthetic biology and technology assessment (http://rafiusa.org/).

TWN: Third World Network (TWN) controls Third World Network Features dealing with Third World affairs and development issues. The features service which was inaugurated in 1985 covers a wide range of issues and topics which include economics, finance, basic needs, environment and culture, as well as political developments. The emphasis of the features is on providing background analysis of structures which form the basis of many current problems facing the Third World today. The information network maintained by TWN retain upto date knowledge of all meetings and decisions regarding biodiversity. It keeps updating about the transactions occurring based on technology based products, patents and biopiracies from all across the globe (http://thirdworldnetwork.net/).

IPBN: The Indigenous Peoples' Biodiversity Network (IPBN) is a coalition of thirty indigenous peoples groups from around the world. An informal network of indigenous scientists, lawyers, community educators, conservation practitioners and activists, IPBN has facilitated an open-ended and ongoing discussion among indigenous peoples concerning the opportunities within the Convention on Biological Diversity for promoting, preserving and protecting their rights to manage, control and benefit from their own knowledge and resources. IPBN has also played an important role in educating governments, multilateral agencies, and non-governmental organisations about the links between cultural and biological diversity and the necessity of addressing indigenous peoples' rights if the Convention is to succeed in its goals (http://povertyandconservation.info).

Although implementation of these laws and act have not been taken on a sufficient scale to address the pressures on biodiversity in most places but attempts are being made.

All these hurdles are being managed by a multi faceted approach i.e. awareness by conducting conference and workshops, generations of public online libraries to encourage knowledge building and development of databases to curate the available knowledge. In addition to conserve the biodiversity it has now become important to ensure the protection of the habitat and other underlying drivers like demographic, socio-political and cultural pressures.

Conclusion

Maintenance of biodiversity may be almost defined as life support system and has become the mandatory means of sustenance of life. The management of biodiversity is a complicated and requires involvement of many different partners ranging from governmental organisations to private companies, NGO's and volunteers. Only then a holistic conservation of diversity of earth may be attained.

References

1. "Countries currently subject to a recommendation to suspend trade". *cites. org. CITES*. Retrieved 6 Jan, 2016.

2. "India's Forests: Forest Policy and Legislative Framework, Chapter 3-5". Ministry of Environment and Forests. 2009.

3. "Summary of the Trans-Pacific Partnership Agreement". USTR. Retrieved on 14 December 2015 [http://www.international.gc.ca/].

4. Aichi Biodiversity Targets retrieved on Jan 9, 2016 https://www.cbd.int/sp/targets/.

5. Arora, S. National report 3 Ministry of Environment & Forest, Government of India (https://www.cbd.int/reports/analyzer.shtml).

6. Basu, S. (2013). Wetlands and lakes will now be protected under single new conservation programme Down to earth retrieved on Jan 9, 2016 (http://www.downtoearth.org.in/tag/wetland-conservation).

7. Biodiversity, access, indigenous knowledge and IPR http://www.twn.my/access_7.htm retrieved on Jan 10, 2016.

8. Convention on International Trade in Endangered Species of Wild Fauna and Flora (Text of the Convention – 1) retrieved from https://www.cites.org/sites/default/files/eng/disc/CITES-Convention-EN.pdf

9. Convention Text retrieved from http://www.cms.int/en/legalinstrument / cms) on Jan 2, 2016.

10. Destructive Insects and Pests Act, 1914 ACT NO. 2 OF 1914 1 [3rd February, 1914.] retrieved on Jan 5, 2016.

11. Ehrlich, P.R. and Wilson, E. (1991). Biodiversity Studies: Science and Policy Science. *Science*. 253(5021): 758-762.

12. EU Biodiversity Strategy to 2020 – towards implementation retrieved on Jan 9, 2016 http://ec.europa.eu/environment/nature/biodiversity/comm2006/2020.htm.

13. Forest Right Act 2006 retrieved on Jan 10th, 2016 from http://fra.org.in/.

14. Gazette of India, Part II, Section III sub section ii, Nov 2010 government of india, ministry of environment and forest.

15. Global Biodiversity Strategy Guidelines for Action to Save, Study, and Use Earth's Biotic Wealth Sustainably and Equitably (1992) Kathleen Courrier (Ed) WRI, IUCN, UNEE All rights reserved.

16. Government of India Ministry of Law and Justice Right to Information Act, 2005 (Act No. 22 of 2005) [As modified up to 1st February, 2011].

17. GRAIN Programme https://www.grain.org/pages/programme retrieved on Jan 10, 2016.

18. Hosetti, B.B. (2013). Concepts in Wildlife Management Daya Publishing House. 3rd Edition. 167pg, Karnataka, India.

19. Indigenous Peoples' Biodiversity Network (IPBN) retrievd from http://povertyandconservation.info/en/org/o0147 retrieved on Jan 10, 2016.

20. International harmonization is essential for effective plant variety protection, trade and transfer of technology UPOV Position based on an intervention in the Council for TRIPS, on September 19, 2002 retrieved from http://www.upov.int/export/sites/upov/about/en/pdf/international_harmonization.pdf.

21. Krugman, P. TPP at the NABE. (2015). The conscience of the liberal: The Opinion pages. New York Times.

22. Marine Products Export Development Authority Act,1972, No. 13 OF 1972 [20th April, 1972], retrieved on Jan 5, 2016.

23. Mast, R.B., Mittermeier, C.G., Mittermeier R.A., Rodriguez-Mahecha J.V. and Hemphill, A.H. (1997). Megadiversity: Earth's Biologically Wealthiest Nations. Ecuador. In: Mittermeier, R.A, P. Robles Gil, and C.G. Mittermeier (Eds.). Cemex, Mexico. 108-127.

24. Mission and current focus retrieved from http://www.etcgroup.org/mission retrieved on Jan 10, 2016.

25. Nair, M.D. (2011). Opinion. TRIPS, WTO and IPR: Biodiversity Protection – A Critical Issue. *Journal of Intellectual Property Rights.* 16: 519-521.

26. Nath, K. (1992). National conservation strategy and policy statement on environment and development. *Ministry of Environment and Forests.*

27. National Biodiversity Strategies and Action Plans (NBSAPs) retrieved on Retrieved 18 December 2015. .

28. National Communication to UNFCCC; http://moef.nic.in/sites/default/files/introduction-csps.pdf (retrieved on Jan 9, 2016).

29. National environmental policy: The Five Year Plan 2002-2007, Volume II, Chapter 1. Planning Commission, Government of India.

30. NNRMS Programs overview retrieved from http://www.nnrms.gov.in/ on Jan 2, 2016.

31. Plant patent http://www.uspto.gov/patents-getting-started/patent-basics/types-patent-applications/general-information-about-35-usc-161.

32. Repositories of Biodiversity Act. Retrieved 18 December 2015 (http://nbaindia.org/uploaded/pdf/Repositorites_of_BDAct.pdf).

33. Rules of Manufacture, use/import/export and storage of hazardous microorganisms/ genetically engineered organisms or cells, 1989 G.S.R. 1037(E). MoEF Notification December 1989 retrieved on Jan 7, 2016.

34. Study on the relationship between the agreement on trips and biodiversity related issues Final Report CEAS Consultants (Wye) Ltd Centre for European Agricultural Studies in association with Geoff Tansey and Queen Mary Intellectual Property Research Institute September 2000.

35. Tansey, G. (1999). Trade, Intellectual Property, Food and Biodiversity Quaker Peace & Service, London, February 1999. Also available in electronic format and downloadable from the QUNO web site: http://www.quaker.org/quno ISBN 0-85245-311-6.

36. The Food Safety and Standards Act, 2006 Act 34 of. 2006 retrieved on Jan 7, 2016.

37. The Indian Fisheries Act, 1897 ACT NO. 4 OF 1897 1 [4th February, 1897.] retrieved on Jan 5, 2016.

38. The Patents Act, 1970 No. 39 OF 1970 retrieved on Jan 6, 2016.

39. TRIPS: Annex 1c: agreement on trade-related aspects of intellectual property rights; Final Act of the 1986–1994 Uruguay Round of trade negotiations.

40. United Nations: Decade of Biodiversity. The first internationally recognized certificate of compliance is issued under the Nagoya Protocol on Access and Benefit-sharing Press Release (https://absch.cbd.int/).

41. USAID initiatives (What We DO) retrieved on Jan 9, 2016 https://www.usaid.gov/what-we-do.

42. Water (Prevention and Control of Pollution) Cess Act, 1977 No.19 OF 2003 retrieved on Jan 6, 2016.

43. Water Prevention and Control of Pollution Act (1974) No. 6 OF 1974. retrieved on Jan 6, 2016.

44. What is CITES?".*cites.org. CITES*. Retrieved 6 Jan, 2016.

45. Wildlife (Protection) Act, 1972 No. 53 of 1972). (9th September, 1972). retrieved on Jan 5, 2016.

46. World Animal Health Organisation (OIE) and CITES agree to collaborate on animal health and welfare issues worldwide to safeguard biodiversity and protect animals Joint press release retrieved from World Animal Health Organisation (OIE) and CITES agree to collaborate on animal health and welfare issues worldwide to safeguard biodiversity and protect animals _ CITES.html. Retrieved 15 December 2015.

47. World Conservation Monitoring Centre of the United Nations Environment Programme (UNEP-WCMC), 2004. Species Data (unpublished, September 2004).

48. WWF: Our mission retrieved from http://www.worldwildlife.org/initiatives retrieved on Jan 10, 2016.

24

Biodiversity Utilization and Human Well-Beings: Current State and Trends

S. K. Patel, Rinku, A. Sharma and G. S. Singh*

*Institute of Environment and Sustainable Development,
Banaras Hindu University, Varanasi-221 005, Uttar Pradesh, India
E-mail: gopalsingh.bhu@gmail.com*

Abstract

Biodiversity is a natural wealth which plays a significant role to support the livelihoods of human-beings. The human population is constantly dependent on biodiversity for food, fuel, fiber, clean water, pollination, and fertility of soil, nutrient management, cultural aesthetics and even more benefits. The presence of the life on the earth surface is the inimitable character of earth combine with existent of diversity in life. The overexploitation and loss of biodiversity is responsible for change in abundance and distribution of biodiversity may leads to thoughtful consequences for the well-being of global society. The objective of this paper is to review the current status of utilization of biodiversity for human well beings at global level. The services provided by the biodiversity include provisioning, regulatory, supportive and cultural services. The consumption of the biodiversity for varieties of purposes reaches the biodiversity at the threshold levelthat reflect for viable solution for conservation. The increase in ecosystem services mainly come from expands of habitat which leads to reduction in species abundance and species extinction ultimately alarming biodiversity loss. This alarms us to sustainable use of ecosystem services for conservation of biodiversity and associated landscape. It also alerts us to promote the management of ecosystem that sustains a large and balanced ecosystem services for present and future scenario.

Keywords: Biodiversity, ecosystem services, utilization, and conservation

Introduction

Biodiversity is a natural wealth which plays a significant role to support the livelihoods of human-beings (Baul*et al.*, 2015). Biodiversity is defined as the variability not only in the species of plant, animal and micro-organism but also the variation in genes of species and level of ecosystem in which they interact with their physical environment (Global Biodiversity Outlook,2010). The biodiversity provided essential life-support services that underpin the civilization of human being but apart from this, some people rely that services deliver by the biodiversity is free of cost, so there are no or very little valuewere given to these services (Summer *et al.*, 2012).The human population is constantly dependent on the biodiversity for their subsistence livelihood and rising population and demand resources imposing threat to the biological diversity. The distributions of biodiversity on the earth are not uniform, and the certain regions are found rich in biodiversity called as megadiversity region. The IUCN has recognized 234 biodiversity rich regions around the world these are mostly localize in central America, tropical Africa, north western south America, eastern Mediterranean region and south Asia. The tropical forest gets worldwide attention due to their rich biodiversity and great devastation of habitat. Further, the tropical forest contain 7% surface of the earth but 70% of the species of all organism of the earth (Singh *et al.*, 2014). The presence of the life on the earth surface is the inimitable character of earth combines with extraordinary feature to existent of diversity in life, within the 9 million typesof animals, plant, fungi and protists life form with 7 billion of human population (Cardinale*et al.*, 2012). The world ecosystem sustain1399189 fauna with 5416 mammalian, 1020007 insects, 6771 amphibian, 9230 reptiles (ZSI, 2014); and 317950 plant species (algae 40800, bryophyte 14500, pteridophyte12000, gymnosperm 650, angiosperm 250000) (BSI, 2009).India also biologically recognized as one of the megadiverse nation of the world because of its diverse climate. It is comprised of 2.4% of world land areas and contains nearly 8% of world biodiversity with 45000 of plant and 91000 of animal species with 10 biogeographical zones. Out of the 34 biodiversity hotspot in the worldwide 4 hotspot is recognized in India namely Himalaya, Indo-Burma, Western Ghats and Sundaland (MoFE, 2014).

The some of the prominent services provided by the biodiversity include food, medicine, and fuel as a provisioning services; climate regulation, and pollination,as a regulatory; nutrient cycle, and soil formation as supportive and ecotourism, cultural diversity as cultural serviceswhich support the economy, health, and climate for the human well-being (MA, 2003; de Groot *et al.*, 2002). Examples of the economic benefits are more visible as some of the records pointed that the production of fisheries is approximate US$ 58 billion/year, anti-cancer agent from marine water organism fetches about US$ 1billion/year, herbal medicine of global market surfaced US$ 43 billion/year as provisioning services and pollination by honeybees in agriculture contribute US$ 8 billion/year as a regulating services and coral reefs fisheries support ecotourism to contribute about US$ 30 billion/ year as cultural services (Emerton and Bos, 2004; FAO, 2004; MA, 2005b;Nabhan and Buchmann, 1997; UNEP, 2006; WHO, 2001; GEO, 2007).Although, nearly 7000 species of food crops have been forming the requirement of human food, but only 30 types of crops provided 95% of foods energy and fisheries has delivers about

20% of protein and employ for 200 million of people (CBD, 2015). In pharmaceutical application about 25% of drug is obtained from 120 plant species (MoFE, 2014).

Cumulativelyfive direct and indirect factors areresponsible for change in biodiversity and its ecosystem services at regional as well as global scale. These are economical, demographical, sociopolitical, cultural and religious, scientific and technological growths which havedirect visible effects on habitat loss, climate change, excessive load of nutrient and pollution, over exploitation of natural resources and invasive species create more pressure on biodiversity whichultimately leads to biodiversity loss (MA, 2005a).This change responsible in abundance and distribution of biodiversity may leads to thoughtful consequences for the well-being of society of human.The thought of ecosystem services getting more importance because thedemand of society mainly dependent on ecosystem services (Daily, 1997; de Groot *et al.*, 2002).The modernization of the services of biodiversity is stared in late 1970s; it begins with forming of serviceable product of ecosystem that attracts the public interest in conserving the biodiversity (Westman, 1977; Ehrlich and Ehrlich, 1981; de Groot, 1987; Kuniyal*et al.*, 2004). Modernization coupled with population growth recognized the indirect causes of biodiversity loss,as the population increases the demand for food, fuel and other life support resources which led degradation of biodiversity. It is also evidenced that area with rapid growth of population to have more number of threatened diversity, maximum loss of habitat with rich endemic species then the average population density region of the world(Population Action International, 2011).The degradation of global ecosystem services has many defects on the society such as change in function of organization and policy, gap in scientific knowledge, disaster event and other factor. In most of cases we do not clear about a policy instrument succeeded or fail (Carpenter *et al.*, 2008). There are examples to modify the natural ecosystem to increase the production of different kind of biodiversity resources e.g. addition of fertilizer, agroforestry, and creation of dam (Ehrlich and Ehrlich, 1992; Daily *et al.*, 1997; Rodríguez*et al.*, 2011). This chapter will highlight about the various use, status and trends of biodiversity and associated ecosystem services in current scenario.

Man-Nature Interaction

Men have always been interacting to nature and its component since immemorialand the human dependencies on biodiversity are more in virgin or undisturbed forest(Singh, 1999; Baul*et al.*,2015). Commonly, man interacts with nature for four prominent components for example economy, security, health and environment which support the existence of human being on earth (Fig. 1). Human beingsare obtaining various products like food, water, fiber, and fuel wood from nature and these resources played a significant role in the evolution of human being.

The nature is strongly linked with human health in form of use of natural medicine that obtained from various plant species and the protection and functioning of the ecosystem is responsible for existence of life on the earth (di Giulio and Monosson, 1996).The global economy is greatly dependent on the different product of ecosystem which also affectsthe economy of the world (Hancock, 2010).

Ecosystem degradation may flavor to various environmental crisis for example air and water pollution increase the carcinogenic, cardiological and respiratory disease in human (McMichael*et al.*, 1996); promote threats,floods, drought storm, climate regulation, and natural disaster (Sims, 2006;Wenning*et al.*, 2007; MA, 2005a). These crises triggered the society and scientific community to make batter information system (Apitz, 2007;de Sherbinin*et al.*, 2007).The environmental interaction of human for the services of nutrient cycling, physiological protection and aesthetic value that provide us clean water, air and minimize the exposure of contamination to the acceptable label, and maintain the ecosystem condition to its sustainability level (Louv, 2008).

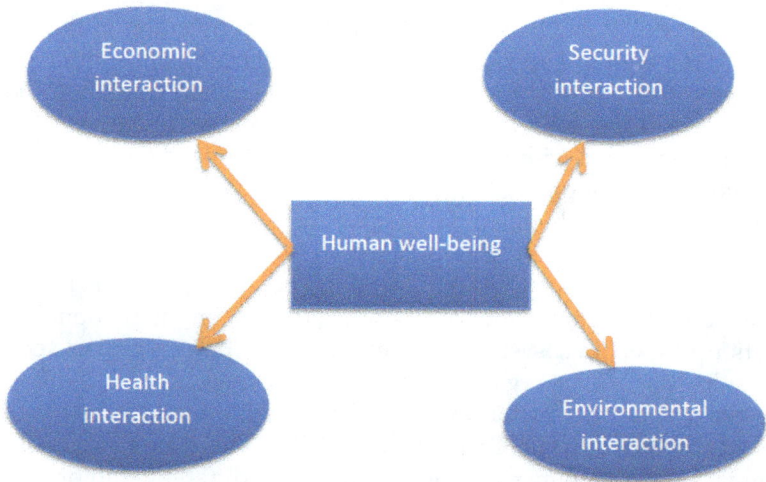

Figure 1: Interaction of human with nature (Source: drawn from Hancock, 2010).

Ecosystem and Biodiversity Services

The ecosystem services commonly addressed as "the benefit obtainsby the human from ecosystem"(MA,2005a).Biodiversity is counted one of the valuation services of ecosystem in different way. In certain casesthe word biodiversity and ecosystem services are interchangeable and if biodiversity manage then ecosystem service are automatically manage for example conservation of the wild species can alsoconserve forest ecosystem services (Mace *et al.*, 2012).

The term biodiversity arises by Walter and Rosen in the National Forum that organized by National Research Council under the title of "Biodiversity" (1986) (Wilson, 1988). But the term is used frequently after 1990, and three division of biodiversity (genetic, species, and community/ecosystem diversity) has been initiated in "Convention of Biodiversity" at the occasion of "Earth Summit" 1992 and biodiversity define as "the variability among the living organism of all sources containing inter alia, terrestrial, aquatic and other organism of biosphere" (UNEP, 1992; CBD, 1992).

Biodiversity service is important for society and some of them are inimitable such as the supply of food is not possible except the natural system (soil, crop and pollination); the availability of drinking water is not succeed without water cycle which is mostly depend on the plant and microorganism. So the conservation of ecosystem service is very necessary for the survival of human society (Fitter*et al.*, 2010). But despite of these profit the conservation strategies does not equal to the benefit obtain from the ecosystem (Pearce, 2006). To understanding the significant of biodiversity for global society the national and international levels afford have been adapted by world to conserve the biodiversity and its ecosystem services. Some of the prominent conservation effortswere addressed by world highlighted below (Table 1).

Table 1: Various global and national efforts for conservation of biodiversity.

Organization/Conference	*Venue*	*Conservation Strategies/Goal*
IUCN-The World Conservation Union (1948)	Fontainebleau, France.	Encourage and promote the global society to conserve the integrity and diversity of nature and sustainable use of natural resources.
Ramsar Convention on wetland (2 February 1971	Ramsar Iron	It is an intergovernmental treaty that provides the framework for national action and international cooperation for the conservation and wise use of wetlands and their resources.
Wildlife (Protection) Act, 1972	India	Aim for the management of National Parks and Wildlife Sanctuaries, protection to Scheduled Species, Conservation Reserves.
CITES-The Convention on International Trade in Endangered Species of Wild Fauna and Flora (1973)	Washington U.S.A.	It aims to ban international trade in wild in worldwide.
WRI-World Resource Institute (1982)	Washington U.S.A.	Societies live in a ways that protect Earth's environment and built the capacity to provide the needs and aspirations of current and future generations. It works to achieve the six critical goals in one decade (e.g. Climate, energy, food, forest, water and cities and transport).

Organization/Conference	Venue	Conservation Strategies/Goal
CMS-The Convention on Migratory Species (1983)	Bonn, Germany	Conservation of migratory species of avian in terrestrial and aquatic system.
CBD-Convention on Biodiversity (1992)	Rio de Janeiro, Brazil	Conserve the biological diversity with sustainable use and equable sharing of benefit arising from biodiversity utilization.
Millennium Developmental Goals (MDG) 2000	United Nation U.S.A.	It originated in Millennium summit of United Nation in 2000 with eight international goals in to achieve /eradicate extreme poverty and hunger; achieve universal primary educationand other many.
Biological Diversity Act (2002)	New Delhi, India	Sustainable use of biodiversity, benefit sharing of biodiversity,declaration of Biodiversity Heritage Sites
TEEB-The Economics of Ecosystems and Biodiversity (2007)	Geneva, Switzerland	Global initiative focused on "making nature's values visible". Its principal objective is to mainstream the values of biodiversity and ecosystem services into decision-making at all levels.
The REDD+-Reducing Emission from deforestation and Forest degradation (2007)	Germany	It introduced by UN Framework Convention on Climate Change(UNFCCC) in 2007 and procedure is finalized in 2013 to reduce emission by deforestation and forest degradation and conservation of carbon stack.
National Biodiversity Action Plan (November, 2008)	New Delhi, India	Strengthening and integration of *in situ*(on-farm)and *ex situ*(off-farm) conservation.
10th COP-(Conference of Parties) (CBD, 2010)	Nagoya, Japan	The United Nation Assembly at its 65th session declares 2011-2020 decade on biodiversity with 5 goals and 20 targets.

Organization/Conference	Venue	Conservation Strategies/Goal
Conference of Parties (COP-11) (CBD, 2012)	Hyderabad, India	This conference encouraged the world to develop commitment to sustainable development. It signifies a historic step forward in the conservation of biodiversity, the sustainable use of its constituents, and equivalent distribution of the fair benefits rising from the use of genetic resources.
2nd International Conference on Biodiversity & Sustainable Energy Development (2013)	North Carolina, USA	Offers a platform for scientists, engineers, directors of companies and students in the field of Biodiversity to meet and share their knowledge.
The IUCN World Parks Congress November (2014)	Sydney	Giving three 'P' concept; PARKS – Valuing and conserving nature, PEOPLE – Effective and equitable governance of nature's use and PLANET-Deploying nature-based solutions to global challenges
ICCB-27th International Congress for Conservation Biology (4th European Congress for Conservation Biology) August 2015	Montpellier, France	It is joint meeting of world community to conservation professionals to address conservation challenges and present new findings, initiatives, methods, tools and opportunities in conservation science and practice.
1st annual International Conference on Biodiversity Nov. (2015)	Colombo Sri lanka	Future of Biodiversity Conservation within the Sustainable development Goals.
5th International Conference on Biodiversity March, 2016	Madrid, Spain	It also a global platform to discuss and learn about Ecology and its associated fields, Biodiversity management, fauna & flora, Biodiversity & food security, conservation of Endangered species of marine.

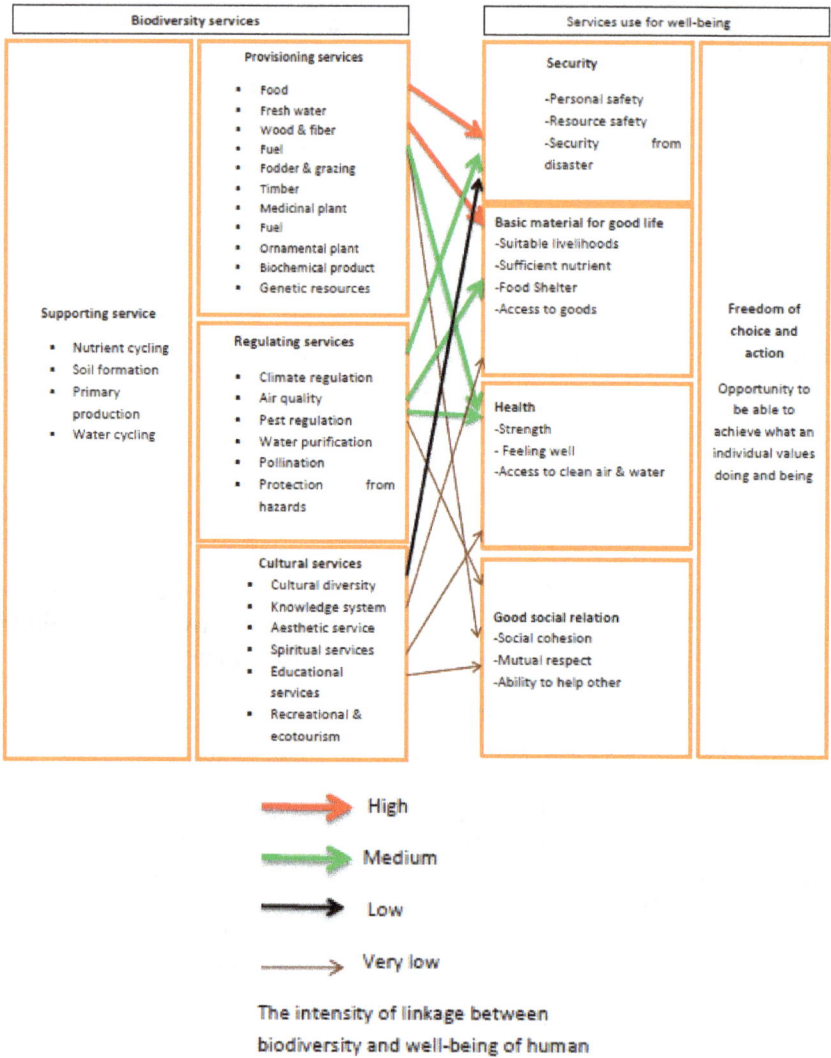

Figure 2: Interaction of Biodiversity with society and human well-being (Source: MA, 2005).

The Millennium Ecosystem Assessment (2005) categories the ecosystem services into fourcategories. The main objective of MA to aware people and state the scientific method to improve the management and sustainable utilization of ecosystem services (Biodiversity) (MA, 2005a).It presents an outline model for categorizing services of ecosystem into provisioning, regulating, cultural and supportive services(Fig. 2). Which reveals that alteration in services of ecosystem leads to direct impact on well-being of people that also directly responsible to change in the security, basic material for good life, health, and social relation of the human being andcollectively these elements are responsible for change in the freedom of choice and action of people. The change in relationship between human well-being and ecosystem services alter the present and future dimensionin terms of short and long-term effects. For example, overexploitation of biodiversity leads temporarily rise of material for well-being and decline poverty for short time which show unsustainable utilization that ultimately reduction in material of well-being and leads growth in poverty (MA, 2003).The MAreflect human well-being iscomprise of five main constituentsfor life that includes needs of basic material for a good life, health, good social relations, security, and freedom of choice and action. Human well-being is the product of various factors, some of them leads to biodiversity and ecosystem services in directly and indirectly form and many of them is independent (MA, 2005b).

Provisioning services

Provisioning services are a process to produce renewal resources (Cardinale*et al.*, 2012). The natural resources included food, medicine, timber, fuel, and biochemical obtain from the biological resources which is providing services to well-being of society (de Groot *et al.*, 2002).These provisional biodiversity services are very essential for global society and which used by human population very rapidly. The rate of the useof provisional biodiversity services are more than the globalgrowth of population and lower than the economic growth during second half of twenties century. There are different provisioning biodiversity services as summaries below.

Foods

The population of the world projected to be touched the nine billion till 2050 and the foodrequirement for the world will be increases about 60%. About 7000 species of plant have been forming for the requirement of human food. Currently,nearly 30 types of crops provided 95% of foods energy to human, with 5 cereal crops such as (millet, rice, wheat, millet and sorghum) that provided 60% of energy requirement for the world population. The foods security of the world depends upon the four dimension of the society such as physical availability of food, economics and physical access to food, food utilization and stability of other thee dimension over

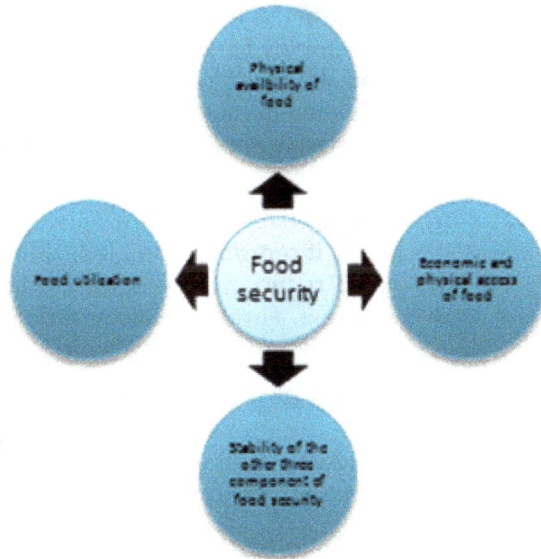

Figure 3: Four dimension of food security (Source: CBD, 2015).

time as illustrated in the fig. 3 (CBD, 2015).

The existence of life is completely dependent on foods (primary producer), a wider range of species diversity potentially underpin the important services of health, nutrition and livelihood that finally support ecological sustainability of society(Jaenicke andHoschle-Zeledon, 2006). The food service obtained by human population in different aspects of life including food, livestock, and fisheries were mentioned in Table 2.

Table 2: Different use of biodiversity as food.

Foods categories	Uses	Source
Crops	Crops have great nutrition importance for growth and development of human. The world cultivation mainly covers 12 grain, 23 vegetable with 35 fruit and nut species. other example of crop use as foods are rice, wheat, vegetable, fruits, maize, sugarcane, linseed, mustard etc.	Fowler and Mooney, 1990.
Livestock	It is very essential part of human diet and provided the employment for the rural and landless people. It has great source of protein carbohydrate fat and other vitamin. E.g. milk, butter, chicken, meat, wool, honey and fuel etc.	5th Indian national report 2014.
Fisheries	The fisheries bring 20% of protein to the 3 billion and employ for 200 million of people. The 30% of food stack of the oceanoverexploited which cause reduction in ecological and biological potential of the marine water.It is great source of protein for human.	CBD, 2015; Nelson, 1994; Global Biodiversity Outlook, 2010
Aquaculture	It is a globally significant source of foods in last 50 years. It contributes about 27% fish production. It could reduce the consumer demand on the wild fish. E.g. shrimp, prawn, carp breeding, Trout breeding, *Alevin* culturing etc.	MA, 2005
Wild plant and animal product	Bushmeat of wild birds, mammals, and reptile, insect can provide 85% of protein to the human. Provision of these foods also decline due to decrease of natural habitat. E.g. different plant and animal that is used as foods.	CBD, 2015

Timber

Timber has most important and valuable product of the forest ecosystemwhich is reflected in many timber based industries as raw materials (Table 3). The wood based industries provided the income and employment for the people particularly in rural areas (FSRI, 2010). Material contains woods, jute, cotton, hemp, silk and wool counted in fiber categories, whereas the hard woody product of forest called timber. The global timber production has increased 60% in last 40 years due to its economic value, butreplantation provided a growth of 35% in the global timber in harvested region to 2000 (MA, 2005a). Timber forest have also important social function, because their trees are goods consumer of carbon dioxide which is a greenhouse gas, it also protected from soil erosion and provided the pleasant of nature with purification of air, water and goods habitat for large wild animal and plant (Federal Ministry of Food, Agriculture and Consumer Protection, 2010).

Table 3: Use of timber product in deferent industries.

Timber based industries	Uses	Reference
Saw Mills	It hasthe largest consumerof timber, timber is used as 62% housing, 8% sleeper, 6% packing, 7% furniture, 7% vehicle industries, 4% ship building.	Pandeyand Rangaraju, 2008
Plywood, veneer and composite wood	It is a wood product& very popular because of its application. Federation of Indian Plywood and Panel Industries (FIPPI) reported 62 large 2500 small scale plywood industries is operating in India.	Forest sector report India, 2010.
Pulp and Paper	It also one of the largeconsumers of forest raw material. In India 660 pulp and paper mill out of which 25 wood and bamboo based and rest is agro and recycle fiber based mill.	Kulkarni*et al.*, 2006.

Medicinal plant

Biodiversity is the rich source of pharmaceutical thatvaluable for human health.Approximately 80% of population of certain African and Asian nations depended on the traditional medicine for their health (WHO, 2003).At the global scale 10000-20000 species of plant is used as traditional medicine but25% of drug is obtained from only 120 plant species. India is also one of the richest sources of medicinal plant in the world from ancient time, which included 6560 species of medicinal plant that underpin and regenerative for health of local society. It also cater 80% of Ayurvedic, 46% Unani and 33% of Allopathic medicine requirement, India has different medical system that included 900 species ofAyurveda, 800 species of Siddha and 700 species of Unani(FSRI, 2010; MoFE, 2014). Different plant derive substances that valuable for the drug or raw material for pharmaceutical and diverse type disease is summerised in Table 4.

Table 4: Use of some valuable substances that derive from plant in various human diseases.

Substances	Plant sources	Uses	Extracted part
Jalap	*Exogoniumpurya*	Used as a purgative	Root
Valerian	*Valerianaofficinalis*	Nervous afflictions, coughing and hysteria	Root
Quinine	*Cinchona spp.*	Malaria	Bark
Ephedrine	*Ephedra sinica. E. equisetina*	Colds, asthma,hay fever	Stem, wood
Belladonna	*Atropa belladonna*	Relieve pain, perspiration and coughs	Leaf
Santonin	*Artemisia cina*	Remedy forintestinal worms	Flower
Colchicine	*Colchicum autumnale*	Antigout	Root

(Source: http://www.faculty.ucr.edu/~legneref/botany/medicine.htm)

Fuel

The use of wood as fuel energy is getting more importance in developing countries, and they obtain more than half of the energy by wood but in the industrial countries such as Sweden and United States obtain their total fuelenergy requirement by wood only17% and 3% respectively. But some developing countries of Africa such as Rwanda, Tanzania and Uganda used 80% of wood energy for their total energy consumption. Rural sector of developing countries consume 95% energy as firewood, whereas in urban areas 85% as charcoal (MA, 2005b). Requirement of energy is anticipated to grow at least 53% till 2030 (IEA, 2006) and the use of biomass and waste product as an energy sources would expected to reached 10% of global scale till 2030.The biofuel is the main option of the energy in future, so many countries is investing in these energy source. The production of biofueldone on the 1% arable land of the worldwhich supports 1% transport demand which is expected to increase to 4% till 2030 (IEA, 2006). Some major biofuel producing countries are given in Table 5.

The fuels obtain from biodiversity included different source, it ranges from millennium year old energy source such dung of livestock, wood fire to modern sources bioenergy from crop, vegetation replace fossil fuel (Webb, and Coates, 2012). Biological diversity has excellent potential of energy production, due to rapid consumption of fossil fuel the human approach towards wood and non-timber forest product for energy source increases, jatropha, fermentation product of grains,aquatic plant and algae is the good source of ethanol and methanol which is used as good energy source in present time (MA, 2005).

Table 5: Use of biodiesel and bio-ethanol in different countries of the world.

Countries	Biodiesel (million/liter)
Germany	1920
France	511
United states	290
Italy	270
Austria	83
Brazil	16500
United state	16230
China	2000
European	950
India	300

Source: Worldwatch Institute, 2006

Biochemical product

Many plants secrete different type of biochemical product as a result of metabolic activity which is used as medicinal importance because of its importance for metabolite, pharmaceutical, nutraceutical, protection against pest, cosmetic and other different raw material for manufacturing the product in industrial sector (Fitter *et al.*, 2010). The reaping of biochemical product from plant cause undesirable effect on the biodiversity and these leads over overexploitation and extinction of resource (Fitter *et al.*, 2010). Resin has also source of rosin and turpentine which is widely used making of soap, paint, paper, varnish and pharmaceutical e.g. *Pinusroxburghii*(FSRI, 2010).

Genetic resources

The genetic resources provision gene and genetic material for vegetation and animal and also to biotechnology technique (Fears, 2007).The different biodiversity product which ever obtains from the wild resources of animal and plant but currently it obtained from the cultivated plant and housetrained animal by the use of genetic resources. However numerous cultivated plants are not able to sustain their marketable position without the genetic support to its wild relatives. The production of these species conserve by the modification of define qualities such as test, smell, resistance to pests and adaptation to change in environmental situationwhich brought by variability in gene, change in product quality as a result by in the genetic materialfrom their wild primitive to modern domesticated product and it very important for their economic position (Oldfield, 1984).There are manyInstitutes that work together tostore and conserve these genetic materials for*in-situ* (crops cultivation) and *ex-situ* (gene bank, seed bank) for universe. The progress in science also introduces the new methods that using molecular marker that provision new precision in signified the biodiversity (Fears, 2007).

Regulating services

Certain component of biological diversity such as dominant species of terrestrial plant work to regulate the climate by inspiring the capacity of carbon sequestration to regulate climate at regional and global level (Diaz *et al.*, 2005). The regulating services of biodiversity is an important ecological activity and provision the life support system by the biogeochemical cycle and other different ecosystem process to sustaining health of natural biosphere which useful to human in the form of clean air and water and climate regulation (de Groot *et al.*, 2002). CO_2 consumption by plant provided indirect feedback services to biomass and soil organic carbon which regulate climate and productivity of soil. The climate also be regulated by different other ecosystem process i.e. albedo, temperature, evaporation, fire, transpiration and land use and large surface areas covers with biodiversity (Diaz *et al.*, 2005). The regulating services of biodiversity perform different other regulating service of ecosystem as given below.

Climate regulation

The climate regulation isthe function role of ecosystem to sustaining the different factor of climate change such as greenhouse gases in the atmosphere (Fitter at al., 2010). The weather and climate of a region has determined by a compound interface of different regional and global air circulation and also with longitude latitude, topography, vegetation cover, albedo and conformation that including river, lake and bays. Some greenhouse gases (CO_2, CH_4 and NOx) absorbing IR radiation are important in climate regulation (de Groot *et al.*, 2002). The present change in climate is largely effected by the trace gases in the atmosphere which is result of the Industrialization, land use change and combustion of fossil fuels. CO_2 is the major GHG gas and marine system is important for their regulation by photosynthesis of aquatic microorganism and algae (Fitter at al., 2010). The different climate regulation performs by the ecosystem is given in Table 6. Currently people perception of climate change and its impact which flavoured sporadic study, is prime focus of the study (Tripathi and Singh, 2013) however adaption and mitigation data were more limited (Tripathiet al., 2016).

Table 6: Climate regulations by biodiversity.

Regulation services	Mechanism	Effect on climate	Reference
Disease and pest control	The disease and pests of an ecosystem is regulated by action of predator and parasite and defense ability of prey. The evidence is say the spread of pathogen is less in diverse ecosystem.	The ecosystem is protected by disease and pests.	Fitter *et al.*, 2010;

Regulation services	Mechanism	Effect on climate	Reference
Carbon sequestration	Capturing and long term storage of CO_2 in the earth crust. Afforestation and cover crop and poster land is mostly used technique of carbon storage.	Decrease global warming and also promote biomass production and greeneries.	Delagrange*et al.*,2008;Laikree*t al.*, 2013.
Disturbance prevention	Biodiversity directly responsible to prevention of the natural hazards (storms, drought and floods) by tree and surface conformation and storage capacity for e.g. Coral reef can reduce wave of sea in coastal region and prevent weathering of coastal region.	Regulating natural hazards.	de Groot *et al.*, 2002.
Trophic level	In any ecosystem composition of producer consumer is constant, if it change ecosystem will collapse.	The maintenance of the ratio of trophic level.	Dıaz *et al.*, 2005.
Nitrogen fixation	Fixation of atmospheric nitrogen by bacteria, which is use full for plant as nutrient.	Increase the productivity of soil.	Fitter *et al.*, 2010
Microbial decomposition of organic waste	Microorganism decompose organic waste into its elemental component that used by plant as nutrient.	Clean environment and increase productivity of soil.	de Groot *et al.*,2002.

Air quality

The atmosphere has ability to clean itself from different air pollutant in pre-industrial period, but increase of industrialization the self-cleaning ability of atmosphere is decreases. The actual fraction decrease of the air pollutant by plant is not known, but different plants and microbes are able to sink atmospheric pollutants(O_3NOx, SO_2, PM and CH_4) andwork to clean the atmosphere (MA, 2005a). Filtering importance of particulate dust by tree in urban ecosystem not consideredduring plantation of tree in olden time, but in the modern housing project a prospect to control atmospheric particulate dust has achieved by plantation of tamarind (*Tamarindusindicus*) thathave small and compound leaves which has great ability to remove dust pollutant than large leaf plant (Kumar *et al.*, 2013). The shape of leaf of Ashoka(*Polyalthealongifolia*), Mango (*Mangiferaindica*), Umbrella (*Thespepsiapopulnea*) and Pongamia (*Derris indica*) have more ability to capturing particulate dust than the other plant (Shetye and Chaphekar, 1989).The filtering effect of ever green plant is more than deciduous plant (Dochinger, 1973).

Water quality

The quality regulation of water is the process ofpurification and decontamination of water by plant and animaldiversity that manage pollutant of water which present in the form of impurity of inorganic and organic waste and that supply for consumption of animal and human. The water is decontaminated by various microbes present in soil and vegetation roots, trees also determine the flow of the water current which checks the erosion of soil (Fitter *et al.*, 2010). The timing and degree of runoff, flooding, aquifer recharge are mainly regulated by the plant that also responsible fordefinite alteration of the water quality (MA, 2005a). The quality of water is degrading by the human activity. The water quality maintain by different ecological process that regulate ecosystem (Kumar *et al.*, 2013). The different water quality regulating by plant is described in the Table 7.

Table 7: Ecosystem services regulated by water quality.

Water quality regulating process	Mechanism
Pollutant-filtering mechanisms of vegetative filter strips.	The water is filter by three zone of vegetation cover as vegetation surface layer zone, roots zone and sub-soil horizon zone.
Removal of water pollutants (Pb, Cu, Cd, Fe, hg and Cr) by Aquatic plants	Hydrillaverticillata; Spirodelapolyrrhiza; *Bacopamonnierii; Phragmiteskarka; Scirpuslacustris;* Water hyacinth (*Eichhorniacrassipes*); Pennywarth (Hydrocotyle umbellate; Duck weed (*Lemna minor;* Water velvet (*Azollapinnata*) remove water pollutant.
Rhizofiltration	It similar to remediation but it only remove contamination of water by adsorbed or absorbed by the plant roots in rhizophere region of plant.
Phytostabilisation	It is the process of immobilization of soil and water contaminant by the root of plant by absorption or adsorption.
Rhizodegradation	It breaks down the organic matter into its component by microbe.
Phytovolatilization	It the process of uptake of contaminated water and spread it into the atmosphere by transpiration.

Source: Based on Kumar et al., 2013

Pollination

Pollination is the natural process that regulates most important ecosystem services for increase in productionand creates diversity in nature. Pollination is process to transfer the pollen between flowers, without the pollination the sexual reproductions are not succeeded.In addition, pollination is the interaction of plant and animal thatimportant function of ecosystem to develop ecosystem services (Buchmann and Nabhan, 1996; Allen-Wardell*et al.*, 1998). It was observing the annual economics of pollination vary widely from $120 billion per year for all the ecosystem services (Costanza*et al.*, 1997) to $200 billion per year for the pollination by global agriculture (Richards, 1993).The pollination is conveyed by the 100000 of insect species and bird and animal, whichactively participate in the pollination of two-third of food plant and support 35% of the production of the crops (CBD, 2015). Almost 65% of plant pollination done by animals, and study of 200 countries also show that 72% of crop species which is important for food production depend on animal pollination mainly insect and 40% depend on wild pollinator (Klein *et al.*, 2007)

Pest control

The control of pest by the biodiversity is a significant regulation process of biodiversity. It is proof by the study that, the spread of pest is low in the diversity rich ecosystem it also evidences that a soil system with rich microbial community can help to reduce the losses of production of crop by the disease and pest generated by soil based (Wall and Virginia, 2000). The effect of the pest (parasite) on the plant is also destroying by the increase in the trophic level in the soil ecosystem and it's also affectsthe dynamics of the nutrient by plenty of the intermediary consumer microbes (Fitter *et al.*, 2010).

Protection from hazards

The protection from hazards is helpful services of biodiversity to reduce the effect of natural force on the global society and their environment. Many of the hazards is generated by people which interfaces with the natural ecosystem and responsible for the change in the environment. The severe use of fossil fuel together with great summer temperature and forest fire (cause by drought) and interferences of human cause air pollution in the environment (Fitter *et al.*, 2010). Ecosystem honesty essential to protect from this hazard, but it is no so effective to geological hazards such as hazards generated by the earthquake and eruption of volcano. This activity restricted to a few susceptible area i.e. in alpine region (Quetier*et al.*, 2007). The microbial diversity of soil also plays a significant role in check erosion of soil by upsetting roughness of surface and porosity (Lavelle *et al.*, 2006). The increase in vegetation cover is also support the protection against soil erosion and rock fall in hilly areas (Dorren*et al.*, 2006). The extensive growth in urbanization and more change in land use also diminish capability of ecosystem to reduce this extreme action (Fitter *et al.*, 2010).

Supporting services

These type of services is essential for the creation of all different ecosystem services, it also differ from the provision, regulation, and cultural services in respect to their impact on people which occurs indirectly way in long duration of time, or impact may occurs in other categories directly in short time on the people. Some regulation services can be characterized as both regulatory and supporting service depending on the duration of time. These services included-

Soil formation

The dependency of provisioning services on the soil fertility, soil formation is characterized into the categories of supportive ecosystem services thataffect the human well-being in many away (Dıaz, *et al.*, 2005). It is the Consecutiveprocess in all terrestrial ecosystem and mainly significantto early stage when land surface is active afterexposing (fallowing glaciation, river etc.) and it is the function of the parent material, topography and climate (Fitter *et al.*, 2010).The inland biodiversity is the key cause of formation of soil and loss of soil biodiversity (bacteria, fungi, algae, leguminous plant and deep rooted plant species) leads the reduction in soil formation, from theexperimental it is evidence that only soil biota is not important but composition of biodiversity is more significant for the soil formation. The soil in Northern Europe in the beginning of the post-glacial is often healthy for intensive agricultural (Newman, 1997), but in Mediterranean climate the soil is so old and not suitable for production and it faced more damage from erosion (Poesenand Hooke, 1997). In the alpine region the high damage is balanced with high rate of soil formation (Fitter *et al.*, 2010).

Primary production

Accumulation of chemical and sun energy into the food energy by the use of CO_2 inthe process of photosynthesis is called primary production. The production at primary level is very important for all different type of ecosystem services. The primary productivity is generally high in young and fertile soil with favorable climatic condition. The productivity is low in very cold region (Arctic and alpine), dry region (Mediterranean region) and with extremely polluted and degraded environment. The primary production in term of economy is most important and it highly depended on biodiversity and much evidence is explaining the relationship between primary production and biodiversity are more complex. The maximum production is generally attain in intensively maintain ecosystem with very low diversity and great requirement of input resources, while the high productivity is sustain by biodiversity rich system without high level of input resources. But due to the increases environmental pressure with land use, climate change and extremely increasing pollution are destroy the potential of biodiversity which leads loss of primary productivity (Fitter *et al.*, 2010;Ciaiset *al.*, 2005).

Nutrient cycle

About 20 nutrients are needed for life, it mainly includesnitrogen, phosphorus, and carbon distribute in the ecosystem at different concentration by nutrient cycle. The nutrient cycle affects ecosystem services and nutrient availability in ecosystem from local to global level (Vitouseket *al.*, 1997; Galloway *et al.*, 2004). The two most important nutrients that limit biological diversity's production in agro-ecosystem are nitrogen and phosphorus and they are also added as fertilizer in agro-ecosystem(Vitouseket *al.*, 1997).Different practices such as cover cropping, intercropping and microbial assimilation of nitrogen is useful for vulnerable loss of nitrogen. The additional nutrient management is done by variation in nutrient source, increasing biological nitrogen fixation and solubilizing the phosphorus and also by crop rotation and integrated management of biogeochemical that control cycling of other nutrient example carbon can decrease the need of extra nutrient in agriculture (Drinkwater andSnapp, 2007). Current study of nutrient cycling anticipate that change in the nitrogen and phosphorus cycle of soil may impact in many ecosystem functioning till 2050 and which further leads the disappearing of nutrient cycle in ecosystem and imbalance in nutrient (Bouwman*et al.*, 2009).

Water cycle

Water cycle essential for distribution of water in ecosystem and maintain the water balance and regulate ecosystem services as rainfall, humidity, transpiration and others (MA, 2005a). The delivery of appropriate amount of clean water is necessary for ecological services provided by ecosystem, the agro-ecosystem use about 70% of global water (FAO, 2003). The infiltration rate of forest soil more than any other soil system, forest has also the capability to cut off the flow and floods and sustain the base flow (Maes*et al.*, 2009). The availability of water and nutrient to the plant and other species of deep rooting system is maintained by hydraulic lift and vertical uplifting. The slow rate of soil erosion generates less pollution which helps respectable quality of water. In rapidly growing plantation forest may be exception to this generalization that they regulate ground water recharge but adversely the cut the flow of stream, salinization and acidity of soil (Jackson *et al.*, 2005; Power, 2015).

Cultural services

Culture ecosystem services define as "the non-material services obtain by society from ecosystem in the form of spiritual enrichment, cognitive development, reflection, ecotourism, recreation, and aesthetic experiences" (MA, 2005). The cultural services divided into different classes according to qualitative and quantities aspect of cultural diversity, spiritual and religious values, knowledge system, educational values, inspiration, cultural heritage values, recreation and ecotourism, aesthetic values, social relation and sense of place (MA, 2003).

Cultural diversity

The human civilization linked with nature from thousands of generation (Balée, 1994; Norgaard, 1994; Denevan, 2001; Toledo, 2001; Maffi, 2001; Gunderson andHolling, 2002; Harmon, 2002; Heckenberger*et al.*, 2007). This civilization is reproduce as culture in long historical regime and rule to protect natural place whichestablishesas sacred sites, national park, nature reserves and community conserved areas and mostly dominating in the mountain areas which directly related to biodiversity (Negi, 2010; Pretty *et al.*, 2009).The separation in nature and culture arenot common and it born in society from our need to manage thenature (Pretty *et al.*, 2009). Cultures are complex and changeable with post of time, in the all situation biological diversity support resilience of natural system where as cultural diversity has ability to grow resilience of social system (Maffi, 1998; Singh, 2000; Gunderson andHolling, 2002; Harmon, 2002). The main role of cultural activity to conserve nature near-pristine which regarded as sacred and if the resource is managed by community and its traditional practices is termed as community-based conservation and location manage in this condition is called community conserved areas or indigenous and community conserve areas(ICCAs) (Callicottand Nelson, 1998; Pretty *et al.*, 2009).

Knowledge system

Nature adoration has been an intrinsic force that is working in the approach towards conservation and sustainable utilization of biodiversity. Conservation of natural resources by traditional system is the integral part of different indigenous community of the world. Much of the knowledge system based traditional conservation practices are transfer from generation to generation by indigenous people around the world and that protecting shrubs, herbs and small patched of forest that has important for them and dedicated them to the local deity or combing them with evil spirits and these work is greatly support the conservation and protection of biological diversity (Ramakrishnan, 2009). Various communities in India fallow the nature-worship based that all creation on the planet has to be protected; they believed a close ritualistic relationship with many plant and vegetation to grow them around house. These sacred plants are mostly seen in the form of garden in fresh surrounding(Phurailatpami andNongthombam, 2014). The knowledge system is considered in two form (1) text-book based knowledge which is based on biophysical value not of the human dimension on the other hand (2) the empirical knowledge that originated as the result of human experiential process and strongly related with human dimension (Ramakrishnan, 2009). The knowledge has divided in two forms.

Formal knowledge

The formal knowledge system is theoretical process which is born by biophysical knowledge of landscape and deliberate and extended over the century and become more superior after every interval. This knowledge system always talks about complete conservation of ecosystem in the industrialized world (Ramakrishanan, 2009).

Traditional knowledge

The formal knowledge is not sufficiently explain the community pertaining ecosystem resources and management while this system is related with traditional knowledge and social discrimination (Ramakrishnan, 1984), which also associated with formal knowledge system. The natural resource conservation is subjected to introduce the socio-economic and cultural knowledge of different tribal and tradition community which transfer in the form of direct or indirect in the surrounding. Traditional knowledge is the fundamental of realization that human and nature form, the whole world associated with each other in symbiotic manner, the ecological opinion of the traditional community mainly related to plant, animal, river and whole earth. The advantage obtain from the traditional knowledge in the form of economic benefits by traditional crop variety, wild plant& animal species and medicinal significance, whereas the ecological and social benefit obtain form subspecies and landscape form, and ethical benefit support cultural, spiritual and religious belief which pointed on the sacred specie sacred groves and sacred landscape (Ramakrishanan, 2009). The different plant species which has spiritual importance and conserve by society by religious aspect are given in Table.

Table 8: Some religious important plant species.

Botanical name	Local name	Religious aspect
Aeglemarmelos	Bael	Lord Shiva
Ocimum sanctum	Tulsi	Lord Krishna
Buteamonosperma	Palash	Lord Brahma
Elaeocarpusganitrus	Rudraksha	Tears of lord shiva
Ficusreligiosa	Peepal	Buddha
Acacia catechu	Kaachu, Khadira	Gayathri Devi, Vastakarve Devi
Achyranthesaspera	Uttarani	IndraDeveIruthanth
Calotropisgigantean	Jilledu	Vinaayaka, Shiva Sangraha
Musa paradisicea	Arati or Kela	Sri satyanarayana Swami mahathya
Nelumbiumskeciosum	Kamal	Sri Laxmi
Piper betle	Paan	Vedaas

Source: (Phurailatpami andNongthombam, 2014; Singh, 2015)

Recreation and ecotourism

The recreation and ecotourism aresignifying to generate opportunity of employment and develop relationship between people and ecosystem (Sievänen*et al.*, 2009). It comprises the development of community that provided benefit and protection to ecosystem, and recreational deed such as camping walking and natural study to feel the services of ecosystem openly provided a chance to empathies the value of ecosystem directly (de Groot *et al.*, 2005; Daniel *et al.*, 2012). This activity mainly related to the people that living in the cities areas and direct interaction of people to the natural environment is not happen. In the field of conservation of ecosystem the recreation and tourism degrade the natural ecosystem such as wildlife disturbance and fragmentation of natural habitat or the adverse effect of other indirect source such as growth of infrastructure for economic gain from tourism (Reed and Merenlender, 2008; Liddle, 1997; Weaver, 2006; Krippendorf, *et al.*, 1989). The recreational and ecotourism give real impressions of physical exercise, experience of aesthetic, intellectual inspiration and different other experience of physical and psychological that is important for human well-being (Chan *et al.*, 2011).

Ecotourism is the small representative of a natural ecosystem (nature tourism), nature tourism is a kind of travel that is undisturbed or without contamination natural habitat and create about 15% of all tourism (WWF, 1995). The IUCN (1990) to promote the ecotourism and conservation of biodiversity hasdeclared different protected areas categories to promote ecotourism and conservation of biodiversity as given in the Table 9.

Table 9: Categories of protection used as ecotourism

Categories of protected areas	Activity for ecotourism and biodiversity conservation
Scientific Reserve/Strict Nature Reserve	Protected mainly for science or wilderness protection, for scientific purposes, tourism not allowed
National Park	Mainly for ecosystem protection and recreation purpose Tourism is dominated.
Natural Monument/Natural landmark	These protected areas mainly for conservation of specific natural features of ecosystem and tourism.
Nature Conservation Reserve/ Managed Nature reserve/ Wildlife Sanctuary	Manage for conservation through management interference. Tourism and hunting both are permitted.
Protected Landscape / Seascape	Conservation and recreation of landscape and seascape with tourism.
Multiple Use Management Area/ Managed Resource	Tourism permitted with hunting
Biosphere Reserve and World Heritage Site	For tourism

Source: IUCN, 1990; IUCN, 1994

Aesthetic services

The scenery of natural areas and landscape always attract the people, most of the people like to live in aesthetically attractive region. Aesthetic rich area has significant importance in term of economic (house or hotel in region of national park and natural scenery of ocean beach is more costly then house and hotel of other region) (Costanza*et al.*, 1997). Demand of aesthetically and attractive natural landscape has increased in wake of natural ecosystem is degradation, this reduction in availability of natural areas may cause unfavorable effect on public health and it economics, to fulfillment aesthetic requirement many people create different form of ecosystem as creation of park, scenic drive (trip to hill station), also creation of home garden (MES, 2005). In the scientific research the qualitative feature of aesthetic is mostly measured perceptual survey but the quantitative feature of aesthetic is determined for the landscape is by choice of averaging or rating or other different statics methods (Daniel *et al.*, 2001).

Major Challenges

The maximum global threat to the biodiversity is climate change, habitat loss and invasive species (Ewers andDidham, 2006; Thomas*et al.*, 2004; Sax and Gaines, 2003). The extinction debt is a process that ignore by community but extinction rate of a more prone endangered species can be highlighted and reduce by conservation process (Hanski and Ovaskainen, 2002). The Convention on Biological Diversity (CBD, 2002 in sixth Conference of Parties) target to get a significant reduction in recent rate of loss of biodiversity at regional, national and global scale that subsidizegrowth in the economy and increase of well-being of life on the earth and other number of indicator set by the CBD to achieved till 2010 (Chettri*et al.,* 2012), But it was not achieved by the most of the nation. So there are need to create new target by decision-maker for conservation of ecosystem services,in these sequencethe Global Outlook 3 target the dire need of an improved and combine information of biodiversity and ecosystem services and also construct scientific capability at regional, national and international scale by the CBD's 2011-2020 strategic plan, conformed at COP-10 in Nagoya by attendance large number of stakeholders (Chettri*et al.,* 2012). There is demand to improve production of food need to be increases 60% till 2050 for expected population growth will reach nine billion (CBD, 2015).

In wake of emerging crisis on biodiversity, CBD has proposed some challenges for conservation of biodiversity as discussed below (CBD, 2012): (i) extended worldwide organization to manage ecosystem service within the developmental sector and wider growth planning frameworks; (ii) develop technology that is use full to restoration of degraded ecosystem and distribute to worldwide; (iii)enhance the useof information (use information from degrading traditional knowledge); (iv) economic interventions (elimination of process that encouragesextreme use of biodiversity goods and services(fisheries or agriculture);(v) delivery of basic requirement of human well-being for long time is necessary to sustain the health of ecosystem and its productivity; (vi) initiate the legislation for justifiable sharing of benefit access from ecosystem; (vii)introduce market protocol and financial inducement at all scale of society to develop green economy concentrating on pro-poor growth; (viii)introduce the protocol that is combination of traditional knowledge and scientific technique to create ability of local people to sustain biodiversity.

Global initiatives

Knowing the importance of biodiversity utilization for mankind different international initiative promotesvarious conservation applications for biodiversity. The Global Biodiversity Outlook (GBO) is an intermittent report that published by CBD. The third edition of this reportvaluating the movement of the 2010 biodiversity target for significant fall in the rate of loss biodiversity at worldwide scale and it was a vital source of information to create the strategic plane for biodiversity 2011-2020 and Aichi Biodiversity Target. The strategic plan for Biodiversity 2011-2020 with Aichi Biodiversity Target was implemented by parties at the Convention on Biological Diversity (CBD) October 2010. The plan has five strategic goals and targets to getting in theme of "a world living in harmony with nature and where by 2050, the biodiversity is valued, preserve, restore, wisely used, maintaining services of ecosystem and sustaining a healthy planet that provided essential services to people well-being". In this series Millennium Developmental Goals (MDG) originated with eight international goals in Millennium summit of United

Nation in 2000 to achieve /eradicate extreme poverty and hunger;achieve universal primary education;promote gender equality and empower women;reduce child mortality;improve maternal health;combatHIV/AIDS, malaria and other diseases;ensure environmental sustainability and develop a global partnership for development by 2015 (MDG, 2015). The completion of the MDG on 14 July 2014 UN general assembly opens a working group of Sustainable Development Goals forward blue print of SDG for assembly with 17 goals and 169 targets that contain different aspect of sustainable development such as poverty, health, education, energy, economy, infrastructure, climate, ecosystem and sustainability. The theme of SDG is "Transforming Our World" for 2030 agenda for sustainable development (SDG, 2015).

Conclusion and discussion

The consumption of biodiversity as food, fiber and energy is main cause for the reach the biodiversity at the threshold level and conversion of natural system to semi-natural ecosystem to forming or urban areas, and overexploitation of biological resources. The loss of the biodiversity and reduction of the capacity of ecosystem services is bringing often to respond in similar ways to become a common driver; the relationship among them is not same but it differs for different aspect of biodiversity. The increase in ecosystem services mainly come from expands of habitat loss which leads to reduction in species abundance and species extinction. The increase of the small range of these provisioning services in the short time naturally negative impact of biodiversity and significant change in the sustaining, regulating and cultural ecosystem services appear critical. This alarming us to sustainable use of ecosystem services for conservation of biodiversity. It also alert to promote the management that sustain a large and balanced ecosystem services with present ecosystem services as similar to focusing on the provisioning services(for example biodiversity scenarios: projections of 21st century change in biodiversity and associated ecosystem services). Biodiversity has various importance, it give the ecosystem services as a regulator of ecosystem services that regulate climate, water, air, pollination etc. it also work as a provisioning services that provide us food, fuels, medicine, that are the essential needs of our life. Biodiversity has ability to support the ecosystem in term of soil formation nutrient cycle and water cycle that give pillar support to sustaining the ecosystem. So biodiversity has enormous importance value without this possibility of the life is not possible. The earth is mega network of different ecosystem landscape and biome, if any link of this chain is break the all network are collapse. So we need to save it, control use of these treasures in sustainable way.

References

1. Acar, C., Acar, H. and Eroglu, E. (2007). Evaluation of ornamental plant resources to urban biodiversity and cultural changing: A case study of residential landscapes in Trabzon city (Turkey). *Building and Environment.* 42(1): 218-229.

2. Allen-Wardell, G., Bernhardt, T., Bitner, R., Burquez, A. and Cane J. (1998): The potential consequences of pollinator declines on the conservation of biodiversity and stability of crop yields. *Conservation Biology.* 12: 8–17.

3. Altieri, M.A. (1999). The ecological role of biodiversity in agro ecosystems. *Agriculture, Ecosystems & Environment.* 74(1): 19-31.

4. Apitz, S.E. (2007). Conceptual frameworks to balance ecosystem and security goals.In *Managing Critical Infrastructure Risks.* Springer Netherlands. 147-173.

5. Baul, T.K., Rahman, M.M., Moniruzzaman, M. and Nandi, R. (2015). Status, utilization, and conservation of agrobiodiversity in farms: a case study in the north western region of Bangladesh. *International Journal of Biodiversity Science, Ecosystem Services & Management.* 11(4): 318-329.

6. Balée, W. (1994).Footprints of the Forest: Ka'aporEthnobotany.NewYork, NY, USA: Columbia University Press.

7. Beattie, A.J., Barthlott, W., Elisabetsky, E., Farrel, R., Kheng, C.T., Prance, I., Rosenthal, J., Simpson, D., Leakey, R., Wolfson, M. and Ten Kate, K. (2005). *Ecosystems and Human Well-Being, Volume 1, Current State and Trends; Millennium Ecosystem Assessment,* ed. R. Hassan, R. Scholes and N. Ash, Island Press, Washington. 1-273.

8. Bharti, R., Shrivastava, A., Choudhary, J., Tiwari A. and Soni N. (2013). Ethno Medicinal Plants used by Tribal Communities in Vindhya region of Rewa and Sidhi District of Madhya Pradesh, India. *IOSR Journal of Pharmacy and Biological Sciences.* 23-28.

9. Boo, E. (1990). *Ecotourism*: The Potentials and Pitfalls, vol. 1 and 2. WWF, Washington, DC.

10. Bouwman, A.F., Beusen, A.H.W. and Billen, G. (2009). Human alteration of the global nitrogen and phosphorus soil balances for the period 1970–2050. *Global Biogeochemical Cycles.* 23(4).

11. Brown, K., Turner, R.K., Hameed, H. and Bateman, I.A.N. (1997). Environmental carrying capacity and tourism development in the Maldives and Nepal. *Environmental Conservation.* 24(4): 316-325.

12. BSI. (2013). Official communication from Botanical Survey of India, Kolkata, India.

13. Buchmann, S.L. and Nabhan G.P.(1996).*The forgotten pollinators.* Island Press, Washington, D.C. 1-312.

14. Budowski, G. (1976). Tourism and environmental conservation: conflict, coexistence, or symbiosis? *Environmental conservation.* 3(1): 27-31.

15. Convention on Biological diversity. (2015). Biodiversity for food security and nutrition.

16. Cardinale, B. J., Duffy, J. E., Gonzalez, A., Hooper, D. U., Perrings, C., Venail, P. and Kinzig, A.P. (2012). Biodiversity loss and its impact on humanity. *Nature.* 486(7401): 59-67.

17. Carpenter, S. R., Mooney, H. A., Agard, J., Capistrano, D., DeFries, R. S., Díaz, S. and Perrings, C. (2009). Science for managing ecosystem services: Beyond the Millennium Ecosystem Assessment. *Proceedings of the National Academy of Sciences.* 106(5): 1305-1312.

18. Callicott, J.B. and Nelson, M.P. (Eds.). (1998). *The great new wilderness debate.* University of Georgia Press.

19. Ciais, P.H., Reichstein, M., Viovy, N., Grainer, A., Ogee, J., Allard, V., Aubinet, M., Buchmann N., Bernhofer Chr., Carrara A., Chevallier F., de Noblet N., Friend A.D., Friedlingstein, P., Grunwald, T., Heinesch, B., Keronen, P., Knohl, A., Krinner, G., Loustau, D., Manea, G., Matteucci, G., Miglietta, F., Oureival, J.M., Papale, D, Pilegaard, K., Rambal, S., Seufert, G., Soussana, J.F., Sanz, M.J., Schutze, E.D., Vesala, T. and Valentini, R. (2005). *Nature.* 437: 529.

20. Carpenter, S.R., Mooney, H.A., Agard, J., Capistrano, D., DeFries, R. S., Díaz, S. and Perrings, C. (2009). Science for managing ecosystem services: Beyond the Millennium Ecosystem Assessment. *Proceedings of the National Academy of Sciences.* 106(5): 1305-1312.

21. Chan, K.M.A. (2011). Cultural services and non-use values. The Theory and Practice of Ecosystem Service Valuation in Conservation, edsKareiva P, Daily G, Ricketts T, Tallis H, Polasky S (Oxford Univ Press, Oxford). 206–228.

22. Chan, K.M., Goldstein, J., Satterfield, T., Hannahs, N., Kikiloi, K., Naidoo, R. and Woodside, U. (2011). Cultural services and non-use values. *Natural Capital: Theory & Practice of Mapping Ecosystem Services. Oxford University Press, Oxford, UK*, 206-228.

23. Chettri, N., Sharma, E. and Zomer, R. (2012).Changing paradigm and post 2010 targets: Challenges and opportunities for biodiversity conservation in the Hindu Kush Himalayas. *Tropical Ecology.* 53(3): 245-259.

24. Costanza, R. and Daly, H.E. (1992). Natural capital and sustainable development.*Conservation biology.* 6(1): 37-46.

25. Costanza, R., d'Arge, R., de Groot, R., Farber, S., Grasso, M., Hannon, B., Limburg, K., Naeem, S., O'Neill, R.V., Paruelo, J., Raskin, G.R., Sutton, P. and Van-der Belt, M. (1997). The value of the world's ecosystem services and natural capital. *Nature.* 387: 253–260.

26. Daily, G.C., Alexander, Ehrlich, S.P.R., Goulder, L., Lubchenco, J., Matson, P.A, Mooney, H.A., S. Postel, S., Schneider, S.H., Tilman, D.G. and Woodwell, G.M. (1997). Ecosystem services:benefits supplied to human societies by naturalecosystems. *Issues in Ecology.* 2:1–16.

27. Daily, G.C. (1997). Nature's Services: Societal Dependence on Natural Ecosystems. *Island Press,* Washington, DC.

28. Daniel, T.C., Muhar, A., Arnberger, A., Aznar, O., Boyd, J.W., Chan, K.M. and Grêt-Regamey, A. (2012). Contributions of cultural services to the ecosystem services agenda.*Proceedings of the National Academy of Sciences.* 109(23): 8812-8819.

29. Daniel, T. C. (2001). Whither scenic beauty? Visual landscape quality assessment in the 21st century.*Landscape and urban planning.* 54(1): 267-281.

30. De-Groot, R. and Ramakrishnan, P.S. (2005). Cultural and Amenity Services. Ecosystems and Human Well-Being: Current State and Trends. Findings of the Condition and Trends Working Group. Millennium Ecosystem Assessment (Island Press, Washington, DC). 455–476.

31. De-Groot, R.S., Wilson, M.A. and Boumans, R.M. (2002). A typology for the classification, description and valuation of ecosystem functions, goods and services.*Ecological economics*. 41(3): 393-408.

32. De-Groot, R.S. (1987). Environmental functions as a unifying concept for ecology and economics. *Environmentalist*. 7(2): 105-109.

33. De-Groot, R.S., Wilson, M.A. and Boumans, R.M. (2002). A typology for the classification, description and valuation of ecosystem functions, goods and services.*Ecological economics*. 41(3): 393-408.

34. Delagrange, S., Potvin, C., Messier, C. and Coll, L. (2008). Linking multiple-level tree traits with biomass accumulation in native tree species used for reforestation in Panama. *Trees*. 22(3): 337-349.

35. De-Sherbinin, A., Chen, R. and Levy, M. (2007). What does climate change mean for the hazards community? *Natural Hazards Observer*. 31(6): 11-13.

36. Denevan, W.M. (2001). *Cultivated landscapes of native Amazonia and the Andes*. Oxford University Press on Demand.

37. Di-Giulio, R.T. and Monosson, E. (1996). Interconnections between human and ecosystem health: opening lines of communication. In *Interconnections between Human and Ecosystem Health*. Springer Netherlands. 3-6.

38. Dıaz, S., Tilman, D., Fargione, J., Chapin III, F.S., Dirzo, R. and Ktzberber, T. (2005). Biodiversity regulation of ecosystem services. *Trends and Conditions*. 279-329.

39. Dochinger, L.S. (1973). *Miscellaneous Publication No.1230*.USDA, Forest Service, Upper Darby, Pa. 1-22.

40. Dorren, L.K.A., Bergera, F., Imesonb, A.C., Maiere, B. and Reya, F. (2004). *Forest Ecol. Manag.* 195: 1-165.

41. Drinkwater, L.E. and Snapp, S.S. (2007). Nutrients in agroecosystems: rethinking the management paradigm. *Advances in Agronomy*. 92: 163-186.

42. Ehrlich, P.R. and Ehrlich, A.H. (1992). The Value of Biodiversity. *Ambio*. 21:219-226.

43. Ehrlich, P.R. and Ehrlich, A.H. (1981). *Extinction: the causes and consequences of the disappearance of species*. New York: Random House. 72-98

44. Ewers, R.M. and Didham, R.K. (2006). Confounding factors in the detection of species responses to habitat fragmentation. *Biological Reviews*. 81(01): 117-142.

45. FAO. (2004). The State of the World's Fisheries and Aquaculture 2004. Food and Agriculture Organization of the United Nations, Rome.

46. Fears, R. (2007). *'Genomics and Genetic Resources for Food and Agriculture'*, CGRFA, Background Study Paper No. 34. FAO, Rome.

47. Federal Ministry of Food, Agriculture and Consumer Protection. (2010). *Conservation of Agricultural Biodiversity, Development and Sustainable Use of its Potentials in Agriculture, Forestry and Fisheries*.

48. Emerton, L., Bos, E. and Value, E. (2004).Counting ecosystems as an economic part of water infrastructure.*IUCN, Gland, Switzerland and Cambridge, UK*.

49. Fowler, C. and Mooney, P. (1990). *Shattering: Food, Politics and the Loss of Genetic Diversity*.University of Arizona Press, Tucson, AZ. 1-278.

50. Galloway, J.N., Dentener, F.J., Capone, D.G., Boyer, E.W., Howarth, R.W., Seitzinger, S. P. and Karl, D.M. (2004). Nitrogen cycles: past, present, and future. *Biogeochemistry*. 70(2): 153-226.

51. George, P. and Arekar, C. (2011). Biodiversity survey of trees and ornamental plants in KarunyaUniversity, Coimbatore, India. *International Journal of Biodiversity and Conservation*. 3(9): 431-443.

52. Goodwin, H. (1996). In pursuit of ecotourism. *Biodiversity & Conservation*. 5(3): 277-291.

53. Gossling, S. (1999). Ecotourism: a means to safeguard biodiversity and ecosystem functions? *Ecological Economics*. 29(2): 303-320.

54. Global Environmental Outlook (GEO 4). *Environment for development*, (2007). United Nations Environment Programme.

55. Gunderson, L.H. (2001). *Panarchy: understanding transformations in human and natural systems*. Island press.

56. Hancock, J. (2010). The case for an ecosystem service approach to decision-making: an overview. *Bioscience Horizons*. 3(2): 188-196.

57. Hanski, I. and Ovaskainen, O. (2002). Extinction debt at extinction threshold. *Conservation biology*. 16(3): 666-673.

58. Holmlund, C.M. and Hammer, M. (1999). Ecosystem services generated by fish populations. *Ecological economics*. 29(2): 253-268.

59. Harmon, D. (2002). In light of our differences: Washington: Smithsonian Institute.

60. Heckenberger, M.J., Russell, J.C., Toney, J.R. and Schmidt, M.J. (2007). The legacy of cultural landscapes in the Brazilian Amazon: implications for biodiversity. *Philosophical Transactions of the Royal Society of London B: Biological Sciences*. 362(1478): 197-208.

61. IEA. (2006). *World Energy Outlook 2006*. International Energy Agency, Paris.

62. Indian Council of Forestry Research and Education. (2010). *Forest Sector Report India*. P.O. New Forest, Dehradun (Uttarakhand).

63. Jackson, R.B., Avissar, R., Roy, S.B., Barrett, D.J., Cook, C.W., Farley, K.A. and Murray, B.C. (2005). Trading water for carbon with biological carbon sequestration. Science. 1-310.

64. Jaenicke, H. and Hoschle-Zeledon, I. (2006). *Strategic Framework for Underutilized Plant Species Research and Development: With Special Reference to Asia and the Pacific, and to Sub-Saharan Africa*. BioversityInternational.

65. Klein, A.M., Vaissiere, B.E., Cane, J.H., Steffan-Dewenter, I., Cunningham, S.A., Kremen, C. and Tscharntke, T. (2007). Importance of pollinators in changing landscapes for world crops. *Proceedings of the Royal Society of London B: Biological Sciences*. 274(1608): 303-313.

66. Krippendorf, J. (1989). *FüreinenanderenTourismus*: Probleme, Perspektiven, Ratschläge (Fischer–Taschenbuch, Frankfurt am Main, Germany.

67. Kumar, S.R., Arumugam, T., Anandakumar, C.R., Balakrishnan, S. and Rajavel, D.S. (2013). Use of Plant Species in Controlling Environmental Pollution-A.*Bull. Env.Pharmacol. Life Sci.* 2: 52-63.

68. Kuniyal, J.C., Vishvakarma, S.C.R. and Singh, G.S. (2004). Changing crop biodiversity and resource use efficiency of traditional versus introduced crops in the cold desert of the north western Indian Himalaya: a case of the Lahaul valley. *Biodiversity and Conservation.* 13: 1271-1304.

69. Laikre, L., Schwartz, M.K., Waples, R.S., Ryman, N. and Ge M Working Group. (2010). compromising genetic diversity in the wild: unmonitored large-scale release of plants and animals. *Trends in ecology & evolution.* 25(9): 520-529.

70. Lavelle, P., Decaëns, T., Aubert, M., Barot, S., Blouin, M., Bureau, F. and Rossi, J.P. (2006). Soil invertebrates and ecosystem services.*European Journal of Soil Biology.* 42: 3-15.

71. Liddle, M. (1997). *Recreation ecology: the ecological impact of outdoor recreation and ecotourism.* Chapman & Hall Ltd.

72. Kulkarni, A.G., Mathur, R.M., Thapliyal, B.P. and Dixit, A.K. (2006). Current Status of Wood Supply to Industries Based on Agro Forestry, Specific Reference to Pulp and Paper Sector, Paper presented at the workshop on Rural Development through Integration of agro-forestry and wood-based industry by the *Commonwealth Forestry Association of India,* 11-12 October, New Delhi.

73. Louv, R. (2008). The right to walk in the woods. http://www. Childrenandnature.org/blog. Accessed 27 June 2008.

74. Maes, W.H., Heuvelmans, G. and Muys, B. (2009). Assessment of land use impact on water-related ecosystem services capturing the integrated terrestrial– aquatic system. *Environmental science & technology.* 43(19): 7324-7330.

75. Mace, G.M., Norris, K. and Fitter, A.H. (2012). Biodiversity and ecosystem services: a multi-layered relationship. *Trends in ecology & evolution.* 27(1): 19-26.

76. Maffi, L. (2001). *Onbiocultural diversity: Linking language, knowledge, and the environment.* Washington, DC: Smithsonian Institution Press.

77. McMichael, A., Haines, A., Slooff, R. and Kovats, S. (1996). Climate Change and Human Health, Geneva: World Health Organisation Middleton B.A., (2013). Rediscovering traditional vegetation management in preserves: Trading experiences between cultures and continents. *Biological Conservation.* 158: 271–279.

78. Millennium Ecosystem Assessment. (2003). *Ecosystems and Human Well-being: A Framework for Assessment.* World Resources Institute, Washington, DC

79. Millennium Ecosystem Assessment. (2005a). *Ecosystems and Human Well-being: Synthesis.* Island Press, Washington, DC.

80. Millennium Ecosystem Assessment. (2005b). *Ecosystems and Human Well-being: Biodiversity Synthesis.* World Resources Institute, Washington, DC.

81. Ministry of Environment and Forests Government of India. (2014). *India's fifth national report to the convention on biological diversity.*

82. Ministry of Environment and Forests, Government of India. (2009). *India's Fourth National Report to the Convention on Biological Diversity.*

83. Nabhan, G.P. and Buchman S.L. (1997). Services provided by pollinators. In Daily G. E. (ed.) Nature's Services – Societal Dependence on Natural Ecosystems. Island Press, Washington, DC.

84. Negi, C.S. (2010). Traditional culture and biodiversity conservation: Examples from Uttarakhand, Central Himalaya. *Mountain Research and Development.* 30(3): 259-265.

85. Nelson, J.S. (1994). *Fishes of the world.Wiley*, New York.

86. Newman, E.I. (1997). Phosphorus balance of contrasting farming systems, past and present. Can food production be sustainable? *Journal of Applied Ecology.* 1334-1347.

87. Oldfield, M.D. (1984). The value of conserving genetic resources/US Dep. of the interior. Nat. park service.

88. Pandey, C.N. and Rangaraju, T.S. (2008). India's industrial wood balance. *International Forestry Review.* 10(2): 173-189.

89. Kontoleon, A., Pascual, U. and Swanson, T.M. (2007). *Biodiversity economics.* UK: Cambridge University Press.

90. Perrings, C., Folke, C. and Maler, K.G. (1992). The ecology and economics of biodiversity loss: the research agenda. *Ambio.* 201-211.

91. Phurailatpam, A.K., Singh, S.R. and Nongthombam, R. (2014). Conservation of medicinally important plants by the indigenous people of Manipur (*Meiteis*) by incorporating them with religion and nature worship. *Current science.*109(1).

92. Population Action International. (2011). "Population Action International." *Publications.*11 Feb. 2011. Web.

93. Poesen, J.W.A. and Hooke, J. M. (1997). Erosion, flooding and channel management in Mediterranean environments of southern Europe. *Progress in Physical Geography.* 21(2): 157-199.

94. Power, A.G. (2010). Ecosystem services and agriculture: trade-offs and synergies. *Philosophical Transactions of the Royal Society of London B: Biological Sciences.* 365(1554): 2959-2971.

95. Pretty, J., Adams, B., Berkes, F., De Athayde, S. F., Dudley, N., Hunn, E. and Sterling, E. (2009). The intersections of biological diversity and cultural diversity: towards integration. *Conservation and Society.* 7(2): 1-100.

96. Quétier, F., Lavorel, S., Thuiller, W. and Davies, I. (2007). Plant-trait-based modelling assessment of ecosystem-service sensitivity to land-use change. *Ecological Applications.* 17(8): 2377-2386.

97. Ramakrishnan, P.S. (2009). Linking knowledge systems for socio-ecological security. In *Facing Global Environmental Change.* Springer Berlin Heidelberg. 817-828.

98. Ramakrishnan, P.S. (1984). The science behind rotational bush fallow agriculture system (jhum). *Proceedings: Plant Sciences.* 93(3): 379-400.

99. Reed, S.E. and Merenlender, A.M. (2008). Quiet, non-consumptive recreation reduces protected area effectiveness. *Conservation Letters.* 1(3): 146-154.

100. Richards, K.W. (1993). Non-Apis bees as crop pollinators. *Revue suisse de Zoologie.* 100(4): 807-822.

101. Rodriguez, J.P., Beard, T.D., Bennett, E.M., Cumming, G.S., Cork, S.J., Agard, J. and Peterson, G.D. (2006). Trade-offs across space, time, and ecosystem services. *Ecology and society.* 11(1): 1-28.

102. Sax, D.F. and Gaines, S.D. (2003). Species diversity: from global decreases to local increases. *Trends in Ecology & Evolution.* 18(11): 561-566.

103. Secretariat of the Convention on Biological Diversity (2010). *Global Biodiversity Outlook 3.*Montréal. 1-94.

104. Shetye, R.P. and Chaphekar, S.B. (1989). Some estimation on dust fall in the city of Bombay, using plants. *Prog.Ecol.* 4: 61-70.

105. Sievänen, T., Arnberger, A., Dehez, J. and Jensen, F.S. (2009). Monitoring of forest recreation demand.*BELL, S.; SIMPSON, M.; TYRVAINEN, L.* 105-133.

106. Singh, G.S. (1999). Utility of non-timber forest products in a small watershed in the Indian Himalayas: The threat of its degradation. *Natural Resources Forum.* 23: 65-77.

107. Singh, G.S. (2000). Traditional society and bio-cultural value in western Himalaya. *J. Social Science.* 4:95-100.

108. Singh, P.S. (2015). Van Sangyan. *Tropical Forest Research Institute, Jabalpur, MP, India.* 2(8): 25–30.

109. Singh, J.S., Singh, S.P. and Gupta, S.R. (2014). Biodiversity: Magnitude function and services threatened species and conservation. *Ecology Environmental Science and Conservation 665-673.* New Delhi, S. Chand.

110. Sims, B. (2007). 'The day after the Hurricane': Infrastructure, order, and the NewOrleans police department's response to Hurricane Katrina. *Social Studies of Science.* 37(1): 111-118.

111. Summers, J.K., Smith, L.M., Case, J.L. and Linthurst, R.A. (2012). A review of the elements of human well-being with an emphasis on the contribution of ecosystem services. *Ambio.* 41(4): 327-340.

112. Toledo, V.M. (2001). Indigenous peoples and biodiversity.*Encyclopaedia of biodiversity.* 3: 451-463.

113. Thomas, C.D., Cameron, A., Green, R.E., Bakkenes, M., Beaumont, L.J., Collingham, Y.C. and Hughes, L. (2004). Extinction risk from climate change. *Nature.* 427(6970): 145-148.

114. Tripathi, A. and Singh, G.S. (2013). Perception, anticipation, and responses of people to changing climate in the Gangetic plain of India.*Current Science.* 105: 1673-1684.

115. Tripathi, A., Tripathi, D.K., Chauhan, D.K., Kumar, N. and Singh, G.S. (2016). Paradigms of climate change impacts on some major food sources of the world: A review on current knowledge and future prospects. *Agriculture, Ecosystems and Environment.* 216: 356-373.

116. UNEP. (2006). Marine and coastal ecosystems and human well-being: A synthesis report based on the findings of the Millennium Ecosystem Assessment. DEW/0785/NA. United Nations Environment Programme, Nairobi.

117. US Environmental Protection Agency, *Bioengineering for Pollution Prevention through Development of Biobased Materials and Energy, State-of-the-Science Report*. US EPA Washington, DC, 2007.

118. Vitousek, P.M., Aber, J.D., Howarth, R.W., Likens, G.E., Matson, P.A., Schindler, D. W. and Tilman, D.G. (1997). Human alteration of the global nitrogen cycle: sources and consequences. *Ecological applications*. 7(3): 737-750.

119. Raven, P.H. (2000). *Nature and human society: the quest for a sustainable world*. National Academies.

120. Weaver, D.B. (2006). *Sustainable tourism: Theory and practice*. Routledge.

121. Webb, A. and Coates, D. (2012). Biofuels and biodiversity. *CBD Technical Series*. (65): 1-69.

122. Wenning, R.J., Apitz, S.E., Baba, A., Citron, M., Elliott, K., Al-Halasah, N. and Rutjes, R. (2007). Understanding environmental security at ports and harbors. In *Managing Critical Infrastructure Risks*. Springer Netherlands. 3-15.

123. WHO, (2001). Herbs for Health, But How Safe Are They? In WHO news. *Bulletin of the World Health Organization*. 79(7): 1-691.

124. Western, D. and Henry, W. (1979). Economics and conservation in third world national parks. *Bioscience*. 29(7): 414-418.

125. Westman, W.E. (1977). How much are nature's services worth? *Science*. 197(4307): 960-964.

126. Wilson, K.A., McBride, M.F., Bode, M. and Possingham, H.P. (2006). Prioritizing global conservation efforts. *Nature*. 440(7082): 337-340.

127. Worldwatch Institute. (2006). *Biofuels for transportation, global potential and implications for sustainable agriculture and energy in the 21st century*. WorldwatchInstitute, Washington, DC.

128. WWF (World Wildlife Fund). (1995). *Ecotourism*: conservation tool or threat? *Conserv.* 2(3): 1–10.

129. Zhang, W., Ricketts, T.H., Kremen, C., Carney, K. and Swinton, S.M. (2007). Ecosystem services and dis-services to agriculture. *Ecological economics*. 64(2): 253-260.

130. Zhasa, N., Hazarika, P. and Tripathi, Y.C. (2015). Indigenous Knowledge on Utilization of plant Biodiversity for Treatment and Cure of diseases of Human beings in Nagaland, India: A case study. *International Research Journal of Biological Sciences*.4(4)**:** 89-106.

131. ZSI. (2014). Official communication from Zoological Survey of India, Kolkata, India.

25

Biodiversity Business: New Facet of Revenue Generation

Manali datta and S. L. Kothari

Amity Institute of Biotechnology,
Amity University Rajasthan, Jaipur-302001, Rajasthan, India
E-mail: slkothari@jpr.amity.edu

Abstract

Biodiversity or diversity in species has become a means of sustenance. Biodiversity has enabled the supply and maintenance of food security, dietary health and livelihood sustainability. Affluent presence of species in an area has generated ecological networks and functions. Many species tend to be endemic in nature due to environmental effects. Thus this region specific variability has created an innovative field of business opportunity and thus, major source of revenue generation in the world. Biodiversity business is defined as business activities which are commercially viable and deliver net positive effects to biodiversity. The business based on biodiversity may be direct, passive or indirect in nature. As numerous types of ecosystems are present, some specific products become the forte of particular region. The business includes trade of raw materials, generation of finished products, tourism and energy trade.

Key Words: Biodiversity, Business, food security.

Introduction

Biodiversity, a contraction of "biological diversity," generally refers to the variety and variability of life on Earth. This can refer to genetic variation, species variation, ecosystem variation and functional variation (based on goods and services) within an area, biome, or planet.

Biodiversity may seem to have little relevance to business and the area of supply chain management, but is crucial for the functioning of ecosystems, which in turn provide essential goods and services on which people, business, and global economies rely.

Biodiversity business is defined as: 'commercial enterprise that generates profits via activities which conserve biodiversity, use biological resources sustainably, and share the benefits arising from this use equitably. Many industries thrive majorly on raw materials obtained directly from natural origin. For example, Amazon rainforest described as the "Lungs of our Planet" contributes to 20% of the world. 80% of the developed world's diet originates in the tropical rainforest. Approximately 121 prescription drugs sold worldwide have originated from plant-derived sources. 70% of plants capable of generating cancer therapeutics, identified by U.S. National Cancer are found in the rainforest. Twenty-five percent of the active ingredients in today's cancer-fighting drugs come from organisms found only in the rainforest (Taylor, 2004). Consequently, loss in biodiversity is a significant threat to long term economic sustainability. This chapter discusses the role of biodiversity in business development and revenue generation and its role in economic sustainability.

Biodiversity: Source of Business

The dynamic analysis of relationship between people and biodiversity has shown an intrinsic dependency on the ecological system (Figure 1). This is because, in general, the greater the biodiversity, the greater is the resilience of the ecological system. Hence, with an increased variance in biodiversity, humans have better security against the impacts of ecological threats (Benett, 2003).

As far as an economist's outlook is towards the impact of biodiversity, they have termed it "Anthropocentric, or people centric, utilitarian or aimed at improvement of well-being of people and marginalist, in that it involves the consideration of the impact on people that will occur when a specified change is imposed".

Biodiversity as a business can be categorized in three categories:-

1. **Direct use** where benefits arising from marketed commodities are impacted by the intensity and diversity of biological resource and tourism and recreation activities are directly correlated to biological variability.
2. **Passive use** like life support services such as nutrient removal, flood control, climate stabilization etc.
3. **Non use** where the ethical considerations are made to maintain the sustainability of species or "existence value".

Ecosystems have been classified in three major types:-
1. Freshwater Ecosystems
2. Terrestrial Ecosystems
3. Marine Ecosystems

The terrestrial ecosystems have been further classified in seven categories namely:-
1. Tropical rain forests
2. Savannahs
3. Deserts

4. Prairies and Pampas

5. Deciduous forest

6. Coniferous forest

7. Tundra

Each of these types and subtypes has become a major source of revenue generation.

Marine ecosystems constitute 71% of Earth's cover and contain 97% of out planet's water. The different divisions of the marine ecosystem are:

a. Oceanic: Shallow part close to the continental shelf.

b. Profundal: Bottom water.

c. Benthic: Bottom substrates.

d. Inter-tidal: The place between low and high tides.

e. Estuaries

f. Coral reefs

g. Salt marshes

h. Hydrothermal vents where chemosynthetic bacteria make up the food base.

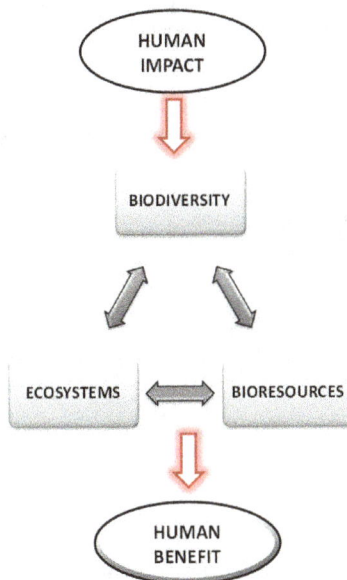

Figure 1: Hierarchy of business of biodiversity

Ecosystems induce changes in energy, water, and carbon balance of the atmosphere at local and larger scales and thus impinge on climate as a whole. Climate change threatens ecological systems at every manner throughout the world. Efforts have been made to mitigate climate change through carbon sequestration

Marine Ecosystem as Source of Revenue

Marine ecosystem provide multiple avenues like fisheries, shipping, mineral resources, energy tourism and bioprospecting for business [Oliver, 2009].

Marine reserves in addition to protecting biodiversity locally provide potential economic benefits including enhancement of local fisheries, increased tourism, and maintenance of ecosystem services.

Coral reefs (Figure 2) have the greatest biodiversity of all marine ecosystems. Tropical reefs shelter one quarter to one third of all marine species. The reefs present a greater occurrence of rare species (38%) and low spatial overlap (81% found in only one locality) and localization of diversity is biogeographically restricted and consistent (Indo-West Pacific>Central Pacific>Caribbean) (Plaisance, 2011). In a typical reef we can encounter species of corals, snails, clams, sponges, anemones, crabs, worms, starfish, shrimps, lobsters, sea cucumbers, sea lilies, fish such as groupers, snappers, breams, surgeonfish, damselfish, butterflyfish, parrotfish, clownfish and a number of other highly coloured ornamental fish, sharks, turtles, dolphins, green algae, brown algae, red algae, sea grasses (Wafar and Wafar, 2001). Surprisingly it has also been found that calcium carbonate skeletons of the reefs have capability of absorbing and harnessing photosynthetically active radiation (PAR). Strong UVR absorbance by the skeleton enables the successful growth of the corals under UVR levels that are damaging to most marine life (Reef, 2009).

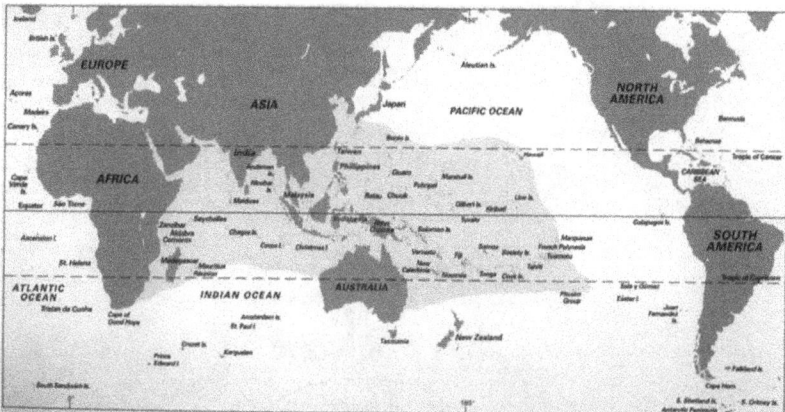

Figure 2: Incidence of coral reefs in the temperate regions of the world especially Red Sea, Persian Gulf or our Gulf of Kachchh. [http://www.nio.org/].

Belize Barrier Reef Reserve System (BBRRS), one of the largest reef complex and world heritage Site is comprised of seven protected areas; Bacalar Chico National Park and Marine Reserve, Blue Hole Natural Monument, Half Moon

Caye Natural Monument, South Water Caye Marine Reserve, Glover's Reef Marine Reserve, Laughing Bird Caye National Park and Sapodilla Cayes Marine Reserve. BBRRS comprise 12% of the entire Reef Complex (Cooper, 2008). Reef based tourism in Belize contributes 12-15 percent of gross domestic product. In 2007, activities generated about $4.4 billion in local sales, almost $2 billion in local income, and 70,400 full and part-time jobs. Currently the reefs also had an asset value of $8.5 billion (UNESCO).

India has a rich resource of coral reefs in the Palk Bay, Gulf of Mannar, Gulf of Kutch, Andaman and Nicobar Islands and Lakshadweep. Indian Ocean, has 199 species consisting of 37 genera, from India, which includes both hermatypic and ahermatypic corals. These reefs serve has a repository for variety of fishes. Hence the extensive areas of reefs serve as a catchment area for fisheries.

Another source of income from reef is the practice of reef gleaning. It serves as a major source of income for women on the islands. Edible shell fish, octopus, ornamental shells and cowries are the main types of materials which are of interest and are each of these pieces are sold for a price value of Rs 25-45/piece (Kumar, 2014).

Marine coastline of India stretches upto 7517 kilometers with 3,827 fishing villages, and 1,914 traditional fish landing centers. Fresh water resources consists of 195,210 kilometers of rivers and canals, 2.9 million hectares of minor and major reservoirs, 2.4 million hectares of ponds and lakes, and about 0.8 million hectares of flood plain wetlands and water bodies. As of 2010, the marine and freshwater resources offered a combined sustainable catch fishing potential of over 4 million metric tonnes of fish (www.fao.org). India is a major supplier of fish in the world. Fishing in India employs about 14.5 million people. The export of marine products has steadily grown over the years-from a mere Rs.3.92 crore in 1961-62 to Rs. 16597.23 crore in 2011-12(Fig 3b). Frozen Shrimp continued to be the major export value item accounting a share of 51.35% of the total US $ earnings.. The giant tiger prawn (Penaeus monodon) is the dominant species chosen for aquaculture, followed by the Indian white prawn (Fenneropenaeus indicus).

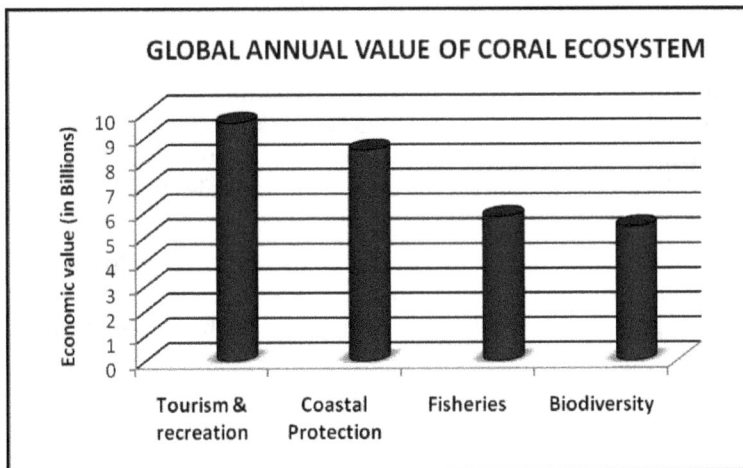

GLOBAL ANNUAL VALUE OF CORAL ECOSYSTEM

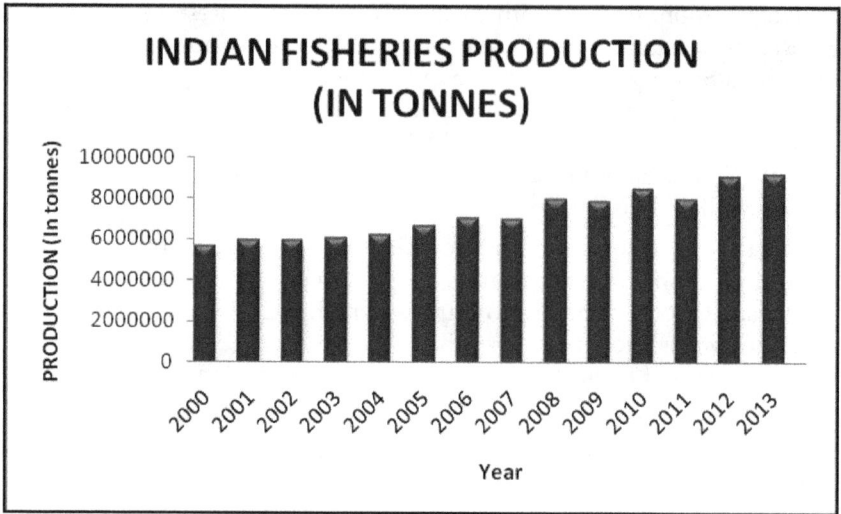

Figure 3: (a) Breakdown of component values that contribute the the global annual value of coral ecosystems (b) Graphical representation of Fisheries production in India. http://coralreef.noaa.gov/ aboutcorals/values/

As per the 2012-13 reports, the revenue generated by fish export to south east Asia alone, the largest importer, was 811 million US dollars and an approximate of 3511 million US dollars from total fish exports (MPEDA, Cochin). The income has been a steady source of economic growth for the country. Indian seafood exporter Seven seas, exports mackerel, cuttlefish sardines and shrimps make 80% of business

Maintenance of fish population has enables a sustainable supply of seafood, jobs, infrastructure, and tax revenue leading to key social and economic benefits. Attainment of productivity and accomplishment from fish businesses at major seafood buying companies to mid-level suppliers to independently owned restaurants depends solely on the sustained availability of fish supply For example in USA, Magnuson-Stevens Fishery Conservation and Management Act (Magnuson-Stevens Act) has played a pivotal role in moving U.S. fisheries toward greater achievement. The act basically added 200 nautical miles into US jurisdiction so that accessibility of other fishing ships was restricted (www.nmfs.noaa.gov). National Marine Fisheries Service projects aims to rebuild U.S. fish populations, indirectly will lead to a$31 billion increase in annual sales and support for half a million new U.S. jobs (Jeans, 2015).

Similarly in India, adoption of this act has resulted in inclusion of 2 million square kilometers. In the mid-1980s, only about 33 percent of that area was being exploited. The potential annual catch from the area has been estimated at 4.5 million tons (www.fao.org).

Offshore petroleum reserves are another source of income being generated (Fig 4a). Although the cause of oil reserves formation in area is decided by plate tectonics and presence of anoxic water, it is the dead microorganism prevalent in the area which generates the oil and gas (Schmidt, 2004).

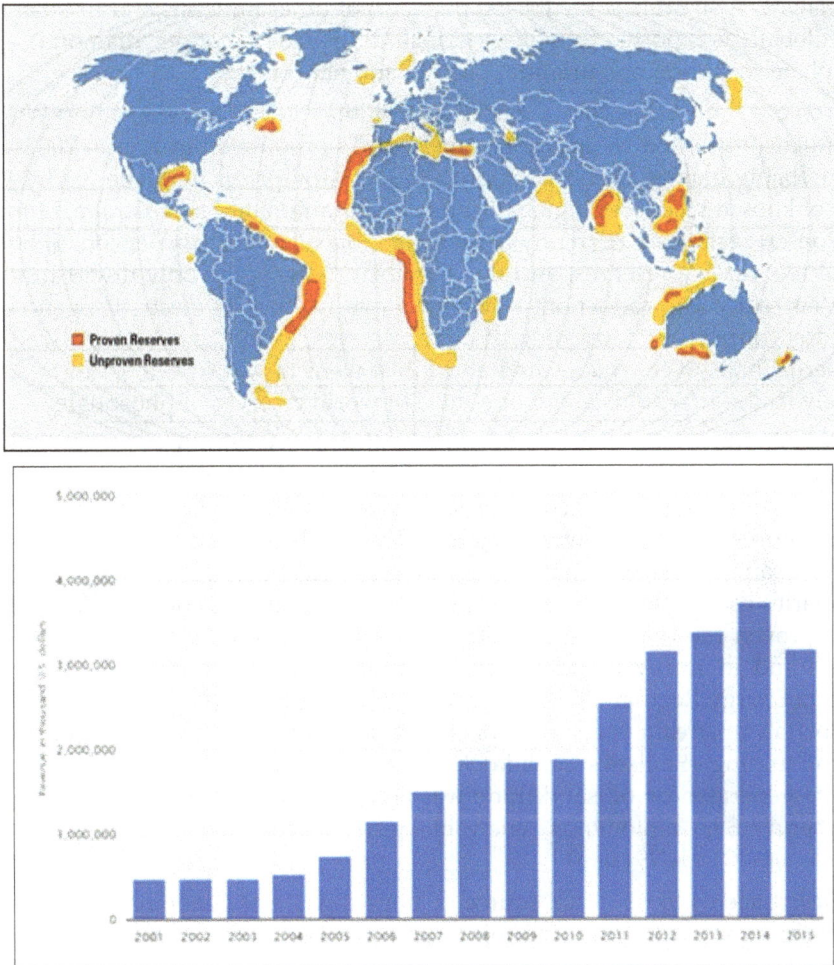

Figure 4: (a) Five most coveted offshore petroleum reserves (https://www.quora.com/). (b) Helmerich & Payne's revenue from 2001 to 2015 @Statista 2016.

H&P is a leading global drilling contractor with activities in the U.S., Latin America, Africa and in the Gulf of Mexico. In 2006, the company generated revenues of around 1.1 billion U.S. dollars (Figure 4b).

Desert Ecosystem as Source of Revenue

Most dust particles in the global atmosphere originate from the deserts of northern Africa (50–70 per cent) and Asia (10–25 per cent). Nutrients carried by desert dust, such as phosphorus and silicon, enhance growth in oceanic phytoplankton by increasing the productivity of some marine ecosystems, and also of nutrient-poor tropical soils, as observed from Saharan dust deposited in the Amazon basin.

Water-soluble salts, such as gypsum, borates, table salt, sodium and potassium nitrates have been historically a product of deserts. Evaporite minerals, such as soda, boron, and nitrates, are common in deserts and are not found in other ecosystems. A sizeable share (30–60 per cent) of other minerals and fossil energy used globally is exported from deserts, including bauxite, copper, diamonds, gold, phosphate rock, iron ore, uranium ore, oil, and natural gas

Iron ore contributes 40 per cent of Mauritania's export income; desert Western Australia contributed 16 per cent of the world's production of iron in 2003. Both will probably maintain their position, although iron prices fluctuate wildly. One-third of known recoverable global reserves of uranium are in Australia, but none of its desert reserves is currently mined. Namibia has about six per cent of known global recoverable reserves, and the Namibian mine is the only desert uranium mine currently in production. A global move to nuclear electricity generation would encourage the reopening of other reserves, as in Kazakhstan, Niger, and northern Chad (over which Chad and Libya went to war in 1987). North Africa (largely its deserts) holds about one-third of world reserves of phosphate.

Another upcoming source of income is oil sand, a naturally occurring mixture of sand, clay or other minerals, water and bitumen. Bitumen is so viscous that at room temperature it acts much like cold molasses. Alberta's oil sands have the third largest oil reserves in the world, after of Venezuela and Saudi Arabia. Alberta's three oil sands areas of Athabasca, Cold Lake and Peace River have regulatory boundaries established by the Alberta Energy Regulator. As of 2014, Alberta's oil sands proven reserves were 166 billion Barrels (bbl). Total oil sands production (mined and in situ) reached about 2.3 million barrels per day (bbl/d) in 2014 (Alberta Energy Regulator (AER)). Approximately 1.3 lack people were employed in Alberta's upstream energy sector, which includes oil sands, conventional oil, gas and mining. (Statistics Canada, Survey of Employment, Payrolls and Hours). Royalty generated by oil sand trade was about $5.2 billion or almost 55 per cent of Alberta's $9.6 billion non-renewable resource revenue (http://www.energy. alberta.ca/OilSands/791.asp).

Desert seems to be a rich source of trees capable of generating oil. *Citrullus colocynthcus* or bitter apple and jojoba both natives of desert, are the two trees contributing to the market of oil.

Argania spinosa, the source of "Liquid Gold" is a tree endemic to the southwest of Morocco and the Tindouf region of Algeria. The tree has long, tap root systems that go deep into the thin soil in search of the water table and hence can survive in desert areas which have low levels of water table. The depth of the root system means the trees are firmly anchored to the ground and resistant to strong winds, which in turn protects the soil from erosion. The fruit is the economically most useful part of the tree. The fruit generates raw material for pottery products, cosmetics and food industry (Cherki, 2006). The oil used for cosmetic purposes is approximately 2 times richer in tocopherol than olive oil (620 mg/kg vs 320 mg/kg). The fat composition of the argan seed kernel oil is 45% monounsaturated fatty acid, 35% polyunsaturated fatty acid, and 20% saturated fatty acids (Dressi, 2004).

A herb from the desert known as the Devils claw has internationally been traded for more than 50 years. This plant, *Harpagophytum*, which is native to Africa, contains chemicals that might decrease inflammation and swelling and resulting pain. About 600-700 tonnes of this herb is traded producing a revenue of US$ 100 million (Ameye, 2006).

At present, Moroccan Argan groves cover an area of approximately one million hectares (2.47 million acres) in the country's southwest; between the Atlas Mountains and the Atlantic coast (Wilson 2014). Over the time, Argan oil has become a high-end commodity in the lucrative global cosmetics market. Today, Argan oil is one of the world's most expensive selling wholesale for €28 ($36) per liter (CIFOR, 2004). Another well known and traded product from the Arabian Peninsula is red dates or Phoenix dactylifera. The cultivation of dates, the fruit of the date palm, began more than 5,000 years ago in regions of the Arabian Peninsula, Middle East, and North Africa. Phytochemical investigations have revealed the beneficial effects and highlighted its potential medicinal importance for humans all around the world (Vayalil, 2012). Currently Egypt is largest producer of date with a production of nearly 1570 metric tones (http://www.perfectinsider.com) contributing 19% of total production in 2011 followed Saudi Arabia at 15% and Iran (14%) (Ali, 2014).

Sand and gravel constitute the largest volume of solid material traded globally. The process of erosion regerates this raw material used in many construction areas (John, 2009). 47 and 59 billion tonnes of material is mined every year (Steinberger *et al.*, 2010) and the fastest extraction increase (Krausmann *et al.*, 2009). In the United States the price of sand has remained very stable, fluctuating from US$4.50 to US$6.7 a tonne between 1910 and 2013 (UNEP, 2014).

Mountain Ecosystem as Source of Revenue

Hindu Kush-Himalayas (HKH) is a dynamic landscape region hosting parts of the four Global Biodiversity Hotspots; namely, the Himalayas Hotspot, the Indo-Burma Hotspot, the Mountains of South-West China Hotspot, and the Mountains of Central Asia Hotspot. Himalayan range is a rich repository of biodiversity and the eastern part is known as "Cradle of Flowering plants" and has more than 400 species of medicinal plants (Sharma *et al.*, 2008). An estimated 10,000 plant species are found in the Himalaya out of which about 3,160 are endemic. Himalayas also houses the largest family of flowering plants in the hotspot, Orchidacea with 750 species. The Himalayan medicinal plants play an imperative role in the local culture and economy of the region. The whole of Himalayan region abounds in medicinal plants that are used by the doctors as well as the local inhabitants for preparing various kinds of medicines, especially the eastern part (bsienvis.nic.in). Search for wonder drug in the Himalayas has led to a latest discovery of Rhodiola by DIHAR scientists, a wonder plant having immunomodulatory, adaptogenic and radio-protecting abilities due to presence of secondary metabolites and phytoactive compounds unique to the plant (TOI, 2014).

A spring is a water resource formed when the side of a hill, a valley bottom or other excavation intersects a flowing body of ground water at or below the local water table, below which the subsurface material is saturated with water.

Health-conscious consumers appreciate the high mineral content like calcium carbonate, iron, magnesium and its naturally high pH to help offset modern acidic diets. Mountain spring water has become a requirement for the elite class and thus generates handsome revenue from its trade. Mountain spring water, a spring water bottling sells Mountain Valley Spring Water, Mountain Valley Sparkling Water, and Diamond Spring Water under its flagship and the company has an indicated an income of US$19.9 million in nine months, but it did make a profit of $6.5 million (Taulli, 2011)

Timber is natural product from wood used for construction. Its flexibility and versatility in addition to its aesthetic value, makes it a suitable choice for building purpose. There are almost 150 species of timber producing trees alone in India. About hundred countries account for 99.7% of the value of all timber shipments for 2014 with China topping the list by contributing 10.3% (approx US$14.5 billion). Overall timber exports totaled US$140.9 billion during 2014, up 32.5% since 2010 (http://www.forestindustries.se).

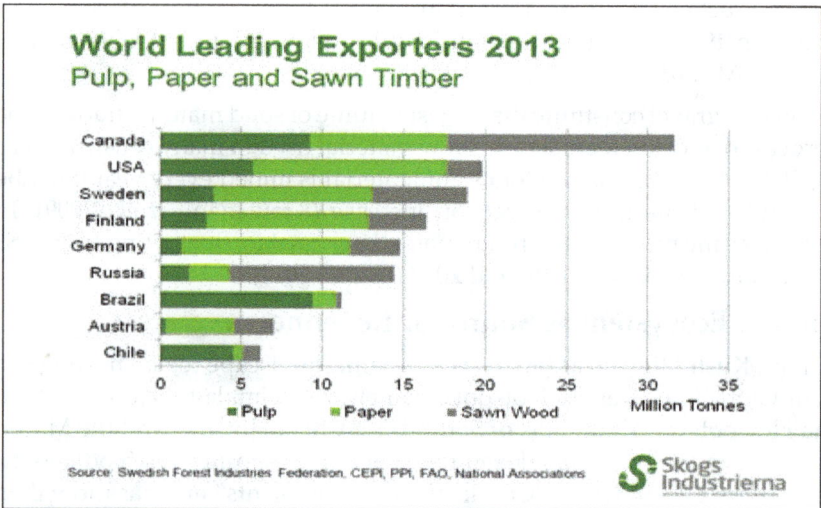

World Leading Exporters 2013
Pulp, Paper and Sawn Timber

Source: Swedish Forest Industries Federation, CEPI, PPI, FAO, National Associations

Figure 5: Top timber producing countries of the world (2013). (http://www.forestindustries.se/documentation/statistics_ppt_files/)

Another upcoming arena of trade is the trade in non-timber forest products (NTFPs). From the economic viewpoint, NTFP are equally important as wood-based products.

About 150 types of NTFP are significant in international trade. They are also increasingly being acknowledged for their role in sustainable development and conservation of biological diversity.

In Nepal up to 80% of the population in developing countries depend on NTFP for subsistence, both economically and for nutrition. It is an important asset especially for people living in Nepal's Sacred Himalayan Landscape contributing significantly in their livelihood. NTFP contributes to about 20% to 40% of the annual income of forest dwellers comprising dominantly of tribals. In Nepal almost 275 million poor rural population (27%) depend on NTFPs for at least part of their subsistence and cash livelihoods (Malhotra & Bhattacharya, 2010; Bhattacharya & Hayat, 2009). Tribals (89 million) of India rely mostly NTFPs for their economic sustenance. Some initiatives have been taken by countries to bring proper revenue to NTFPs trade (Table 1). Roughly 90% of the herbal raw materials from NTFP enter the industry. Of the 7000 plants used in Indian System of Medicine, 960 have been recorded in trade and 178 are traded in high volumes in quantities exceeding 100 metric tonnes per year. As per the study of planning commission, an annual demand of raw botanicals in 2005-06 was around 3, 19,500 million tonnes with corresponding trade value of Rs.1069 crores. Based on the cumulative report of 2005-2009, the Indian share of global medicinal plants trade is increasing at an annual growth rate of 23% and India stands third as exporter of NTFP composed of medicinal herbs after Canada and China (Planning commission Report, 2011). In Cameroon, revenue earned from NTFPs was US$ 1.75 million in 1995.

Table 1: Indian state initiatives for popularization of NTFPs

INITIATIVES UNDER NTFPs IN INDIA

1. **UTTARAKHAND** : Jarie Bootee Mandi
2. **ANDHRA PRADESH** : Girijan Cooperative (Girijan brand rock bee honey, tamarind, and amla)
3. **ORISSA:** Sanjog : Lac cultivation; Hill broom and cashew drupe
4. **CHHATTISGARH** : Chhattisgarh Minor Forest Products Federation
5. **MADHYA PRADESH** : MFP Federation

Crocus sativus or saffron, an exotic spice is a native of Southern Europe and Greece, Turkey, Iran, and in the Indian state of Jammu and Kashmir.. Iran produces 95% of the world's saffron. India, along with Spain, Italy, Greece, Morocco and Ajerbaijan, produce the rest. Kashmir is India's only saffron producing centre (Hill, 2004). A variant of "red gold", Swiss saffron, costs CHF28 ($29) per gram, making it the most expensive saffron in the world. Almost 50,000 flowers are harvested to generated 0.45 kg of dry saffron. Bulk quantities of lower-grade saffron can reach upwards of US$500 per pound; retail costs for small amounts may exceed ten times that rate. Retail price is almost US$1,000 per pound of saffron in western countries.

In 2014, Saffron rates varied from £15-£75 for 0.2 grams in England. The superior 'mongra' variety of Kashmir has a good market in Japan, Australia and Canada because of its aphrodisiac properties. On an average a kilogram of pure saffron can cost Rs.250,000-Rs.270,000 in the domestic market (Gulati, 2014). Iran alone pulls in over $300 million in *revenues* from *saffron (Woodman, 2012)*.

Moschus moschiferus or musk deer is a small, solitary deer that lives in the mountainous forests of Asia and eastern Russia. Six species of musk deer (a topic of dispute among scientists), are found in 13 countries, including Russia, China, India, Nepal, and other Asian nations. Extraction of one kg of musk is feasible only after 160 deer are killed. Musk is worth U.S. $2 to $3 per gram, meaning that the pod of a single stag can easily fetch $70—a huge sum of money to hunters in far-flung regions of Russia or Mongolia (Pickrell, 2004)

John Pickrell (2004) Poachers Target Musk Deer for Perfumes, Medicines National Geographic News.

Biodiversity and Tourism

Influx of tourists to the marine reserves is directly related to the increase in marine life in reserves, in particular large fish as they serve as the main source of attraction for divers. Approximately 7.5 million dives constituting 50% dives in the Caribbean and Pacific coast of Central America, take place within marine protected areas (Green, 2003). In the US, reef-related recreation and tourism account (Figure 3a).for an estimated $364 million in added value to Hawaii's economy each year and its nearshore reefs annually contribute nearly $1 billion in gross revenues for the state. Each square kilometer of reefs of Southeast Asia with tourist potential generate a prospective net benefit of $23,100 to $270,000 (NOAA report).

Desert tourism, another source of investment, has grown quickly. Four million tourists visit Morocco and five million reach Tunisia each year. They contributed six per cent to Tunisian gross domestic product in 1999, and employed over 300 000 people. Desert destinations in both countries outperformed their coasts. There was a 161 per cent increase in tourism to Egypt in 2005. Dubai claims to be the world's fastest growing tourist destination; 100 000 British people have bought homes there, and it is aiming at 15 million tourist visits a year. Baja California is booming. More gambling is said to take place in deserts than in any other global environment. Greater Palm Springs Convention and Visitors Bureau in 2013 confirmed that revenue equivalent to $5.8 billion was generated, a marked hike of 9.4 percent since 2011. Similarly an Economic Impact of Tourism report indicated direct visitor spent approximately $4.5 billion on desert tourism especially during festivals or events (www.desertsun.com). In 2014, desert state of India, Rajasthan generated 6.8% of the foreign tourists visiting India (www.tourism.nic.in). In 2015, India's tourism industry was expected to directly contribute 42.77 billion U.S. dollars to the country's economy, and this figure was forecasted to rise to 85.6 billion by 2025.

Discussion: Loss of Biodiversity and Its Recurpussions

Business based on biodiversity forms a vicious circle in which the viability of the business depends solely on a flourishing diversity of flora and fauna. Rainforest deforestation is leading to loss of 137 plant, animal and insect species every day. Approximately that becomes equivalent to 50,000 species a year. Currently, 121 prescription drugs sold worldwide come from plant-derived sources. With disappearing flora and fauna the possible cures for life-threatening diseases go into oblivion.

NTFPs are used and managed in complex socio-economic and ecological environments. But with shrinking forest and growing human population, the sustainability of many NTFPs is dwindling. For example, as international rattan prices increased in the 1980s and '90s, commercial companies in Asia hired local people to harvest available resources. Widespread over-exploitation resulted in destruction of the resource, affecting the endemic biodiversity and decrease in revenue gained by people (Malleson, 2014).

Considering the current population boom, by 2050, the population is supposed to hit 9 billion mark. This will inadvertly lead to increase in the demand of all products provided by nature. This will eventually lead to overexploitation of the natural resources as the economy will prefer a 'Anthropocentric' outlook. Many law and initiatives have been taken to enable the protection of this thriving source of income as potential ecosystem degradation and biodiversity loss have become central to almost investment decisions. Some legislation in European Union and USA, legislation necessitates compensation for the loss of ecosystems with activities which will enable flourishing of biodiversity.

References

1. Bennett, J. (2003). The economic value of biodiversity: a scoping paper The Economic Value of Biodiversity. Australian Government Department of the Environment and Heritage, and Land & Water Australia.

2. Oliver, J. (2007). Economic Valuation of Large Marine Ecosystems Report 2007 IUCN Global Marine Programme.

3. Sala, E., Costello, C., Dougherty, D., Heal, G., Kelleher, K., Murray, J.H. (2013). A General Business Model for Marine Reserves. PLoS ONE. 8(4): e58799.

4. Lester, S.E., Halpern, B.S., Grorud-Colvert, K., Lubchenco, J., Ruttenberg, B.I. (2009). Biological effects within no-take marine reserves: a global synthesis. Marine Ecology Progress Series. 384: 33-46.

5. Plaisance, L., Caley, M.J., Brainard, R.E. and Knowlton, N. (2011). The Diversity of Coral Reefs: What Are We Missing? PLoS ONE. 6(10): e25026.

6. Reef, R., Kaniewska, P. and Hoegh-Guldberg, O. (2009). Coral Skeletons Defend against Ultraviolet Radiation. PLoS ONE. 4(11): e7995.

7. Wafar, M. and Wafar, S. (2001). 101 Questions on Corals. NIO publications.

8. http://whc.unesco.org/en/list/764

9. http://coralreef.noaa.gov/aboutcorals/values/tourismrecreation/

10. Cooper, E., Burke, L. and Bood, N. (2008). Coastal Capital: Economic Contribution of Coral Reefs and Mangroves to Belize. Washington DC: World Resources Institute.

11. Kumar, J.Y., Geetha, S., Raghunathan, C. and Venkataraman, K. (2014). An Assessment of Faunal Diversity and its Conservation in Shipwrecks in Indian Seas. Marine Faunal Diversity in India: Taxonomy, Ecology and Conservation. 441.

12. Sala, E., Costello, C., Dougherty, D., Heal, G., Kelleher, K. and Murray, J.H. (2013). A General Business Model for Marine Reserves. PLoS ONE. 8(4): e58799.

13. Green, E. and Donnelly, R. (2003). Recreational scuba diving in Caribbean marine protected areas: Do the users pay? Ambio. 32: 140-144.

14. http://www.fao.org/fishery/facp/IND/en#CountrySector-Overview

15. Country wise exports, [Statistics] MPEDA, Cochin http://www.cift.res.in/

16. http://www.nmfs.noaa.gov/sfa/laws_policies/msa/

17. Meghan Jeans Sustainable Fisheries Make Good Business Sense

18. 2015 http://conservefish.org/2015/05/12/sustainable-fisheries-make-good-business-sense/

19. http://www.fao.org/figis/servlet/SQServlet?ds   =   Production&k1   = COUNTRY&k1v = 1&k1s = 100&outtype = html

20. Schmidt, S. and Littke, R. (2004). The petroleum potential of the passive continental margin of South-Western Africa: a basin modelling study. Fakultät für Georessourcen und Materialtechnik.

21. http://www.intrafish.com/free_news/article1412364.ece

22. http://www.nola.com/environment/index.ssf/2014/04/louisiana_commercial_fishery_t.html

23. Events drove $5.8B tourism industry to record levels www.desertsun.com/story/money/tourism/2014/...tourism.../9531745

24. Indian tourists statistics at a glance 2014 http://tourism.nic.in/

25. Drissi, A., Girona, J. and Cherki, M. (2004). Evidence of hypolipemiant and antioxidant properties of argan oil derived from the argan tree (Argania spinosa). Clin Nutr. 23(5): 1159-1166

26. Cherki, M., Berrougui, H., Drissi, A., Adlouni, A. and Khalil, A. (2006). Argan oil: which benefits on cardiovascular diseases? Pharmacol Res. 54(1): 1-5.

27. Matthew, W. (2014). Moroccan argan oil: the 'gold' that grows on trees Financial Times Limited 2016.

28. http://www.energy.alberta.ca/OilSands/791.asp

29. Steinberger, J.K., Krausmann, F. and Eisenmenger, N. (2010). Global patterns of materials use: a socioeconomic and geophysical analysis. Ecological Economics. 69: 1148-1158

30. Krausmann, F., Gingrich, S., Eisenmenger, N., Erb, K-H., Haber, H. and Fischer-Kowalski, M. (2009). Growth in global materials use, GDP and population during the 20th century. Ecological Economics. 68: 2696-2705.

31. John, E. (2009). The impacts of sand mining in Kallada river (Pathanapuram Taluk), Kerala, Journal of basic and applied biology. 3(1&2): 108-113.

32. Sand rarer than one thinks: UNEP 2014. www.unep.org/pdf/UNEP_GEAS_March_2014.pdf

33. Ameye, L.G. and Chee, W.S. (2006). Osteoarthritis and nutrition. From nutraceuticals to functional foods: a systematic review of the scientific evidence. Arthritis Res Ther. 8(4): R127.

34. [The Times of India_Aug 25, 2014_httptimesofindia.indiatimes. comhomescienceIndianscientists-find-a-wonder-herb-in-the-high-Himalayasarticleshow40869492.cms

35. Sharma, E., Tse-ring, K., Chettri, N., Shrestha, A. and Kathmandu, N. (2008). Biodiversity in the Himalayas–trends, perception and impacts of climate change. In Proceedings of the International Mountain Biodiversity Conference Kathmandu.

36. Biodiversity hotspots in India. (2015). Conservation International: www. conservation.org; www.cepf.net http://bsienvis.nic.in/

37. Top Exported Timber Countries http://www.worldsrichestcountries.com/top-exported-timber-countries.html

38. Taulli, T. (2011). Mountain Valley Quenches IPO Thirst A buyout may trump a public offering http://investorplace.com/ipo-playbook/mountain-valley-quenches-ipo-thirst/#.VtSIf_l97IU

39. Riches of the Forest: For Health, Life and Spirit in Africa Citlalli López Binnqüist Patricia Shanley CIFOR, 2004-Forest plants.

40. Ndoye, O., Pérez, M.R. and Eyebe, A. (1997). Rural Development Forestry Network The Markets of Non-timber Forest Products in the Humid Forest Zone of Cameroon.

41. Vayalil, P.K. (2012). Date fruits (Phoenix dactylifera Linn): an emerging medicinal food. Crit Rev Food Sci Nutr. 52(3): 249-71.

42. http://www.perfectinsider.com/top-ten-date-producing-countries-in-the-world/

43. Ali, A.M., Fahad, A.M., El-Habbab, Samir, M. (2014). Saudi Dates Exports Demand In Selected Markets. Intl J Agri Crop Sci. 7(11): 827-832.

44. Woodman, C. (2012). Unfair Trade: The shocking truth behind 'ethical' business

45. Vishal gulati. (2014). Kashmir's meagre saffron production spikes prices (Business Feature) IANS.

46. Hill, T. (2004). The Contemporary Encyclopedia of Herbs and Spices: Seasonings for the Global Kitchen (1st ed.), Wiley, ISBN 978-0-471-21423-6

47. Report Of The Sub-Group-II On Ntfp And Their Sustainable Management In The 12th 5-Year Plan (2011). Planning Commission's Working Group on Forests & Natural Resource Management http://planningcommission. gov.in/aboutus/committee/wrkgrp12/enf/wg_subntfp.pdf

48. Gift of the Himalayas-high value plants and NTFPs (2007) http://wwf. panda.org/wwf_news/?118460 /Gift-of-the-Himalayas-high-value-plants-and-NTFPs

49. Rainforest statistic Facts: http://www.savetheamazon.org/rainforeststats. htm

50. Sustainable Forestry For Food Security And Nutrition-E-Consultation To Set The Track Of The Study http://www.fao.org/fsnforum/cfs-hlpe/sites/cfs-hlpe/files/resources/ SUSTAINABLE%20FORESTRY%20FOR%20FOOD%20SECURITY%20AND%20NUTRITION.pdf

51. Malleson, R., Asaha, S., Egot, M., Kshatriya, M., E. Marshall, E., Obeng-Okrah, K. and Sunderland, T. (2014). Non-timber forest products income from forest landscapes of Cameroon, Ghana and Nigeria – an incidental or integral contribution to sustaining rural livelihoods? International Forestry Review. 16(3): 261-277.